EurographicSeminars

Tutorials and Perspectives in Computer Graphics

Edited by W. T. Hewitt, R. Gnatz, and D. A. Duce

EurographicSeminars
Tutorials and Perspectives in Computer Graphics

Eurographics Tutorials '83. Edited by P. J. W. ten Hagen.
XI, 425 pages, 164 figs., 1984

User Interface Management Systems. Edited by G. E. Pfaff.
XII, 224 pages, 65 figs., 1985

Methodology of Window Management.
Edited by F. R. A. Hopgood, D. A. Duce, E. V. C. Fielding,
K. Robinson, A. S. Williams.
XV, 250 pages, 41 figs., 1985

Data Structures for Raster Graphics.
Edited by L. R. A. Kessener, F. J. Peters, M. L. P. van Lierop.
VII, 201 pages, 80 figs., 1986

Advances in Computer Graphics I.
Edited by G. Enderle, M. Grave, F. Lillehagen.
XII, 512 pages, 168 figs., 1986

Advances in Computer Graphics II.
Edited by F. R. A. Hopgood, R. J. Hubbold, D. A. Duce.
X, 186 pages, 96 figs., 1986

Advances in Computer Graphics Hardware I.
Edited by W. Straßer.
X, 147 pages, 76 figs., 1987

GKS Theory and Practice.
Edited by P. R. Bono, I. Herman.
X, 316 pages, 92 figs., 1987

Intelligent CAD Systems I.
Edited by P. J. W. ten Hagen, T. Tomiyama.
XIV, 360 pages, 119 figs., 1987

P. J. W. ten Hagen   T. Tomiyama   (Eds.)

Intelligent CAD Systems I

Theoretical and Methodological Aspects

With 119 Figures

Springer-Verlag
Berlin Heidelberg New York
London Paris Tokyo

EurographicSeminars

Edited by W. T. Hewitt, R. Gnatz, and D. A. Duce
for EUROGRAPHICS –
The European Association for Computer Graphics
P.O. Box 16, CH-1288 Aire-la-Ville, Switzerland

Volume Editors

Paul J. W. ten Hagen
Stichting Mathematisch Centrum
Kruislaan 413, NL-1098 SJ Amsterdam, The Netherlands

Tetsuo Tomiyama
Department of Precision Machinery Engineering
Faculty of Engineering, The University of Tokyo
Hongo 7-3-1, Bunkyo-ku, Tokyo 113, Japan

This book is a record of the First Eurographics Workshop on "Intelligent CAD Systems" which was held on April 21–24, 1987, at Noordwijkerhout, The Netherlands, sponsored by the European Association for Computer Graphics, and organized by Centre for Mathematics and Computer Science, Amsterdam.

ISBN-13: 978-3-642-72947-8   e-ISBN-13: 978-3-642-72945-4
DOI: 10.1007/978-3-642-72945-4

Library of Congress Cataloging-in-Publication Data.
Eurographics Workshops on "Intelligent CAD Systems" (1st : 1987 : Noordwijkerhout, Netherlands). Intelligent CAD systems I : theoretical and methodological aspects/ edited by P.J.W. ten Hagen and T. Tomiyama. p.cm. – (Eurographic seminars) "Record of the First Eurographics Workshop on 'Intelligent CAD Systems' which was held on April 21–24, 1987, at Noordwijkerhout, The Netherlands, sponsored by the European Association for Computer Graphics" – T.p. verso. Bibliography: p.

1. Computer-aided design–Congresses. 2. Engineering design–Data processing–Congresses. 3. Artificial intelligence–Congresses. I. Hagen, P.J.W. ten. II. Tomiyama, T. (Tetsuo), 1957–. III. European Association for Computer Graphics. IV. Title. V. Series. TA174.E87 1987 620'.00425'0285–dc19 87-27602 CIP

2145/3140-543210

Program Committee of the Eurographics Workshops on Intelligent CAD Systems

(As of June 1987)

Preface

CAD (Computer Aided Design) technology is now crucial for every division of modern industry, from a viewpoint of higher productivity and better products. As technologies advance, the amount of information and knowledge that engineers have to deal with is constantly increasing. This results in seeking more advanced computer technology to achieve higher functionalities, flexibility, and efficient performance of the CAD systems. Knowledge engineering, or more broadly artificial intelligence, is considered a primary candidate technology to build a new generation of CAD systems. Since *design* is a very intellectual human activity, this approach seems to make sense.

The ideas of *intelligent CAD systems* (ICAD) are now increasingly discussed everywhere. We can observe many conferences and workshops reporting a number of research efforts on this particular subject. Researchers are coming from computer science, artificial intelligence, mechanical engineering, electronic engineering, civil engineering, architectural science, control engineering, etc. But, still we cannot see the direction of this concept, or at least, there is no widely accepted concept of ICAD. What can designers expect from these future generation CAD systems? In which direction must developers proceed? The situation is somewhat confusing.

CAD systems tend to be large software systems. This means large investments and long development time. Thus, it is thought that a good system must be equipped with flexibility to absorb the future advances in technology and general applicability for problems that may arise in the future. Unfortunately, introducing knowledge engineering does not always improve the situation. Knowledge-based systems themselves are large and require carefully engineered software as much as do other information systems.

In such a situation, how can one justify one's investment? Can one particular approach survive the long development time and be still applicable? Is it at all possible to guarantee that one's method (in the context of knowledge engineering, e.g. knowledge representation) would not become old-fashioned by the time of delivery? No matter what kind of approach one may take, these must be affirmatively answered.

This book is the record of the First Eurographics Workshop on "Intelligent CAD Systems — Theoretical and Methodological Aspects" held on April 21–24, 1987, at Leeuwenhorst Congress Center, Noordwijkerhout, The Netherlands. There were eighteen paper contributions from nine countries and forty-five participants.

The workshop was organized as the first of a series of three workshops under the same title. In particular, this first workshop was implemented at least to identify the sources of the confusion and to formulate questions to be solved in the future work, focusing at founding a sound and strong theoretical basis for the future development. In 1988, there will be the second workshop concentrating on implementational issues and in 1989, the third one dealing with practical experiences and evaluations. Through these three workshops, we want to clarify the situation and to determine the directions we should take.

We had good variety in the workshop in every aspect; e.g. geographical distribution, application domain, and research background. Reflecting the variety of the speakers, it is not an exaggeration to say that we had as many as eighteen different definitions of *design* and as many as eighteen different approaches, from phenomenologically perceived to vigorously theoretical. Indeed, we found that there was no consent over the concepts *design* or *ICAD*. However, in this book we shall define the term *intelligent CAD systems* as a CAD system built by applying knowledge engineering. We may include in this term the so-called *knowledge-based CAD systems* and *expert CAD systems* as well, because they use advanced techniques of computer science, primarily knowledge engineering, in their kernels.

Summarizing the discussions, we could make the following observations which we hope the readers will validate by reading this book. Our first observation is that we identify three dichotomies that have arisen during the workshop:

- CAD (Computer Aided Design) approach vs. AD (Automated Design) approach.
- Designer's apprentice or assistant vs. Autonomous design system.
- Glass box system vs. Black box system.

The dichotomy of *"CAD approach vs. AD approach"* is a rather philosophical one and very much dependent on the application. If all the design knowledge of the application domain is well-known and clearly defined, we might be tempted to pursue implementing AD systems. On the other hand, if a sufficient amount of domain-specific knowledge is not captured in the system, the design process indispensably involves interaction between the designer and the system, which means we must stick to the CAD approach. These distinctions apply to the second dichotomy, *"Apprentice vs. Autonomous"* as well. An apprentice system consists of less domain-specific knowledge than an autonomous system which is nevertheless good for narrow domains. In this sense, a *glass box* system is transparent to the designer and if something wrong happens one might be able to interfere via interaction which a *black box* system does not allow.

Let us agree to call the CAD/apprentice/glass box approaches *CAD type* approaches, and the AD/autonomous/black box approaches *AD type*. Our second observation is that we could further identify two different approaches within the CAD type approach, i.e., the *"Framework vs. Intelligent tool"* dichotomy. In the *intelligent tool* approach building a set of cooperating expert systems for designing will be examined, whereas in the *framework approach* establishing a basis for incorporating such tools (including existing CAD tools) is emphasized.

The third observation is that, among those three different approaches (i.e. *CAD type-framework*, *CAD type-intelligent tool*, and *AD type*), there are strange differences and commonalities. For instance, the AD type work tries to find a theory on which a design system can perform fully automatic design operations. This idea is quite similar to that of the CAD type intelligent tool approach. The CAD type framework approach insists that there must be a general framework on which individual design subsystems should be built. This framework, therefore, has to be able to represent design knowledge that must be fed into subsystems. This requires the study of *general* design knowledge besides individual domain-specific knowledge. This is a very distinctive issue in the CAD type framework approach.

The workshop was started by the presentation given by A. Bijl (University of Edinburgh, UK) who discussed how to capture design knowledge in the framework of ICAD, taking architecture as an example. B. Veth (Centre for Mathematics and Computer Science, The Netherlands) closed the workshop by answering the questions concerning how to proceed most straightforwardly by showing his model of design processes and design knowledge and his experimental system. He concluded that we need a clean and sound theoretical basis for building ICAD. This view was supported by J.F. Koegel (University of Denver, USA) who showed another model of ICAD from a viewpoint of computer science.

B.T. David (Ecole Central de Lyon, France) presented an architecture where multiple expert systems are cooperating. His architecture too is based on his model of design processes. P. Bernus (Centre for Mathematics and Computer Science, The Netherlands) and Z. Létray (Hungarian Academy of Sciences, Hungary) argued for a fundamental structure and implementation strategy for ICAD. The viewpoints of M. Nadin and M. Novak (The Ohio State University, USA) put strong emphasis on designers. They analyzed how this emphasis influenced the architecture for the future generation CAD systems. A working example of good system architecture was illustrated by H.J. Kahn (University of Manchester, UK). A rather radical view was provided by T. Takala (Helsinki University of Technology, Finland) who suggested a new paradigm which might make the system even innovative.

Knowledge representation and inference techniques are big issues also in the development of ICAD. Both F. Arbab (University of Southern California, USA) and G. Sunde (Center for Industrial Research, Norway) pointed out the importance of geometrical knowledge in design and discussed how to represent and utilize such knowledge.

In the design processes where planning and optimization play crucial roles, knowledge about the design procedures becomes very important. This was discussed by D. A. Hoeltzel and W.-H. Chieng (Columbia University, USA). E. E. Berkhout (Technical University of Delft, The Netherlands) presented a methodology to tackle this problem with examples from architectural designing. Zs. Ruttkay (Hungarian Academy of Sciences, Hungary), R.H. Allen (University of Houston, USA), and B. Laczik (Technical University of Budapest, Hungary) explicitly mentioned their interests in the user interface problem. They relate the user interface issues to the knowledge representation problem.

Some participants were interested in more domain-specific problems, namely how theories can help implementing ICAD. M.P. Fourman and R.M. Zimmer (Brunel University, UK) discussed the use of mathematical formalizations, viz. category theory and temporal logic, in VLSI CAD, while D.R. Genin (Tektronix, Belgium) presented his system based on assumption-based truth maintenance. J. Duhovnik (University Edvarda Kardelja, Yugoslavia) on the other hand discussed how to implement "design methodology" (for mechanical design) in an ICAD environment.

The integration of design and other activities such as manufacturing, assembly, and testing is becoming more and more important. R. Milne (Intelligent Applications Ltd., UK) presented an interesting system for circuit diagnosis using data generated by

CAD systems. Another important issue in knowledge engineering is knowledge acquisition. A methodology and system for this was discussed by J.H. Boose and J.M. Brandshaw (Boeing Advanced Technology Center, USA).

The variety in the workshop elements and differences in the approaches were, however, overcome by the conclusion at Closing Discussion that ICAD must be developed on a theoretical basis. Each presentation showed a methodology to develop an ICAD or even such a system depending on a theoretical basis different from each other. But we finally found a point somehow we could agree upon. The system cannot be built on a weak basis; i.e. we need sound and strong theories. In the future workshops this requirement must be kept in mind and proposed systems must be checked against it.

This book is a collection of eighteen papers. Paper contributions were first published in the draft proceedings of the workshop. After the workshop, authors were requested to revise their papers based on the comments and questions during the workshop. The book also comprises eight session reports, two reports on general discussions, and the closing session. These reports were first drafted by the chairpersons based on notes taken by the co-chairpersons. These were finally rewritten and edited by Varol Akman (Centre for Mathematics and Computer Science, The Netherlands) and the editors to bring uniformity. In addition, this book has a bibliography of all the listed references. We hope this will help the readers' future work.

We would like to thank very much the authors who prepared their valuable contributions in a very short time, the workshop participants who made the workshop intellectually exciting, the chairpersons who worked hard to obtain good results, and the organizers who made the workshop extremely comfortable. The work of the program committee members is much appreciated. They helped us to organize a high quality workshop. Centre for Mathematics and Computer Science, Amsterdam, The Netherlands (Centrum voor Wiskunde en Informatica, CWI), the official organizer of the workshop, allowed us to use its modern computer equipment to edit this book, which made it possible for us to publish this book quickly after the workshop. The workshop itself was partially funded by NFI (Nationale Faciliteit Informatica), a Dutch organization for promoting research in computer science, and CWI. The members of the *Bart Veth IIICAD* group of the Department of Interactive Systems, CWI, helped us with everything. The new idea to have biographies of the workshop participants in this book was realized by Elizabeth Both (Centre for Mathematics and Computer Science, The Netherlands) who was the workshop secretary. Last but not the least, many thanks go to Varol Akman who worked as the *third* editor. Without his help the book should not have been born.

Amsterdam, June 1987

Paul J.W. ten Hagen
Tetsuo Tomiyama

X

Table of Contents

List of Authors

R. H. Allen
Department of Mechnical Engineering
University of Houston
Houston, Texas, USA

Farhad Arbab
Computer Science Department
University of Southern California
SAL 200, MC 0782
Los Angeles, CA 90089, USA

E. E. Berkhout
Department of Architecture
Technical University of Delft
OPM-groep/kmr 13.00B, Berlageweg 1
NL-2628 CR Delft, The Netherlands

Peter Bernus
Center for Mathematics and
Computer Science
Kruislaan 413
NL-1098 SJ Amsterdam, The Netherlands

Aart Bijl
Department of Architecture
University of Edinburgh
20 Chambers Street
Edinburgh EH1, 1JZ, United Kingdom

John H. Boose
Knowledge Systems Laboratory
Boeing Advanced Technology Center
Boeing Computer Services
P. O. Box 24346
Seattle, WA 98124, USA

Jeffrey M. Bradshaw
Knowledge Systems Laboratory
Boeing Advanced Technology Center
Boeing Computer Services
P. O. Box 24346
Seattle, WA 98124, USA

W. H. Chieng
Laboratory for Intelligent Design
Department of Mechanical Engineering
Columbia University
New York, NY 10027, USA

Bertrand T. David
Ecole Central de Lyon, Département
Mathématiques-Informatique-Systèmes
B. P. 163
F-69131 Ecully Cédex, France

Joze Duhovnik
University Edvarda Kardelja Ljubljana
Faculty for Mechanical Engineering
Murnikova 2
61000 Ljubljana, Yugoslavia

N. P. Filer
University of Manchester
Department of Computer Science
Oxford Road
Manchester M13 9PL, United Kingdom

M. P. Fourman
Department of Electrical Engineering
Brunel University, Cleveland Road
Uxbridge, Middx UB8 3PH
United Kingdom

Dominique R. Genin
Manager, AI Research Center
Tektronix
Imec, Kapeldreef 75
B-3030 Leuven, Belgium

David A. Hoeltzel
Laboratory for Intelligent Design
Department of Mechanical Engineering
Columbia University
New York, NY 10027, USA

Hilary J. Kahn
Department of Computer Science
University of Manchester
Oxford Road
Manchester M13 9PL, United Kingdom

J.F. Koegel
Department of Mathematics and
Computer Science
University of Denver
University Park
Denver, CO 80208-0189, USA

B. Laczik
Department of Mechanical Engineering
Technical University
H-1502 Budapest, Hungary

Z. Létray
Computer and Automation Institute
Hungarian Academy of Sciences
P. O. Box 63, XI. Kende u. 13–17
H-1111 Budapest, Hungary

Robert Milne
Intelligent Applications Ltd.
Kirkton Business Centre
Kirklane, Livingston Village
West Lothian EH54 7AY
United Kingdom

Mihai Nadin
Art and Design Technology
The Ohio State University
Cranston Center, 1501 Neil Avenue
Columbus, OH 43201, USA

M. Novak
Art and Design Technology
The Ohio State University
Cranston Center, 1501 Neil Avenue
Columbus, OH 43201, USA

Zsofia Ruttkay
Computer and Automation Institute
Hungarian Academy of Sciences
P. O. Box 63, XI. Kende u. 13–17
H-1111 Budapest, Hungary

Geir Sunde
Center for Industrial Research
P. O. Box 350, Blindern
N-0314 Oslo 3, Norway

Tapio Takala
Helsinki University of Technology
Laboratory of Information
Processing Science
Otakaari 1A
SF-02150 Espoo, Finland

Bart Veth
Interactive Systems
Centre for Mathematics and
Computer Science
Kruislaan 413
NL-1098 SJ Amsterdam, The Netherlands

R.M. Zimmer
Department of Electrical Engineering
Brunel University, Cleveland Road
Uxbridge, Middx UB8 3PH
United Kingdom

Session 1

1. Strategies for CAD

A. Bijl

EdCAAD, University of Edinburgh, Department of Architecture
20 Chambers Street, Edinburgh EH1 1JZ, UK

Abstract: *Assumptions for CAD are discussed, pointing to a distinction between human knowledge and machine representations of knowledge. Implications for future useful CAD systems are considered. A strategy for "mechanistic" symbol processors is presented, employing "mechanisms" of formal logic to manipulate written and drawn expressions of designers' knowledge.*

Keywords: knowledge, notions, representations, formalisms, symbols, drawings, design.

1. INTRODUCTION

We have now experienced two decades of CAD. In that period, the promise of CAD has met with obstacles presented by: firstly, the essentially idiosyncratic nature of design practices, particularly in loosely constrained fields; secondly, the prescriptive nature of conventional computer technology; and, more recently, the assumptions underlying machine intelligence. It is time to review our fundamental assumptions for CAD.

In this paper we will discuss a philosophical point that refers to a basic assumption employed by many AI researchers, and it arises from experience of CAD. I will refer to the distinction between knowledge within people, and its overt representations in the form of symbolic and other expressions outside people. The idea that human knowledge and formal representations of knowledge are not the same should be a fruitful focus for fundamental research contributions from CAD.

Next, we will consider how this starting position impinges on possibilities for future useful CAD systems. It may turn out that we will have to rephrase the intention of AI by characterising future systems as: "intelligent use of dumb machines."

Lastly, a speculative new formalism will be presented, explained by a worked example. This example is not intended to reveal the beginnings of yet another new and better CAD application. Instead, it is intended to illustrate the kind of thinking that needs to be targeted at CAD research in order to reach a better understanding of how designers (and other people) can use computers.

2. REPRESENTATIONS ARE NOT KNOWLEDGE

Figure 1.1 depicts a person as a bounded human being. We attribute properties to these beings, which we indicate by such words as *intelligence* and *knowledge*. We think that these properties are acquired through some combination of physiological "mechanisms" and world experience within people — without explaining how or what, we accept that knowledge does exist within people.

The same figure also shows objects that are made by people, that exhibit some of people's inner knowledge, or intuition. Such objects include text and drawings executed in some medium, on paper or a computer display screen. We read such objects as symbols and symbolic constructs, in the case of text, or more as analogic depictions in the case of drawings. Analogic depictions here refer to objects that exhibit properties which are the same as corresponding properties of other objects which they depict, such as shape properties. Interpretations of symbols and analogues within ourselves then tell us something about the knowledge which we exhibit to each other.

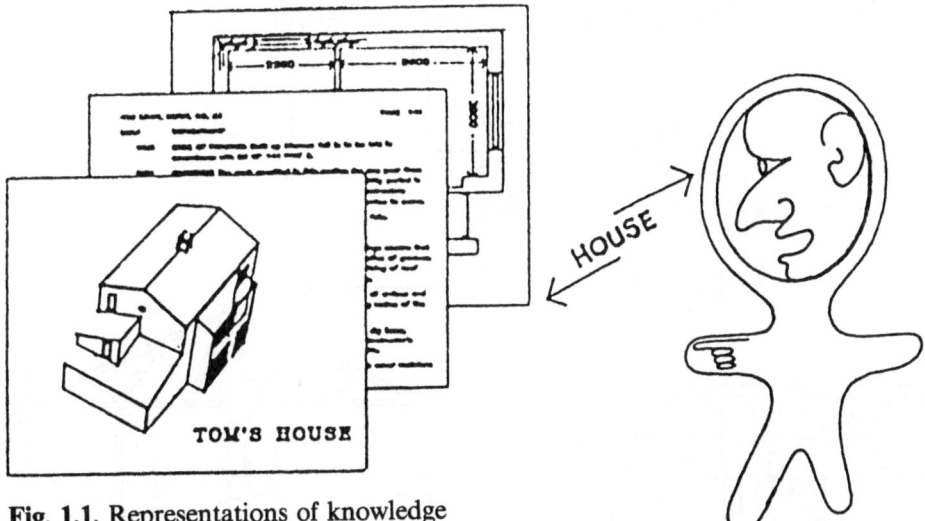

Fig. 1.1. Representations of knowledge

The key point that now needs to be recognised is that all overt expressions are conditioned by formal systems that are employed in their execution. Their formalisms constitute bounded worlds that limit the scope of possibilities for expressive objects. We tend to lose sight of such limitations when humans exercise interpretive ability. We forget we are dealing with objects and think only in terms of interpretations, and somehow we adjust to differences between interpretations within each of us.

We then proceed to do something rather odd. There are people who believe that our interpretations, our knowledge, can somehow exist separately from us within these expressive objects. The oddity of this position becomes apparent when we consider the possibility of symbolic objects being animate and exhibiting design knowledge that you or I might variously know, as is the ambition of AI. Usually researchers depersonalise this position by targeting computers at other people's intelligence.

For certain purposes, this belief in objective knowledge seems to be practical. It is practical when we can satisfy ourselves that overt knowledge includes all that is necessary to produce useful effects. Overt knowledge, as formalised thought, applies when we believe that individual thought from people is no longer required.

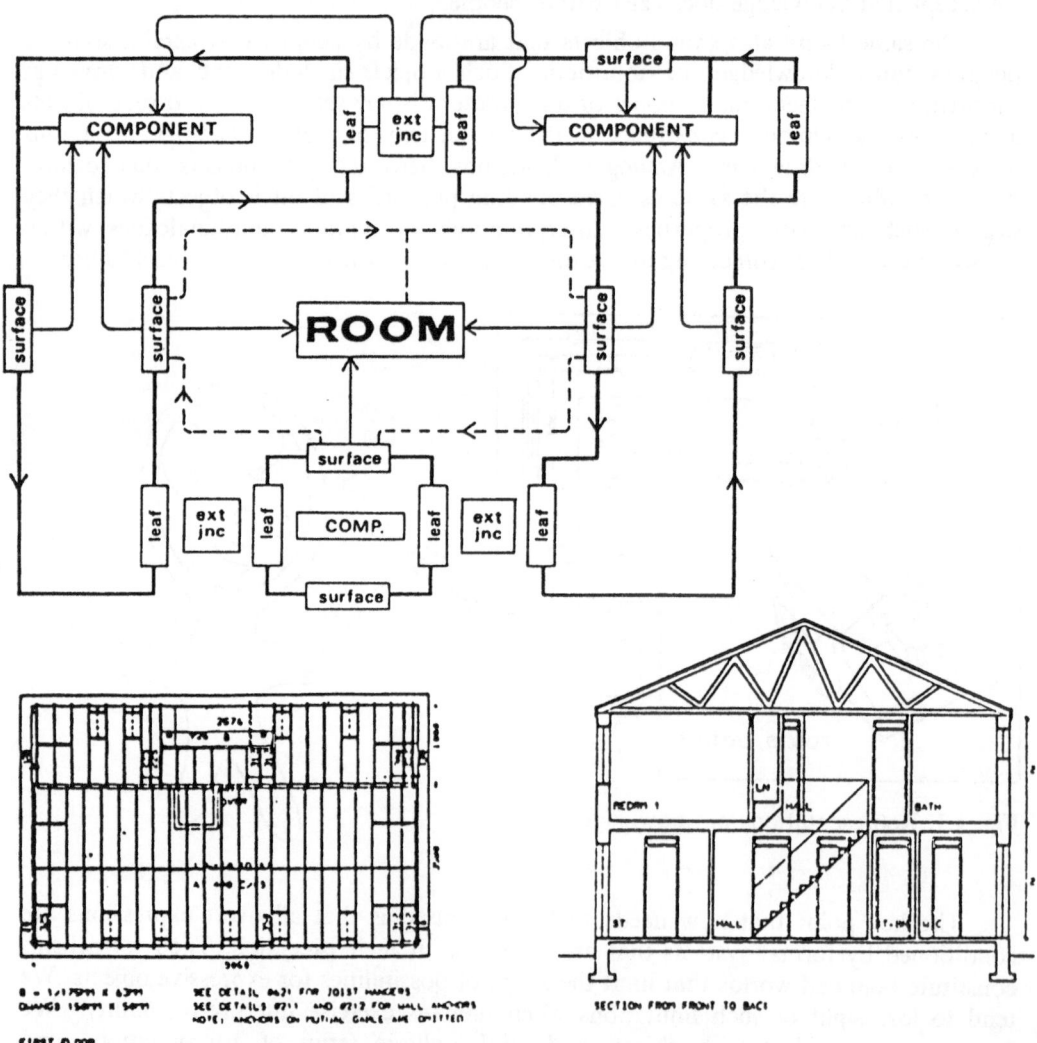

Fig. 1.2. Example of housing & site layout design

3. EXPERIENCE OF INTEGRATED CAD

Figure 1.2 illustrates a large integrated design system that was developed by EdCAAD in the late 60s and early 70s, and was used in practice for over 10 years (Bijl 1979). The system modelled the work-practices of one design office within an organisation that designed, built and managed houses all over Scotland. It represented a particular view of buildings in great detail, and it included abilities to *depict* objects (components and junctions) and to *interpret* drawings (in terms of materials, heating, light and sound), *conditioned* by orthogonality of component arrangements.

The ambition for this CAD system was that designers should be able to draw plans, and that the system should associate other information with the drawings in order to generate descriptions of parts and assemblies, of buildings. The system's representation of parts had to conform to designers' perceptions of buildings. Thus, in the example of materials unit quantities, component surface finishes had to be differentiated between large clear areas and narrow widths (as around doors), and had to include changes in specifications across surfaces to account for different activities to either side of components (to deal with noise transmission). There could not be a one-to-one correspondence between component drawings and other information associated with parts of drawings, and the system was responsible for the necessary further decompositions to generate different views of buildings.

This system focused on junction information between construction components, between components and room spaces, between pairs of houses and between houses and natural ground forms. Junction information was also targeted at roads, footpaths, parking areas and garages, and gardens and communal landscaped areas. Junctions were viewed as further components resulting from the coming together of other things, such as wall and window components, and they propagated changes to the descriptions of these other things. The system did not employ libraries of pre-defined components, but generated descriptions of component instances in the contexts of individual designs.

All information from drawings was targeted at evaluations of design performance, and at detailed quantification of construction materials required to execute designs. These targets were defined by users, conforming to users' own practices for designing buildings.

4. PAST REPRESENTATIONS

Figure 1.3 illustrates the kind of data structuring strategy that was employed for the CAD system, developed in Fortran IV during the late 60s. This strategy referred to locations in memory, resulting in the familiar records and pointers structure. The arrangement of component (and room) types is shown in a master list, with ranges of materials specifications for each type, arranged in three layers; primary, external cladding, and internal cladding. The sort of information which this represents is depicted below, for a typical portion of wall.

This kind of data structuring has rightly been criticised for being prescriptive, being too dependent on the system developer's correct anticipations of users' design practices, and then compelling users to conform to anticipation. This criticism was

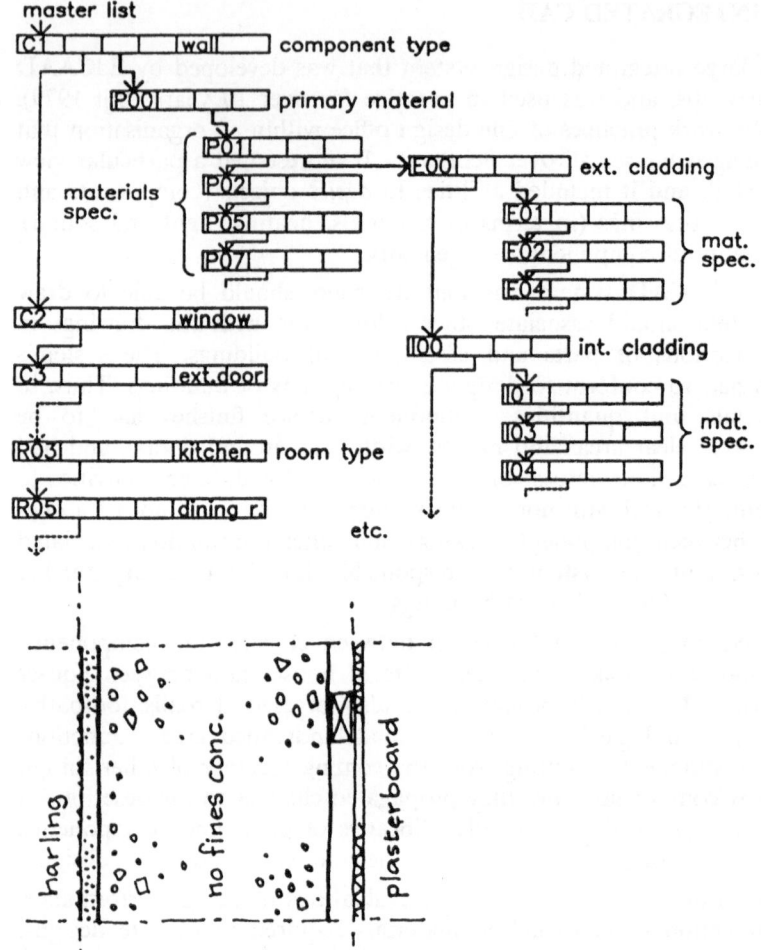

Fig. 1.3. Records and pointers data structure

developed in a published report by EdCAAD for the UK Department of the Environment (Bijl, Stone, and Rosenthal 1979). In the example of wall materials linked to junction information, the three-layered nature of specifications became encapsulated in program code, so that other perceptions of wall materials could no longer be accommodated. Such limitations were attributed to the obscure nature of program code, dealing with precise instructions to machine procedures and hiding the logic of applications as perceived by users. For programs of any size and complexity, it became impractical for subsequent programmers or users to modify a system's representation of a user's world.

In normal practice, designers have to respond to unforeseen demands from other people, as people's perceptions of houses and housing needs change. Design practices have to evolve in response to changing demands. Evolution here refers to changes that occur in imperceptible steps, not directed by prior goals, that become apparent over periods of time — a normal and on-going characteristic of human knowledge. Such evolution could not be accommodated in the system and, eventually, the user-organisation had to decide to scrap the system — a painful decision, after users' practices had become dependent on information held within the system. It is worrying that this decision can take 10 years, during which time the system increasingly inhibits the necessary evolution of design practices.

The question that now needs to be considered is: Can newer software techniques, including those offered by AI, avoid the criticism of prescriptive technology? We have replaced pointers to memory locations with various procedures for indexing by name, and with use of structured terms that refer to symbolic logic super-imposed on machine procedures. These developments can have the effect of masking the earlier prescriptive conditions from users, but they do not remove those conditions. We still do not know how to make formal systems evolve in parallel with human knowledge.

5. WHAT WE CAN DO NOW

Figure 1.4 offers a reassessment of the role of symbolic expressions related to human knowledge and machine processes. Knowledge refers to human notions. Here, notions are thought of as something extending beyond symbols, as part of human being, without implying anything about the actual substance of notions. Notions can be considered to exist as interpretations between symbols and anything else that might occur in the human mind, including states of mind that may be shaped by the totality of human experience (including sight, sound, touch, taste and smell). Domain knowledge then refers to specialised knowledge that people attribute to other people.

Expressions occur as externalised physical manifestations of people's knowledge, as representations of human knowledge. In the normal case of symbolic expressions, prior to computers and in the familiar forms of words and mathematical notations, expressive objects are used to evoke notions, to prod thoughts within people. We should not expect symbolic constructs to possess notions about themselves. This implies, for example, that mathematical notions exist *not* in text books but in the minds of mathematicians and other people who write and read the books. We can make two further general observations. Human validation of knowledge, in the sense of knowledge receiving variously accepted meanings with reference to human notions, cannot be contained wholly within overt expressions. As a consequence of this dependence on interpretations within people, symbolic representations of knowledge do not need to be complete. For example, we do not need an overt description of the syntax and semantics of natural language before we can know how to use it.

Worlds of symbols include relationships and functionalities that can be applied to symbol types, constituted as formalisms for producing compositions and supporting analyses of instances that might represent human knowledge. Processing of symbols then employs the functionality of symbolic logic, with mathematics as a highly developed example. Such processing is commonly executed by people who are literate in the symbolic language. When this processing is executed autonomously within machines, we have computers.

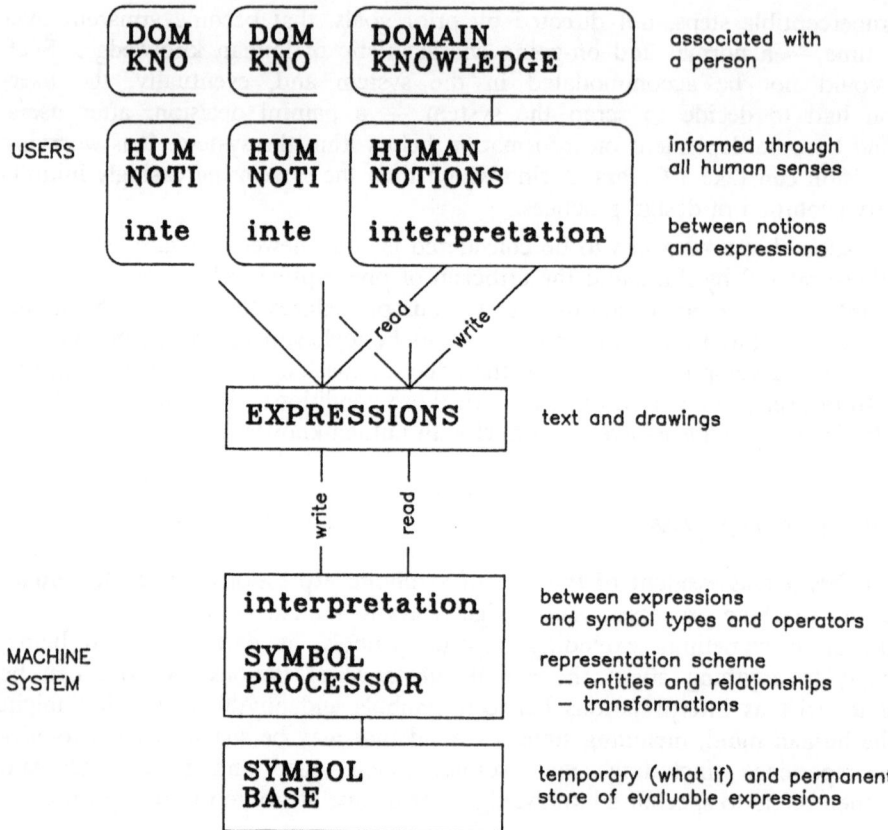

Fig. 1.4. Symbol processors

As a general strategy for computers, we should try to differentiate between knowledge required to process symbols, and people's domain knowledge that might be represented by instances of symbolic expressions. This is not an easy distinction, and it is confused by the use of highly evocative words to indicate abilities of computers, such as *artificial intelligence* and *knowledge engineering*. To redress the balance, we should think of computers more as mechanical devices for operating on symbolic objects. The position that I am advocating appears to be in tune with criticisms of AI by Dreyfus (1979) and Searle (1984) but, in our case, we are not seeking to resolve the debate in AI. Instead we are looking for advances that might be appropriate to CAD. Such advances should employ formal logic to provide the "cogs and levers" for manipulating symbols, as an extension of the kind of functionality provided by "dumb" word processors. What is being suggested here is that we should maintain a clear distinction between the internal structure or syntax of a system and a user's semantics for expressions that are processed by the system. The system's semantics should be limited to interpretations of the user's demands for edits to its current syntactical

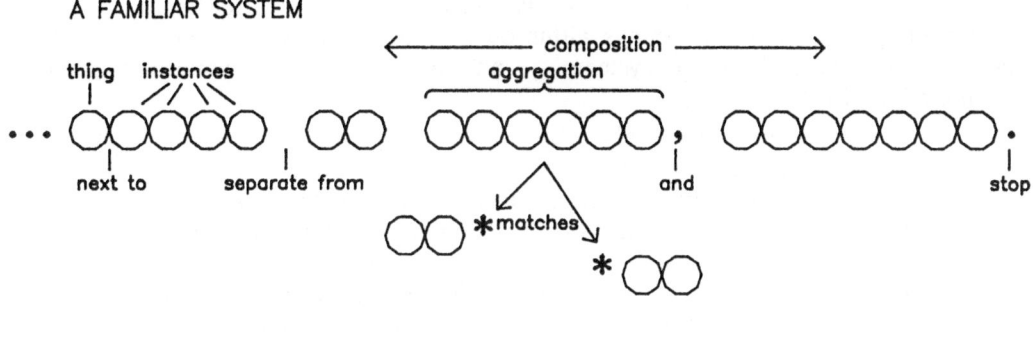

A FAMILIAR SYSTEM

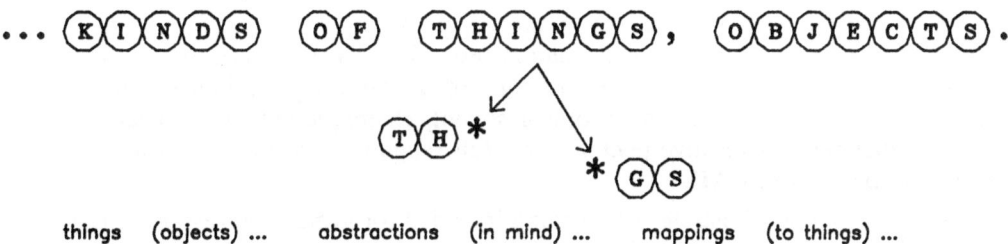

A WRITING MACHINE

things (objects) ... abstractions (in mind) ... mappings (to things) ...

Fig. 1.5. Formalism of a writing machine

representation of the user's expressions. Users, knowing the syntax of the system, might then be able to use it to construct and manipulate expressions of their own design knowledge.

6. WE ALL USE SYSTEMS

The upper part of Fig. 1.5 illustrates a familiar formalism that each of us uses every day. This formalism consists of a symbol type, and aggregations of instances that each have a single uni-directional "next to" relationship. Aggregations are related by empty spaces and special symbols that have meanings like *and* and *stop*, to form compositions. Further relationships can be identified by matching similar instances of symbols.

The lower part of this figure reveals this formalism to be the basis for a "writing machine", or word processor. This is *not* a natural language system; it deals only with those parts-of-speech that occur as parts-of-writing. The basic symbol type is a character and aggregations of instances form written words that approximate to spoken words. Special symbols occur as punctuations only in writing and are not spoken, though they might be implied by intonations or pauses in speech. The ability to translate between spoken utterances and written expressions, plus the ability to make further interpretations into meanings within other forms of knowledge representation, is the goal of natural language systems — an ambitious goal, beyond our present consideration of CAD.

A writing machine deals only with written objects. Thus, in the illustration, THINGS that appears as a character string is very different to the thing that is a character. We can use the writing machine to manipulate characters to produce words to signify different things that are not words. Examples of writing machines are pen and paper, mechanical typewriters and electronic word processors. This spread indicates a progression from artefacts that embody very little knowledge about writing objects, to artefacts that represent quite a lot of knowledge about writing objects. In all cases, they stop short of knowing what is being signified by any instance of writing. This lack of interpretive responsibility, and their obvious dependence on humans for any interpretation of expressions passing between people, makes these machines widely useful to many people.

The virtue of writing machines is that they share a common formalism which consists of few symbol types and relationships. We have learned to employ this formalism to construct partial representations of a vast range of human knowledge, from scientific methodologies and problem solving techniques to the rich ambiguities and contradictions of evocative poetry. This full range is important to all interactions between people, and to CAD.

Writing machines illustrate the meaning intended for symbol processors in Fig. 1.4. The question that now presents itself is: Can we envisage a development of writing machines that will handle more varied and multi-directional relationships between words, and can we link them to drawing machines? Can such a development preserve the generality of these machines and what new kind of literacy will they require from people?

7. AN EMERGING FORMALISM

Figure 1.6 shows the structure of expressions supported by the MOLE logic modelling system (Bijl 1988; Krishnamurti 1987) — a tentative and modest extension of a writing machine which supports multi-directional relationships between symbols. The MOLE system forms part of a theoretical exploration of fundamental CAD strategies which is being undertaken at EdCAAD and is supported by the UK Science and Engineering Research Council.

The concept that is being developed here sees words as symbols that stand for things and parts of things in the minds of people, and sees other symbols as standing for relationships between words, which can be used to model descriptions of things. These relationships have general definitions for users and they have interpretations down into machine procedures, giving the system its ability to manipulate descriptions. The form of expressions generated by the system and its users is the same, with special (non-verbal) symbols alerting users to system functions. All expressions generated by users can be read as being declarative, telling the system what it has to do, and expecting the system to produce intended results.

ENTITIES & RELATIONSHIPS

HIERARCHIES

Kind   slot   Filler

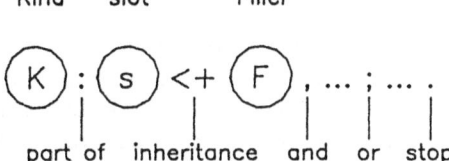

part of   inheritance and   or   stop

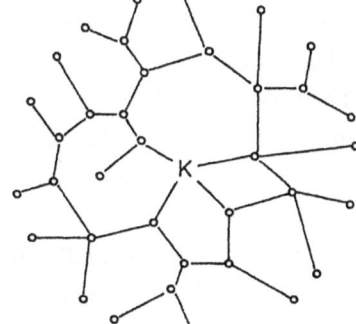

LINKING PARTS

INTERLINKED HIERARCHIES

thing   its part   part instance

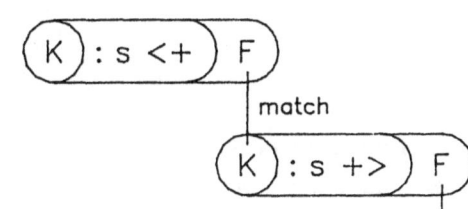

match

Fig. 1.6. Parts hierarchy

Expressions consist of three basic symbol types: kinds (K, for any kind of thing); slots (s, denoting a part of the thing); fillers (F, referring to any other thing that instances a part). This representation has an affinity with frame systems (Minsky 1975), but without any definition of frames as distinct from arbitrary collections of slots; and it also has an affinity with semantic networks (Woods 1975), but without the system reading significance in names of arcs. Any instances of MOLE's symbol types can be indicated by any user-declared words, and the system's functionality is applied only to instances of types, without reading significance in the words that instance them. The further special symbols denote the relationship of a slot as being a *part of* a kind, and a slot's filler as an *inheritance* relationship to something else. Punctuation of composite expressions occurs in a similar manner to normal writing.

A kind name (in upper case characters) refers to something that a user has in mind. A slot label (in lower case) attached to a kind denotes a part of that kind. A filler (may include references to further kinds and slots, as path names) attached to a slot denotes something else that instances a part, which may be matched to another kind name. Matching has the effect of linking expressions, to form virtual hierarchies of parts into composite descriptions of things. Through inheritance relationships, descriptions can include views of parts of other things, resulting in interconnected parts hierarchies.

The central idea here is that we should be able to devise a machine for processing symbols so that expressions can be read by people as being meaningful, and so that expressions make the functionality of the machine visible to users. Words that form parts of expressions remain the responsibility of users, and relationships between words are maintained by the system. We should then be able to make the system do things with words and, if we can satisfy ourselves about mappings of words to other things (such as parts of drawings), we should then expect the system to perform useful tasks on other things.

8. SOME SYMBOLS KNOWN TO THE MOLE SYSTEM

< + slot filled by a new *instance* of F.

+ > slot filled by F.

> > slot filled by *view* of F.

:   whatever follows instances a *part-of* K.

::   expands to a *path* through a sequence of part-of instances of K.

?   denotes queries and identifies *search space* for candidate answers.

{}   denotes queries and sets *conditions* on candidate answers.

| exclusive *not* conditioning answers.

< = *description* of K gets *replaced* by description of F.

This is a fairly comprehensive sample of system-defined symbols included in the system as presently developed. The first three referring to inheritance relationships are the most critical. Each qualifies how a kind sees a part through its slot, where a part is likely to be some instance of a previously declared kind, and each also qualifies how a change to a part may be declared from a kind.

Instance inheritance, < +, makes the part a *new instance* of its former self, which continues to inherit its previous description, but subsequent changes to sub-parts have the effect of masking corresponding inherited sub-parts. Reverse inheritance, + >, makes the part the *same as* the inherited instance (of some other kind), and subsequent changes to sub-parts will be seen by all kinds that are partially described by the same instance. Indirection, > >, serves as a *read-only* instruction, to view an instance of a part of another kind.

The penultimate three symbols are used to express queries to the system, to find out what is already stored in the knowledge base and incorporate answers as parts of further assertions declared by users. Question marks at the end of path names are used to control the extent of search space for answers, and curly brackets contain conditions that have to be matched by answers.

9. WHAT MOLE KNOWS OF PARTS OF DRAWINGS

Expressions can include words which map onto drawing parts, and relationships that represent the general structure of line drawings. They refer to drawing parts as: construction lines with angle values, construction points at the intersections of construction lines, drawing line segments as portions of construction lines delimited by construction points, shapes as chains of segments (open or closed), and compositions as connected shapes (under varying conditions of attachment).

MOLE's symbolic representation of drawings includes no coordinate point values, yet this representation is sufficient to regenerate instances of drawings — a logical representation of the general structure of drawings, attachments and transformations, is being developed as a PhD by Szalapaj (Szalapaj and Bijl 1984). The chosen drawing primitive is a construction line, more in keeping with conventional uncomputerised drawing practice. Coordinate values might be a property of a drawing space (gridded paper or a computer display), in which case they need to come into play only to locate a drawing on a particular drawing space.

A separate drawing machine, a computer (or a person), then passes symbolic representations of instances of drawn objects to MOLE, and these can reflect edits and transformations applied to drawings. Representations of drawings can be linked to further representations of other things that they depict, so that drawings can form parts of descriptions of other things. Similarly, changes made to representations in MOLE can have the effect of driving the drawing machine, to result in changed drawings.

10. RETURNING TO JUNCTIONS

To illustrate how MOLE expressions can be used to make and change descriptions, we can return to earlier experience (Fig. 1.2 and 1.3) and look again at the example of walls and junctions between walls. We might want to remove the condition of orthogonality, as shown in Fig. 1.7, below.

11. DESCRIBING JUNCTIONS

```
JUNC_END:
  [join_pt + > fix_wall:join_pt,
   opp_pt:
     [conline1 + > join_wall:face?:|{conpt?{origin}}:bearer,
      conline2 + > fix_wall:face?:|{conpt?{join_pt}}:bearer].
  junction:
    [dwg:
       [segment:
          [conpt1 + > join_pt,
           conpt2 + > opp_pt,
              style < + DOTTED]]],
  change:
    [join_end + > join_wall:end?:{conpt?{origin}}:
```

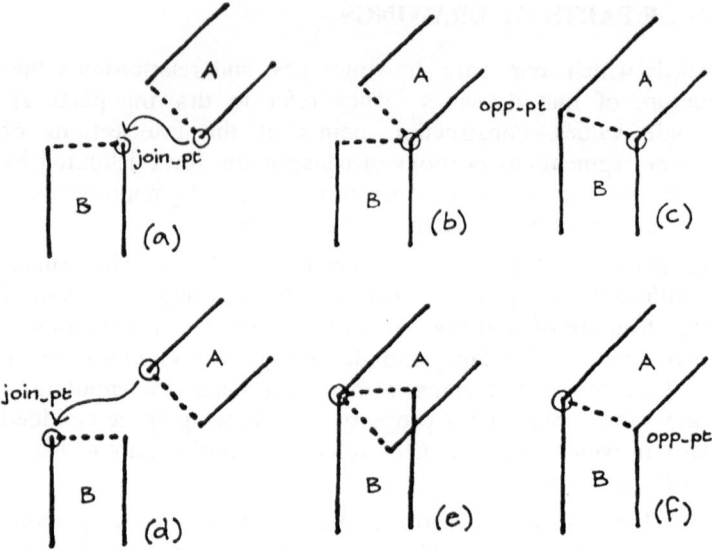

Fig. 1.7. Problem of junctions

[move_pt +> join_end:conpt?|{origin}:
  [move_to +> move_pt<=opp_pt]],
fix_end +> fix_wall:end?:{conpt?{join_pt}}:
  [move_pt +> fix_end:conpt?|{join_pt}:
    [move_to +> move_pt<=opp_pt]]]].

This is a composite MOLE expression which describes the general case of wall end-on junctions. It expects that walls and their drawings, and the fact that two walls are going to take part in a junction, have been or will be described. We then declare a junction to be a kind of thing, giving it any name: JUNC_END. The junction is described by any number of parts, indicated by slot labels. These parts are a *join point* an *opposite point,* a *junction* part, and a *change* part. Each of these parts inherit descriptions of other kinds indicated by the slot fillers.

Thus the *join point* has a reverse inheritance relationship with the kind that fills the *join point* part of the kind that fills the *fixed wall* part of something that is not as yet identified in this description. The filler of this junction description's *join point* is, in effect, a path name to some other kind. The reverse inheritance means that this description can effect changes to the other kind.

The *opposite point* part is filled by an un-named kind that is further described by two *construction line* slots that are filled by kinds that are identified by the following conditions. The first *construction line* is filled by the same kind that fills the *bearer* part of any *face* of a *joining wall,* provided that it does not include within its description any *construction point* that is the same as the *origin* part of the *joining wall.* Notice, once more, that the *joining wall* is a slot label indicating a part of something that is not as yet identified in this description. The second *construction line* is treated in similar fashion.

14

We have now described the *join point* and *opposite point* of any junction as shown in Fig. 1.7.

The *junction* component part identifies the mitre at the junction as a component that is separate to the two wall components — this might not be necessary, but follows the earlier precedent. One might want to attach further information to the *junction*, independently of the walls. The description of the *junction* component says that it has a *drawing* part which has a *segment* part which is described by two *construction points* (the same as the *join point* and *opposite point* described earlier) and its line *style* is dotted.

The *change* part propagates changes to the descriptions of the adjoining walls, to complete the junction. The *change* part has a *joining end* and a *fixed end*. The *joining end* part is filled by the same kind that fills any *end* of the *joining wall*, that includes within its description any construction point that is the same as the *origin* part of the *joining wall*. The *change* part's *joining end's* inherited description then receives a *move point* part which is filled by the same kind that fills any *construction point* of the *joining end* part, provided that the *construction point* is not the same as the *origin* part of the *joining end*. The *move point* part then receives a *move to* part which is filled by the same kind as fills the *move point* part, and the description of that kind gets replaced by the description of the *opposite point* part (declared earlier) using the $<=$ change operator. The *change* part's *fixed end* part gets treated in a similar fashion.

These paragraphs have given a conversational interpretation of the composite MOLE expression describing end-on junctions, using far more words. All these conversational words are needed to express this knowledge about junctions, if we want to do so conversationally. If we then want a machine to represent this knowledge, we have to re-express it in terms of system-defined relationships. Notice that the words included in MOLE expressions can refer to anything that a user might have in mind, to objects, events or tasks. None of the words included in these expressions have any meaning to the system, other than to signify different instances of kinds and slots. Yet, by using system-defined relationships to link words, the system can be made to exhibit behaviour as though it knows what it is doing, by effecting changes to drawings of walls in order to form junctions.

MOLE does things by evaluating descriptions, by using a left-to-right, depth-first search procedure (working on virtual tree structures implicit in composite expressions, as illustrated in Fig. 1.6). As indicated at the beginning of this explanation, this description of junctions includes references to parts that are not yet identified within the description. Something more has to happen before the system can produce instances of junctions.

12. INSTANCES OF JUNCTIONS

JUNC_END AB:
    [join_wall +> WALL A::dwg,
     fix_wall +> WALL B̄::dwg,
     join <+ JUNC_END].

If we want to join two particular walls, say walls A and B, then we have to declare a kind name for this instance, any name: JUNC_END_AB. We can then say that an instance of junction has three parts. These are a *joining wall* part, a *fixed wall* part and a *join* part. The kind that fills the *joining wall* part is declared to be the same as the kind that fills the *drawing* part of wall A. Similarly, the *fixed wall* part is filled by the kind that fills the *drawing* part of wall B. The *join* part then is filled by an instance of the JUNC_END kind, inheriting all the description of that kind. The *fixed wall* and *joining wall* parts referred to in the earlier general description of junctions can now be found as kinds included in the description of this instance of junction.

With this description of an instance of junction, the system is able to evaluate the description and effect changes to the drawings of walls, to produce the completed junction. This approach to describing junctions can then apply to all cases of end-on junctions at any angle and for walls of unequal thickness (with the exception of a straight end-on junction between walls of different thickness) as shown in Fig. 1.8 below.

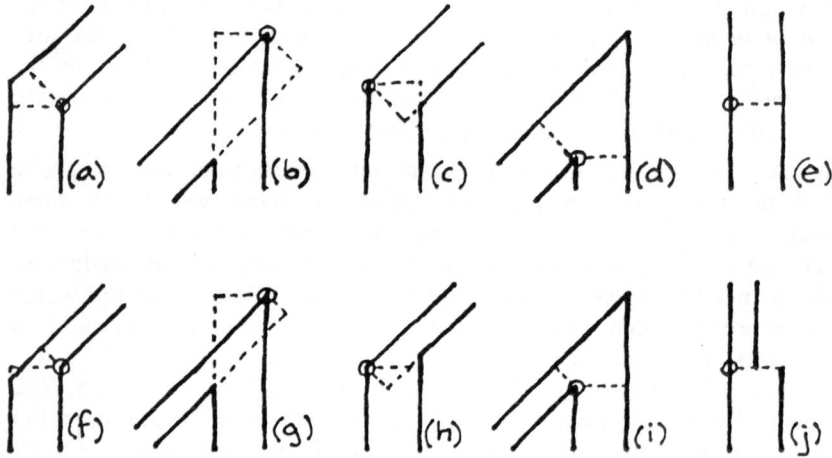

Fig. 1.8. Junction configurations

What has been shown by this example is the use of MOLE to model a perception of parts of buildings (not necessarily a correct perception). This has been done by using words that map onto those parts, plus words that map onto drawings of those parts, and by using system-defined symbols to declare relationships between parts and effect changes to descriptions. More can be added to describe construction materials linked to drawings of parts, or to express anything else about other aspects of a design. All this can be done provided that a user stays within the bounds of formalism of the system (not easy), and provided the user retains responsibility for results.

The example of wall junctions has been chosen because, in previous experience of CAD, it presented severe problems. Previously, we were constrained to orthogonal arrangements of walls, in plans of buildings. This was especially true in those cases where a system was expected to interpret drawings in order to evaluate the effects of junctions on construction materials, and on other aspects of designs for buildings. Now this case of junctions is no longer particularly difficult. The point of the example is that it illustrates a use of a system which does not rely on domain knowledge being held in some prior and separate way within the system.

13. CONCLUSIONS

Figure 1.9 returns to the question of intelligence, knowledge and symbols. When we think of understanding symbols we should draw a distinction between symbols understanding us, and our understanding of symbols. The former suggests fascinating and deep problems, and poses valid goals for AI researchers, but outcomes may never be acceptable to other people who are invited to use computers. The latter appears more in keeping with our established tradition of human knowledge, predating computers, and is likely to prove more acceptable to people.

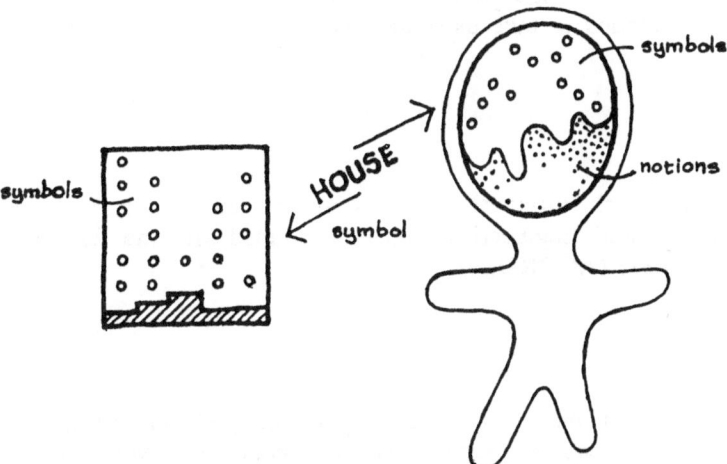

Fig. 1.9. Understanding symbols

By differentiating between human knowledge, symbolic expressions that emanate from such knowledge, and logic systems that support manipulations of expressions, we can then conceive that human knowledge exists only within human beings — not in expressions, nor in logic systems. This assumption admits that knowledge can be represented symbolically, but that a representation is not the same as knowledge within people. A representation can be validated only by people. A representation may come to be accepted by many people, thus acquiring the status of conventional knowledge, but even then it is not the same as human knowledge. It remains subject to validation by people.

The effects of this assumption are not very different to those of the more orthodox assumption that knowledge has an objective existence. It does, however, focus responsibility for knowledge on people. More importantly for CAD, it provides scope for an understanding of symbolic environments and logic systems as devices that are defined in terms of their use by people.

We can then view symbolic expressions as objects that get interpreted into abstract symbols in the minds of people, which in turn get interpreted into human knowledge. Such further interpretations can be regarded as human notions, beyond symbols, that refer to the totality of human experience. Logic systems can be regarded as mechanisms for instancing and manipulating symbolic constructs — without responsibility for correctness, and untroubled by ambiguity. Thus we can expect to create a world of symbols, which itself knows nothing, but which can be used by people to represent their knowledge. We should recognise that symbolic expressions, as in the usual form of words, emanate from the notions of people, and are used to evoke other notions within people.

The goal for CAD, then, is to achieve — within computers — a symbolic logic that will be accessible to users, allow them to formulate *their* expressions of design knowledge, and enable them to call on *their* machine representations of that knowledge to manipulate expressions towards design proposals. An attempt to develop a logic modelling system, which is intended to take us some way towards this goal, has been outlined in this paper.

ACKNOWLEDGMENTS

Research at EdCAAD, which forms the background to this paper, is being supported by the UK Science and Engineering Research Council and the European ESPRIT Programme, project P393 (ACORD).

REFERENCES

Bijl, A. (1979): "Computer aided houses and site layout design," *Proceedings of PARC79 (Planning Architecture & Computer): International Conference on Application of Computers in Architecture, Building Design and Urban Planning,* Berlin, West Germany, pp. 283-292.

Bijl, A., Stone, D., and Rosenthal, D. S. H. (1979): "Integrated CAAD Systems," EdCAAD Report for the Department of the Environment , Edinburgh University, Edinburgh, UK.

Bijl, A. (in preparation, 1988): *Architecture in Mind, Computer Discipline and Design Practice,* Wiley, London.

Dreyfus, H. L. (1979): *What Computers Can't Do — The Limits of Artificial Intelligence (revised edition),* Harper Colophon Books, New York.

Krishnamurti, R. (1987): "Representing Design Knowledge," *(submitted to) Environment and Planning B,* Planning & Design, UK.

Minsky, M. (1975): *The Psychology of Computer Vision,* McGraw-Hill, New York.

Searle, J. (1984): *Minds, Brains and Science (1984 Reith Lectures)*, BBC Publication, London.

Szalapaj, P. J. and Bijl, A. (1984): "Knowing where to draw the line," in *Knowledge Engineering in Computer-Aided Design, Proceedings of the IFIP WG 5.2 Working Conference 1984 (Budapest)*, Gero, J. S. and Gero, J. S. (eds.), North-Holland, Amsterdam, pp. 149-169.

Woods, A. W. (1975): "What's in a link: Foundations for semantic networks," in *Representations and Understanding*, Bobrow, D. G. and Collins, A. M. (eds.), Academic Press, New York, pp. 35-82.

2. A Paradigm for Intelligent CAD †

F. Arbab

Computer Science Department, University of Southern California
Los Angeles, California 90089-0782, USA

Abstract: *An intelligent CAD system is a tool box of automated problem solving aids that allow designers to conceive, evolve, and document their designs. Our research involves CAD systems for mechanical applications where three-dimensional geometric shapes play a significant functional role. In this paper, we consider the design process itself as a problem solving activity. Computer systems suitable for assisting different modes of problem solving must be based on different assumptions about users' intentions and their dialog with the system. We identify three modes of problem solving based on their implications on a system's underlying paradigm. Because of the important role of geometric information in design, a prerequisite for intelligent CAD systems is more explicit representation and manipulation of geometric knowledge. We discuss a paradigm for an intelligent CAD environment specifically suitable for geometric information and its implication on knowledge representation.*

Keywords: intelligent CAD, geometric reasoning, problem solving.

1. INTRODUCTION

The heart of a conventional mechanical CAD system is its geometric modeler[1]. This is only natural. In CAD applications such as VLSI design, geometry plays at most an ancillary role in the functional properties of the design object. In design of VLSI circuits, for instance, one can almost ignore the geometry and concentrate on functionality, which subsequently can induce the actual layout. On the other hand, functionality, aesthetics, and manufacturability of mechanical parts or systems primarily depend on their geometric shape. Furthermore, very little explicit knowledge is

---

† This research is supported in part by the National Science Foundation Grant DMC-8505334.

[1] In this paper we use the term "mechanical" in a very broad sense to include systems of physical entities with loose interconnections, e.g., the art-gallery example in Section 4.

available on how functionality in general induces or even influences the geometric shape of a mechanical object.

Many contemporary CAD systems are simply (two- or three-dimensional) *drafting* systems. Valuable tools as they are, and undeniably useful in communicating the end result of a design, Computer Aided Drafting systems have very little to offer as aid in the process of design itself. Other CAD systems based on various solid modeling schemes (Boyse and Gilchrist 1982; Brown 1982; Fitzgerald, Gracer, and Wolfe 1981; Wesley et al. 1980) go farther than drafting systems in contributing to the activity of design: the integrity of the model as a solid object is guaranteed by the system, mass properties can be obtained, various views and sections can be examined, proper models for finite element analysis can be automatically derived, etc.

An intelligent CAD system is a tool box of automated problem solving aids that allow designers to conceive, evolve, and document their designs. Thus an intelligent CAD system makes a meaningful contribution to the progress of a design activity. In order for a system to make such a contribution, it must have a good understanding of the domain of discourse. In mechanical CAD, this frequently involves geometry. There is now over three decades of experience with computer graphics and geometric modeling (Mortenson 1985; Requicha 1980). But, are present geometric modelers adequate to serve as the core of intelligent CAD systems? To answer this question, we must re-examine the present CAD systems and consider how much they actually contribute to the geometric part of a design activity.

2. INTELLIGENT IN WHAT SENSE?

Our interest is to make CAD systems more intelligent in the domain of geometry. However, present CAD systems can be made more intelligent in at least two (somewhat orthogonal) other ways as well. In most CAD systems of today non-geometric technical information (e.g., material properties, functional requirement specifications, etc.) and administrative information (e.g., standards for common parts, evolution and versions of a design, related families of parts, scheduling, inventory, etc.) are either completely neglected, or at best, treated as almost-meaningless information which passes through for ultimate consumption by other programs or people. One approach to building more intelligent CAD systems is through meaningful integration of non-geometric and administrative information with geometric information into a knowledge base, incorporating real-world and manufacturing constraints into this knowledge base, and devising schemes to use this collection of knowledge to aid designers. Such systems can be called intelligent in the sense that they are more aware of the real-world context in which design problems are to be solved. Thus, they can contribute to a design process by dynamically providing a critique of the design from the organizational point of view, or indicating the more subtle manufacturing-related implications of a designer's decisions.

The issues involved in integration of the three types of information mentioned above presently are not well understood. Even less understood are problems like the interdependence of geometric and non-geometric properties of a part, formal specification of the function of a part and derivation of constraints defining its shape from such specifications, automatic verification of a design against its specification, etc. Until more is learned about how functionality affects geometry in general, the interplay of functionality (and aesthetics) with geometry will continue to be an activity internal to designers. This implies that exchanging explicit geometric information must continue to be an integral part of the interaction between designers and intelligent CAD systems of the future.

The intellectual level of (geometric) communication between a contemporary CAD system and its users is very low: they are primarily passive repositories of geometric information that keep track of the information supplied by a user and perform very specific, well-defined tasks on demand. Examples of such tasks are creating a point at the intersection of two curves, defining a line passing through two points, trimming a line up to an intersecting entity, fileting the intersection of two surfaces, performing a Boolean set operation on two solids, etc. The entities manipulated by the algorithms involved must all be fully defined; these systems generally cannot tolerate partial information. Thus, before a user can tell a CAD system about a point, line, surface, or solid, he must precisely know the attributes (e.g., coordinates, slopes, control points, lengths, etc.) that *uniquely* define that entity. It can be argued that by the time that a concept is broken down to such level of detail, and each detail is known so specifically, the real design, or at least a good deal of it is done. From this perspective, existing CAD systems, including those based on solid modeling, merely aid the *documentation* of a design process which takes place in the mind of a designer. Because design almost invariably involves cycles of revision, such an interactive documentation facility is a valuable tool. Nevertheless, revising a design itself is also an external activity to contemporary CAD systems: they merely record the outcome of a designer's revisions activity. An intelligent CAD system must be an active apprentice in the process of design, not just a passive recipient of information.

We advocate to advance the intellectual level of geometric communication between a CAD system and its users as a way of making CAD systems more intelligent. A CAD system that is more "knowledgeable" about geometry than present systems, is clearly desirable, even if it is still mostly blind to application-domain-dependent, non-geometric, and administrative information. Alternatively, one may reduce the geometric content of the communication between designers and a CAD system. For example, one can limit the scope of a system to special cases where the knowledge about how functionality affects geometry is either available explicitly, or perhaps it can be hard-wired into the system implicitly. With this knowledge at its disposal, the system can communicate with its users at the level of functionality, where fewer and more abstract geometric concepts need to be specified. Such special purpose systems may still deserve to be called intelligent in the sense that they are "experts" in their own domain of application.

3. DESIGN AND PROBLEM SOLVING

Design is at least as much a process of discovery as it is a purposeful manipulation of facts into a contemplated solution. In a design process, crystalization of facts into a coherent problem specification can initiate a more focused activity of problem solving to find its solution. The aspect of design which involves discovery can perhaps never be fully automated. However, we can build systems to aid or automate the problem solving aspect of design.

There is a difference between multiplying two matrices and finding a proof of Fermat's "Last Theorem." One can identify different types of problem solving based on different criteria, for instance, how focused is the activity, what strategy is used, the application domain of the problem, etc. We distinguish between three types of problem solving based on the characteristics of systems that can meaningfully contribute to the problem solving process: *solution instantiation*, *methodical synthesis*, and *solution by refinement*.

Although there are no clear boundaries between these three types of problem solving activity, this classification helps to identify the requirements and responsibilities of the members of a human-computer problem solving team. A computer system is a suitable team-mate if it requires no user intervention in solving the more mundane sub-problems, allowing him to concentrate on the main problem at hand. It is the definition of what is considered to be "mundane" that determines the intelligence level of the system and thus its contribution to the problem solving activity. We consider a CAD system as intelligent if it relieves a designer from the burden of inferring the consequences of his decisions, a task we regard as essential to a design process. This requires a system based on the "solution by refinement" paradigm.

Existing CAD systems are built around geometric modelers that are based on a solution instantiation, or sometimes, on a methodical synthesis paradigm. Consequently, relationships among geometric entities are generally meaningless except as used in the few specialized algorithms that are available in a system. For example, it is usually possible to define a line parallel to another previously defined line, or to inquire whether two given lines are parallel, but it is not possible to use the relationship "parallel" to define the "new" concept of a parallelogram as *a quadrilateral whose opposite sides are parallel*, or to make an inference based on the fact that parallel is a transitive relation.

3.1. Solution Instantiation

The simplest exercises at the end of a chapter of a textbook usually require no more than the ability to "plug-in" values describing a *specific instance* of a problem into a somewhat general *solution procedure* discussed in that chapter. We refer to this type of problem solving as *solution instantiation*. The apparent triviality of this type of problem solving is a bit misleading; there is more involved than meets the eye. Knowing the general solution procedure alone is not enough. How often have we observed a person fail at such a "trivial" task, simply because the problem was not within the context of the end of *the* chapter? In addition to knowing the general solution procedure, one must (1) recognize the problem at hand as an instance where the solution applies and (2) must identify the parameter values that specialize the general solution. For example, consider the problem of finding the intersection point of two given lines l_1 and l_2. Assuming that a canned algorithm exists for finding the solution of a system of simultaneous linear equations, one must first recognize the problem at hand as an instance of the problem solved by this algorithm, *know* about the existence and proper use of this algorithm, and supply the coefficients of the linear equations that describe lines l_1 and l_2.

When the domain of interest is small, the most effective approach is to build a system based on solution instantiation. Most contemporary CAD systems fall into this category of problem solving aids. A system based on solution instantiation paradigm resembles a "desk calculator" with different keys for different operations. Although these keys may be grouped into categories for ease of reference by users, there is no relationship between them. Because in this paradigm tools do not interconnect, neither flow of control nor flow of information is a system's concern. Therefore, control mechanisms and information encapsulation concepts such as types, classes, and objects are not necessary.

When a problem is not (recognized as) an instance of any of those that a system has a solution for, the system may still be used to solve it. In this case, a user must "construct a solution" by using the system to instantiate solutions to a successive sequence of sub-problems. The proper sequence of these instantiations and passing the appropriate information among them is the user's responsibility. As the number of these sub-problems increases, the complexity of this responsibility dominates the problem solving activity. Used in this capacity, a solution instantiation system is a poor substitute for a system based on methodical synthesis.

3.2. Methodical Synthesis

By methodical synthesis we refer to cases where a "canned" solution is not available, but instead a simple methodology for solving the problem is known. As a problem is being analyzed, it is decomposed into a number of smaller problems. Solutions for these simpler problems are usually readily available through instantiation. By interconnecting solutions for these subproblems a solution for the original problem is synthesized. The analysis phase of solving a problem by methodical synthesis is very similar to solving it using a solution instantiation system that provides solutions for its primitive sub-problems only. The difference is that in methodical synthesis, there is the additional complexity of synthesizing a solution recipe. For example, consider the problem of finding the intersection points of two polygons, each represented as an ordered list of vertices. A straight-forward solution scheme involves finding the intersection of all pairs of edges of the two polygons. The methodology used to arrive at this solution is simple:

(1) Some problems involving polygons can be reduced to solving related problems using their edges instead, and

(2) Some problems involving line segments can be reduced to problems involving their corresponding lines.

Our solution can be "constructed" from three "primitive" algorithms using a simple control structure: (1) an algorithm to find the coefficients of the linear equation of the line passing through two given points; (2) an algorithm to find the solution to two simultaneous linear equations of two variables; and (3) an algorithm to classify a point on a line with respect to a line segment.

The paradigm of methodical synthesis is a modular tool kit with special architectural features that allow interconnection of different modules to construct a solution for new problems. A system based on this paradigm is like a programming language environment. The architectural features (control structure) that allow interconnection of individual tools (primitive operations) are as much an integral part of the paradigm as the individual tools themselves. The repertoire of tools in such a system should consist of *robust* solutions to *generic* and preferably *orthogonal* problems. For example, creating primitive geometric entities and finding their intersections are generic problems that a system must have solutions for. Orthogonality allows users to combine solutions to these primitive problems in every conceivable way to solve larger problems. Therefore, it is necessary to make them robust by considering, detecting, and properly reacting to special conditions (e.g., intersection of parallel lines) or degenerate cases (e.g., tangent points as intersections) that may arise. To facilitate flow of information, a methodical synthesis system must support information encapsulation concepts similar to types and instances, or classes and objects.

3.3. Solution by Refinement

When no simple methodology is known for solving a problem, a more sophisticated scheme must be used to refine the problem into more manageable subproblems. The distinction we make between solution by refinement and methodical synthesis is that the latter is a more sharply focused activity. Whereas in methodical synthesis one can pay more attention to the available tools and how best they can fit together to build a solution to the problem at hand, in solution by refinement the distance of the problem from available tools concentrates one's attention on the problem statement. One characteristic of problems that require a solution by refinement approach is lack of a sufficiently detailed problem statement. The starting point is usually a set of (sometimes conflicting) requirements and constraints. Aside from the (sometimes unattainable) solution, an important outcome of the problem solving activity in this case is a precise problem statement which reflects the original set of constraints and their implications. Solution by refinement is more a process of investigating the given constraints and following the chain of their logical implications.

4. A CASE STUDY

As an example, consider the two-dimensional analog of the simple design problem given below. In this section we examine how a designer may attempt to solve this problem using three CAD systems based on the different paradigms presented above. The purpose of this exercise is (1) to draw attention to the division of work between the CAD system and the designer, and (2) to identify the requirements of the CAD system in each case. For reasons that will be explained later, we deviate from the order of paradigms as presented in Section 3, and consider the solution by refinement paradigm first.

> Design Problem:
>
> A square area on the wall of an art gallery (e.g., a painting) is to be illuminated by a spot light. The side of the square is l and the coverage angle of the spot light is α. The light source must be no more than d_{max} away from the area, nor closer than d_{min}. Find an appropriate position for the light source.

Like most real design problems, this problem does not have a unique solution. The tangible final result of a design activity is, of course, *a design*, i.e., a solution, provided that one exists. But, what is perhaps even more important is an understanding of how and why that solution works. To this end, finding a unique solution, or any at all, is somewhat unimportant. As the problem constraints and their implications are better understood and their effect on the space of potential solutions is discovered, the original constraints can be revised to arrive at a solution.

4.1. A Solution by Refinement Approach

Assuming that this is truly a new problem, the natural starting point is to consider the constraints given in the problem statement and "understand" their implications, i.e., the problem itself. This calls for a CAD system based on the solution by refinement paradigm. Indeed, this is why we chose to discuss this paradigm first. The following scenario represents the interaction between a designer and an intelligent CAD system working on our design problem.

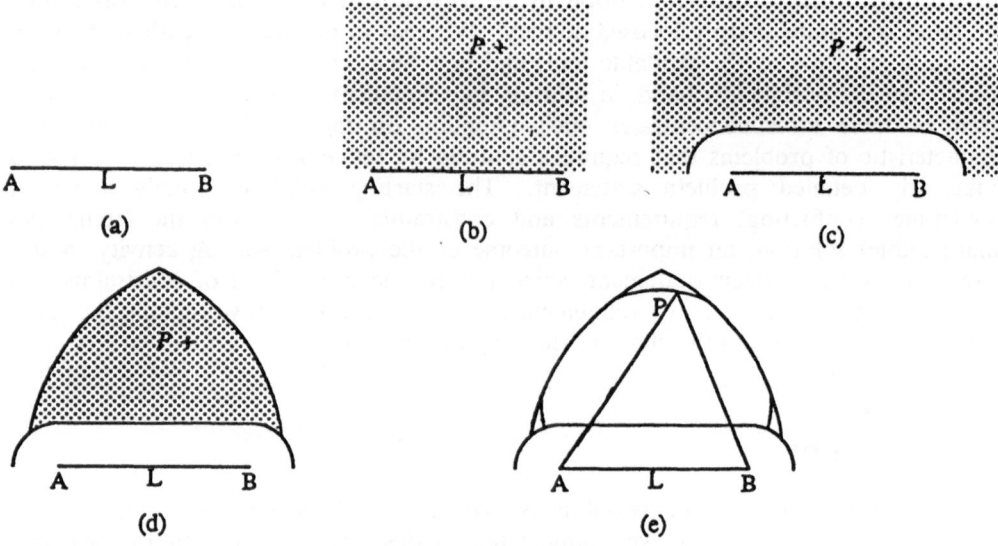

Fig. 2.1. Interactions with an intelligent CAD system

Designer:

There is a line L with length l.

System:

Inferences:

- L is a line segment with length l. [We will refer to the endpoints of L as A and B in this scenario. A real system may eliminate the need for this labeling by allowing designers to use a pick device.]

26

Actions:

- Choose an arbitrary coordinate system.
- Display an arbitrary line segment with appropriate length in this coordinate system (Fig. 2.1.a).

Designer:

Point P is "above" L.

System:

Inferences:

- L is directed.
- P is a point.
- P is on the "above" side of L.

Actions:

- Choose a direction for L and identify the "above" side induced by this direction using shading or color.
- Adjust the display window, if necessary, so that the "above" side of L appears above L on the screen.
- Choose an arbitrary position for P in the identified region (Fig. 2.1.b).

Designer:

The distance from P to L is greater than d_{min}.

System:

Inferences:

- P is outside the oval shape shown in Fig. 2.2.a.
- P is above the dome, as in Fig. 2.2.b.

Actions:

- Display the oval shape induced by the minimum distance constraint.
- Trim the colored region on the screen to the border of the oval shape.
- Adjust the position of P, if necessary, to keep it in the colored region (Fig. 2.1.c).

Designer:

Maximum distance from P to points on L is d_{max}.

System:

Inferences:

- P is in the intersection of the interiors of the two circles centered at A and B both with radius d_{max} (Fig. 2.3.a).
- P is in the intersection of the interiors of these two circles and the colored region on the screen (Fig. 2.3.b).

Actions:

- Display two circles with radius d_{max} centered at A and B.
- Trim the colored region to its intersection with the intersection of the interiors of these two circles.
- Adjust the position of P, if necessary, to keep it in the colored region (Fig. 2.1.d).

Designer:

The angle APB is α.

System:

Inferences:

- P is on the circumference of the two circles that pass through A and B such that the central angle AOB is $2\times\alpha$ (Fig. 2.4.a).

- P is on the intersection of the circumference of these circles with the colored region on the screen (Fig. 2.4.b).

Actions:

- Display the two circles as in Fig. 2.4.a.

- Identify the intersection of the circumference of these circles with the colored region (Fig. 2.1.e).

- Adjust the position of P to keep it in the colored region (i.e., the arcs).

Clearly, an analysis similar to the above scenario is necessary to fill up the gap between the given problem statement and more concrete geometric concepts. Without such reasoning, it is impossible to use a system based on the other two paradigms, or to understand the sequence of actions taken by a designer on such systems. This is the other reason we presented the Intelligent CAD System scenario first.

Observe that in addition to performing very specific and well-defined tasks, such as trimming a line or circle to an intersecting element, the system is performing symbolic reasoning. Indeed, in this simple example, it is the system that is doing most of the reasoning. Some of this reasoning is basically numerical computation, e.g., finding the coordinates of intersection points. But inferring the locus of P from the constraint $APB = \alpha$ involves symbolic reasoning and a declarative form of the corresponding theorem in elementary geometry. Explicit knowledge of geometric theorems in declarative form also enables users to define new concepts using previously defined ones, e.g., *a quadrilateral is a polygon with four sides*.

(a)   (b)

Fig. 2.2. Region with minimum distance d_{min} above L

4.2. A Solution Instantiation Approach

One can easily conceive of a tool built specifically to solve instances of our design problem. It is then trivial to supply values for the parameters l, α, d_{max}, and d_{min} and obtain a solution. In that case a "designer" using this system simply must trust that the solution it provides is appropriate. Aside from deciphering and tracing the actual program code that the system executes, there is no way to follow and understand the system's actions.

It is unlikely that a general purpose CAD system would have a pre-defined solution procedure for our design problem. To achieve generality while maintaining tractability, conventional CAD systems cast their geometric knowledge in terms of solutions to a number of common, well-defined problems: e.g., constructing a line segment given its endpoints, a circle given its center and radius, etc. The following scenario represents the interaction between a designer and a CAD system based on solution instantiation. In this scenario, we assume system capabilities quite similar to what is available in a typical CAD system.

Designer:

Goal:

Define line L with length l.

Inferences:

- Two points are needed to define a line segment.
- If L is horizontal (or vertical) then l can be used as an offset to define the second endpoint.

Actions:

- Choose an arbitrary point A with coordinates A_x and A_y.
- Offset A horizontally by l to define point B.
- Define L using A and B.

System:

In response to the designer's actions, two points and a line segment are created and displayed on the screen.

Designer:

Goal:

Find the locus of points P above L that are at least d_{min} away from it.

Inferences:

- P is outside the oval shape shown in Fig. 2.2.a.
- P is above the dome, as in Fig. 2.2.b.

Actions:

- Offset line L vertically by d_{min} to get line Q above L.
- Draw circle C_1 centered at A with radius d_{min}.
- Trim C_1 to get an arc between Q and the extension of L.
- Draw circle C_2 centered at B with radius d_{min}.
- Trim C_2 to get an arc between Q and the extension of L.

System:

In response to the designer's actions, the dome in Fig. 2.2.b is displayed.

Designer:

Goal:

Find the locus of points P closer than d_{max} to all point on L.

Inferences:

- P is in the intersection of the two circles centered at A and B with radius d_{max}.

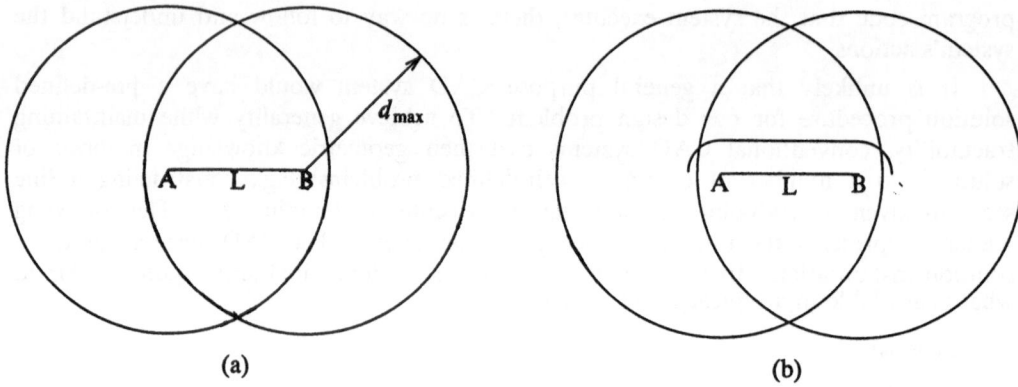

(a)   (b)

Fig. 2.3. Region with maximum distance d_{max} from L

Actions:

● Draw circle C_3 centered at A with radius d_{max}.

● Draw circle C_4 centered at B with radius d_{max}.

System:

In response to designer's actions, two circles are displayed (Fig. 2.3.a).

Designer:

Goal:

Find the locus of points P such that the angle APB is α.

Inferences:

● P is on the circumference of the two circles centered at O_1 and O_2 that pass through A and B such that the central angles AO_1B and AO_2B are equal to $2 \times \alpha$ each (Fig. 2.4.a).

● The triangles AO_1B and AO_2B are isosceles triangles.

● The angles O_1AB and O_1BA are equal to $90° - \alpha$ each.

● The angles O_2AB and O_2BA are equal to $90° - \alpha$ each.

● P is on the segments of the circumference of these circles that are inside both circles C_3 and C_4, and are above the dome.

Actions:

● Draw a half-line M_1 from A, above L, such that the angle between L and M_1 is $90° - \alpha$ (Fig. 2.5.a).

● Draw a half-line N_1 from B, above L, such that the angle between L and N_1 is $90° - \alpha$ (Fig. 2.5.b).

● Find the point O_1 at the intersection of lines M_1 and N_1.

● Draw circle C_5 with center O_1 and radius AO_1 (Fig. 2.5.c).

● Draw a half-line M_2 from A, above L, such that the angle between L and M_2 is $90° - \alpha$ (Fig. 2.5.d).

● Draw a half-line N_2 from B, above L, such that the angle between L and N_2 is $90° - \alpha$ (Fig. 2.5.e).

● Find the point O_2 at the intersection of lines M_2 and N_2.

- Draw circle C_6 with center O_2 and radius AO_2 (Fig. 2.5.f).
- Break and trim circles C_5 and C_6 appropriately.

System:

In response to the designer's actions the proper geometric entities are created and displayed.

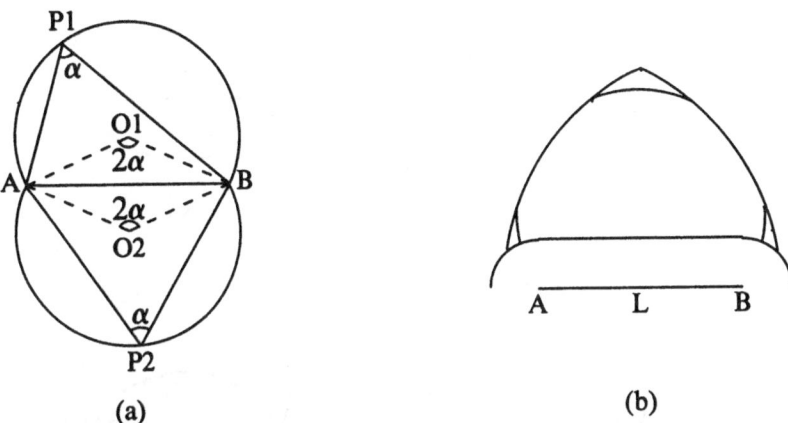

(a)   (b)

Fig. 2.4. Locus of points that see L at α

This rather involved sequence of steps for solving such a simple problem should not be considered as a flaw in existing CAD systems: most of them are capable of solving certain much more complex problems in much fewer steps. For instance, deriving an isometric view of an object from its orthographic projections is a more elaborate geometric problem than our example. However, it is very likely to find a set of pre-defined procedures in a conventional CAD system that simplify isometric construction to a few trivial steps. The complexity of our scenario is a direct result of the inadequacy of the solution instantiation paradigm for design. However, there are more serious problems than increased complexity.

Observe that in this scenario, it is the designer, not the system, that is responsible for higher-level inferences. We can identify two types of inferences: the counterparts of the inferences performed by the intelligent CAD system in Section 4.1, and the type of inferences that the designer makes to identify the necessary actions. The system is completely unaware of this process and of how it determines a *purpose* for its actions. Not only the problem is being solved by the designer alone, he is also hampered by the responsibility for inferring the system-dependent sequence of actions required to instantiate it. What is worse is that there is no separation between the two, and that none of this is recorded anywhere. Are points O_1 and O_2 relevant to the problem or were they constructed only as intermediate results? Every action must result in a fully defined geometric entity. When enough information is not available, the designer must manufacture arbitrary values. This specificity in the designer-system communication is not only irrelevant and unnecessary, it is actually harmful. The system is not aware of the arbitrary nature of, e.g., the location of points A and B, nor it knows about the interdependency of individual elements. Must the two circles C_1 and C_2 always have the same radius? Must they always be tangent to line Q? If any of the values in the problem statement are changed, the system cannot modify the solution on its own.

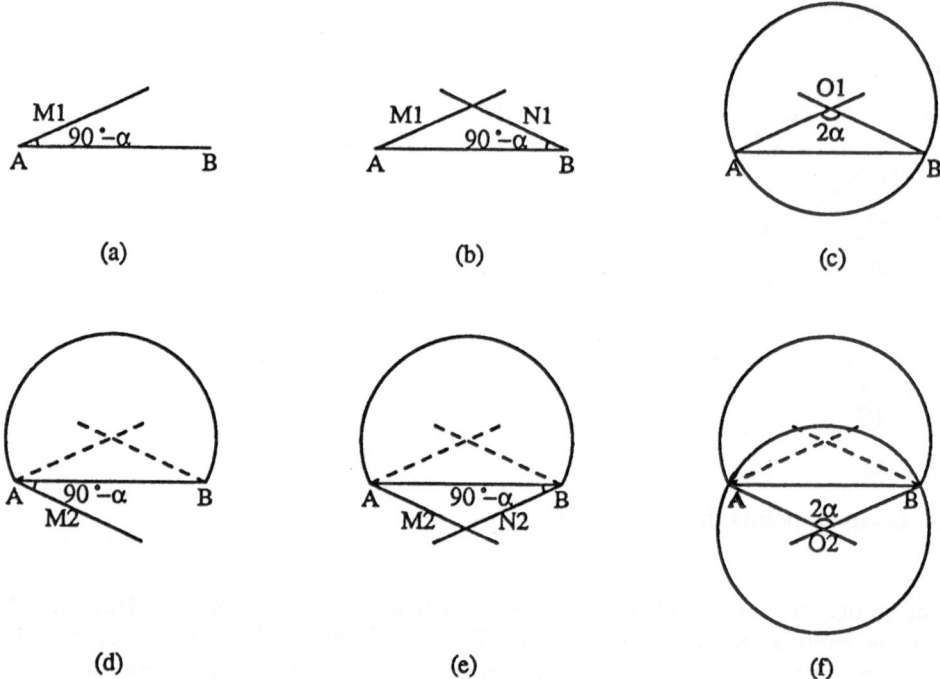

(a) (b) (c)

(d) (e) (f)

Fig. 2.5. Finding locus of points that see L at α

4.3. A Methodical Synthesis Approach

A CAD system based on methodical synthesis must support geometric primitives similar to to those of the system in Section 4.2. Thus, the scenario would be essentially the same, except that here the additional programming environment facilities of the system can be used to alleviate some of the problems in Section 4.2. With even a simple command macro processing capability, it is possible for the designer to document his thought process as a program that reflects some of the semantics and interdependencies of the individual components of his solution. For example, it is easy to state that the circles C_1 and C_2 must have the same radius and that this radius is the same as d_{max}; that the angle O_1AB depends on α; and so on. With slightly more sophisticated control structures, this program can be turned into a parametrized procedure which accepts the values given in the design statement and instantiates a solution automatically. A CAD system may even allow this new procedure to be added to the repertoire of its primitive tools and be used in the synthesis of solutions for other problems. In the following scenario, we assume that the designer wishes to define procedures for some of the more generic sub-problems he encounters.

Designer:

Goal:

Create a procedure to define line segments and use it to define L with length l.

Inferences:

- Two points are needed to define a line segment.
- If L is horizontal (or vertical) then l can be used as an offset to define the second endpoint.

Actions:

- Define procedure *line-seg* with the following steps:
  1. Show the initial point I at a default position and request user to either accept or specify an alternative position for I with coordinates I_x and I_y.
  2. Show the default direction vector $V=(1,0)$ at I, and request user to either accept or specify an alternative direction vector V for the line segment.
  3. Request user to specify the length of the line segment, k.
  4. Offset I horizontally by $k.V_x$ and vertically by $k.V_y$ to define point J.
  5. Define a line segment with endpoints I and J.
- Invoke *line-seg* to define L with length l, with default position and direction.
- Name the endpoints of L, A, and B, respectively.

System:

As the result of invoking *line-seg*, two points and a line segment are created and displayed on the screen.

Designer:

Goal:

Define a procedure to show the locus of points P at a given distance on one side of a line segment. Use this procedure to find the locus of points above L that are at least $d_{\min}$ away from it.

Inferences:

- P is outside the oval shape shown in Fig. 2.2.a.
- P is above the dome, as in Fig. 2.2.b.

Actions:

- Define procedure *Min* with the following steps:
  1. Request user to specify a line segment I. Designate the endpoints of I as E and F.
  2. Request user to specify a distance d.
  3. Show the default direction vector V, perpendicular to I with positive y-component (pointing in the same direction as the positive y-axis). Request user to either accept or inverse the direction of V.
  4. Offset line I horizontally by $d.V_x$ and vertically by $d.V_y$ to get line Q.
  5. Draw circle C_1 centered at E with radius d.
  6. Trim C_1 to get an arc between Q and the extension of I.
  7. Draw circle C_2 centered at F with radius d.
  8. Trim C_2 to get an arc between Q and the extension of I.
- Invoke *Min* with L, $d_{\min}$, and the default direction.
- Name the resulting object *dome*.

System:

As the result of invoking *Min*, the dome in Fig. 2.2.b is displayed.

Designer:

Goal:

Define a procedure to show the locus of points P closer than a given distance to all points on a line segment. Use this procedure to find the locus of points above L, closer than d_{max} to all point on L.

Inferences:

● P is in the intersection of the two circles centered at the endpoints of the line segment with radius d_{max}.

● P is in the part of this intersection that is above L and *dome*.

Actions:

● Define procedure *Max* with the following steps:

1. Request user to specify a line segment I. Designate the endpoints of I as E and F.

2. Request user to specify a distance d.

3. Draw circle C_3 centered at E with radius d.

4. Draw circle C_4 centered at F with radius d.

5. If C_3 and C_4 do not intersect:

5.1. then notify user that there are no solutions and terminate.

5.2. else find the intersection of the closures of C_3 and C_4 and trim the two circles to the boundary of this intersection.

● Invoke *Max* with L and d_{max}.

● Trim the area to L and *dome*.

System:

As the result of invoking *Max*, two circles are displayed (Fig. 2.3.a) and subsequently trimmed to their intersection with L and *dome*.

Designer:

Goal:

Find the locus of points P such that the angle APB is α.

Inferences:

● P is on the circumference of the two circles centered at O_1 and O_2 that pass through A and B such that the central angles AO_1B and AO_2B are equal to $2 \times \alpha$ each (Fig. 2.4.a).

● The triangles AO_1B and AO_2B are isosceles triangles.

● The angles O_1AB and O_1BA are equal to $90° - \alpha$ each.

● The angles O_2AB and O_2BA are equal to $90° - \alpha$ each.

● P is on the segments of the circumference of these circles that are inside both circles C_3 and C_4, and are above the dome.

Actions:

● Request user to specify α.

● Draw a half-line M_1 from A, above L, such that the angle between L and M_1 is $90° - \alpha$ (Fig. 2.5.a).

● Draw a half-line N_1 from B, above L, such that the angle between L and N_1 is $90° - \alpha$ (Fig. 2.5.b).

- Find the point O_1 at the intersection of lines M_1 and N_1.
- Draw circle C_5 with center O_1 and radius AO_1 (Fig. 2.5.c).
- Draw a half-line M_2 from A, above L, such that the angle between L and M_2 is $90° - \alpha$ (Fig. 2.5.d).
- Draw a half-line N_2 from B, above L, such that the angle between L and N_2 is $90° - \alpha$ (Fig. 2.5.e).
- Find the point O_2 at the intersection of lines M_2 and N_2.
- Draw circle C_6 with center O_2 and radius AO_2 (Fig. 2.5.f).
- Break and trim circles C_5 and C_6 appropriately.

System:

In response to the designer's actions the proper geometric entities are created and displayed.

As with other prescriptive specifications, it is generally not possible to make inferences about the properties of solutions in this paradigm. For example, it may not be possible to conclude that circles C_1 and C_2 are always tangent to line segment Q. This information may be useful in a different context. It is also not possible for the system to reason about our procedure and, for example, derive a solution to a similar design problem where line L is replaced by a circular arc.

5. GEOMETRIC REASONING FOR DESIGN

Introduction of a frame of reference for points in space in terms of a coordinate system by Descartes (1954) converted the classical geometry of Euclid (Euclid 1933; Moise 1974) to algebraic manipulation of numbers and structures. This analytic view of geometry has been predominant in computer graphics and CAD since their inception. This view is quite suitable for procedural manipulation of algebraic variables in order to compute specific results. Nevertheless, this is not always the most useful view of geometry. The classical view of Euclid was refined in the late 19th century by Pasch (Behnke et al. 1986) and Hilbert (1965) into what is now called axiomatic geometry. This view of geometry is quite susceptible to making logical inferences about more abstract properties of geometric entities.

To illustrate the relative advantages of each view, consider the theorem in geometry which states that the angle on the circumference of a circle is half the central angle that "sees" the same arc. In analytic geometry, this theorem may be incarnated as a set of algebraic constraints. These constraints involve a minimum of two variables representing the magnitudes of the angles, plus eight more representing the coordinates of the four points (three on the circumference and the center). Thus, given the values for a sufficient number of these variables, a procedure can compute the rest. However, the power of the original theorem goes beyond that of its analytic incarnation. Stated in predicate logic, a system can use this theorem to infer that the locus of points that "see" a line segment through a given angle is a circle. In the case of the example in Section 4.1, for instance, the ability to make such inferences is more important than computing coordinate values.

The type of inferences that an intelligent CAD system such as the one in Section 4.1 is expected to make, requires geometric reasoning (Arbab and Wing 1986). To achieve this, Logic Programming and an explicit knowledge of abstract geometric concepts represented in predicate logic are necessary. Realistically, a CAD system would also need the traditional analytic view of geometry and its numerical computation procedures. Combining these two views of geometry into a single framework raises some interesting issues. For instance, analytic geometry associates specific algebraic definitions to geometric concepts. This is amenable to concrete object representations and procedural manipulation in computer programs. On the other hand, axiomatic geometry never really defines what a geometric entity is. It gives a set of axioms which implicitly define geometric concepts, e.g. a point is simply any entity for which the axioms of point hold. This form of object definition is best suited for making logical inferences about the properties and relationships among objects. How should objects be represented and organized in a framework that accommodates both views of geometry?

Representing objects and the information about them has been one of the core issues in programming languages. .Modern object oriented programming languages (Birtwistle *et al.* 1973; Bobrow and Stefik 1983; Bobrow 1986; Dahl, Myhrhang, and Nygaard 1970; Goldberg and Robson 1983; Liskov *et al.* 1977, 1981; Stefik and Bobrow 1986) properly regard what can be done with an object as information about the object. Thus, they explicitly place all valid operations that can be performed on an object (called *methods* in Smalltalk) in its class definition. Communication between objects is only allowed through the uniform mechanism of *message passing*. In most object oriented programming languages it is difficult or impossible to represent knowledge of declarative nature. An important exception is the information about how objects of a certain class are related to some other classes. This kind of knowledge is represented declaratively in the form of a class hierarchy. The majority of object oriented languages, e.g., Simula (Birtwistle et al. 1973; Dahl, Myhrhang, and Nygaard 1970) and Smalltalk (Goldberg and Robson 1983), use a mechanism called *inheritance* to pass on properties of super-classes to the objects in their sub-classes. This mechanism imposes a rigid structure on classes and the class hierarchy, and makes a clear distinction between classes and objects. Some consequences of this rigid dichotomy are that only objects, not classes, can be operated on; classes are used only to define subclasses, or to *instantiate* objects; every object is an instance of a single class; before an object can be instantiated, the class it is an instance of and its superclass(es) must be defined.

In a "solution by refinement" paradigm, the system must be able to infer object properties (i.e., its class) from a given set of constraints. Moreover, it must do so incrementally and dynamically, as a design develops. This leads to class definitions through predicates that represent the set of constraints on their member objects. Using predicates to define classes is compatible with the axiomatic view of geometry, but it is incompatible with the rigid structure that inheritance imposes on a class hierarchy. For example, assuming a class hierarchy that defines *triangle*, *trapezoid*, and *parallelogram* as subclasses of class *polygon*, it is easy to imagine a case where a generalization of *trapezoid* and *parallelogram* into *quadrilateral* would become necessary. Not allowing this generalization to take place in the system means that a user must first decide whether his object of interest is a parallelogram or a trapezoid. But, this in itself is part of the refinement that the system is supposed to help the user with! In predicate logic this generalization can be written as *quadrilateral(X) implies parallelogram(X) or trapezoid(X)*. Allowing this as a class definition in an environment based on inheritance makes *quadrilateral* an uninstantiable class: given an object, the system can determine whether or not it is a quadrilateral, but it cannot instantiate a quadrilateral object (McAllester and Zabih 1986).

In geometric reasoning, a class is an encapsulation of a set of constraints in the form of predicates. Thus, the distinction between an object and the class (i.e., a set) it belongs to is not as clear as in a language like Smalltalk. It is not clear (and immaterial) whether X in *parallelogram(X)* is a specific object or a set of objects. The predicate *parallelogram(X)* can be used both to define other classes of objects and to operate on individual parallelograms. This view of objects and classes is similar to the notion of *prototype objects* in some more recent languages (Borning 1986; Lieberman 1986b). An alternative mechanism to inheritance, called *delegation*, is used to share properties among classes (Lieberman 1986a).

6. RESEARCH PROBLEMS

In the previous section we discussed a number of issues involving geometric reasoning for intelligent CAD. Mechanisms for incorporating algebraic procedures in a framework based on an axiomatic view of geometry, and representation of objects and classes still need more work. Class definition through predicates allows multiple inheritance (e.g., a right-angled-isosceles triangle is both a right-angled triangle and an isosceles triangle) and specialization (e.g., an equilateral triangle is a triangle with equal sides). It also allows cancellation (e.g., a rhombus is a square *without* right angles) and conflicts (e.g., a triangle with four sides). In view of this, sharing procedural knowledge through delegation among objects can lead to disaster. For example, suppose procedures *area* and *perimeter* are associated with the class of squares to compute their area (as the square of their side) and perimeter (as four times their side), respectively. If a rhombus is later defined as a square without right angles, it is correct to delegate computation of its perimeter to *perimeter*, but it is incorrect to use *area* to compute its area.

7. CONCLUSION

An intelligent CAD system for mechanical applications must be based on a "solution by refinement" paradigm. Because geometry is an important component of the initial constraints in a design problem, it becomes part of the interaction between designers and CAD systems. In order for a system to aid a designer in refining the design problem, both axiomatic and analytic views of geometry must be incorporated in the geometric reasoning component of intelligent CAD systems. The notion of objects and classes in this paradigm is incompatible with the inheritance mechanism that is popular in object oriented programming languages. Further research is necessary to accommodate procedural knowledge in this paradigm as properties of objects defined through predicate logic.

REFERENCES

Arbab, F. and Wing, J. M. (1986): "Geometric reasoning: A new paradigm for processing geometric information," in *Design Theory for CAD, Proceedings of the IFIP W. G. 5.2 Working Conference 1985 (Tokyo)*, Yoshikawa, H. and Warman, E. A. (eds.), North-Holland, Amsterdam, pp. 107-121.

Behnke, H., Bachmann, F., Fladt, K., and Kunle, H. (1986): *Fundamentals of Mathematics Volume II: Geometry*, MIT Press, Cambridge, MA, USA. Translated by S.H. Gould.

Birtwistle, G. M., Dahl, O. J., Myhrhaug, B., and Nygaard, K. (1973): *Simula Begin*, Van Nostrand Reinhold, New York.

Bobrow, D. G. and Stefik, M. J. (1983): "The Loops Manual," Technical Memo No. KBVLSI-81-13, Xerox Palo Alto Research Center, Palo Alto, CA, USA.

Bobrow, D. G. (November 1986): "CommonLoops: Merging lisp and object-oriented programming," *Special Issue of ACM SIGPLAN Notices Notices (Proceedings of Object-Oriented Programming Systems, Languages and Applications '86)*, **21**(11), pp. 17-29.

Borning, A. (November 1986): "Classes versus prototypes in object oriented languages," in *Proceedings of Fall Joint Computer Conference*, ACM/IEEE, Dallas, Texas.

Boyse, J. W. and Gilchrist, J. E. (March 1982): "GMSolid: Interactive modeling for design and analysis of solids," *IEEE Computer Graphics and Applications*, **2**(2), pp. 27-40.

Brown, C. M. (March 1982): "PADL-2: A technical summary," *IEEE Computer Graphics and Applications*, **2**(2), pp. 69-84.

Dahl, O. J., Myhrhang, B., and Nygaard, K. (1970): "Simula 67 Common Base Language," Technical Report No. S-22, Norwegian Computing Center.

Descartes, R. (1954): *Geometry*, Dover Publications, Inc., New York. Translated by D.E. Smith and M.L. Latham.

Euclid (1933): *Elements*, J.M. Dent & Sons, London. I. Todhunter's edition of R. Simson's "Euclid".

Fitzgerald, W., Gracer, F., and Wolfe, R. (July 1981): "GRIN: Interactive graphics for modeling solids," *IBM Journal of Research and Development*, **25**(4), pp. 281-294.

Goldberg, A. and Robson, D. (1983): *Smalltalk-80: The Language and its Implementation*, Addison-Wesley, Reading, MA, USA.

Hilbert, D. (1965): *The Foundations of Geometry*, The Open Court Publishing Company, La Salle, IL, USA. Authorized English translation by E.J. Townsend.

Lieberman, H. (1986): "Delegation and inheritance: Two mechanisms for sharing knowledge in object oriented systems," in *3eme Journees d'Etudes Langages Orientes Objets*, Bezivin, J. and Cointe, P. (eds.), AFCET, Paris, France.

Lieberman, H. (November 1986): "Using prototypical objects to implement shared behavior in object-oriented systems," *Special Issue of ACM SIGPLAN Notices Notices (Proceedings of Object-Oriented Programming Systems, Languages and Applications '86)*, **21**(11), pp. 214-223.

Liskov, B., Snyder, A., Atkinson, R., and Schaffert, C. (August 1977): "Abstraction mechanisms in CLU," *Communications of the ACM*, **20**(8), pp. 564-576.

Liskov, B., Atkinson, R., Bloom, T., Moss, E., Schaffert, C., Scheifler, R., and Snyder, A. (1981): *CLU Reference Manual*, Springer-Verlag, Berlin, Heidelberg, New York, Tokyo.

McAllester, D. and Zabih, R. (November 1986): "Boolean classes," *Special Issue of ACM SIGPLAN Notices Notices (Proceedings of Object-Oriented Programming Systems, Languages and Applications '86)*, **21**(11), pp. 417-423.

Moise, E. E. (1974): *Elementary Geometry from an Advanced Standpoint*, Addison-Wesley, Reading, MA, USA.

Mortenson, M. E. (1985): *Geometric Modeling*, John Wiley & Sons, New York.

Requicha, A. A. G. (December 1980): "Representations for rigid solids: Theory, methods, and systems," *ACM Computing Surveys*, **12**(4), pp. 437-464.

Stefik, M. and Bobrow, D. G. (Winter 1986): "Object-oriented programming: Themes and variations," *AI Magazine*, **6**(4), pp. 40-62.

Wesley, M. A., Lozano-Perez, T., L.Lieberman, Lavin, M. A., and Grossman, D. D. (January 1980): "A geometric modeling system for automated mechanical assembly," *IBM Journal of Research and Development*, **24**(1), pp. 64-74.

Report on Session 1

Chair: P.J.W. ten Hagen †
Cochair: W. Eshuis †
Edited by: V. Akman †

Bijl's main point is that there is a difference between knowledge within people and its overt representations (e.g. visualizations). An illustration of knowledge is less than the knowledge itself. A CAD system has a responsibility to represent the designer's knowledge. This is irrespective of whether such knowledge is factual, unambiguous, etc. Therefore, as a first step, a theory must be developed for an appropriate and consistent (i.e. using the same symbols for the same concepts all the time) representation of knowledge. This theory need not cover more advanced topics such as understanding the knowledge it represents.

Arbab's main premise is to clarify the kinds of intelligence that can be attributed to a CAD system. His work aims at an understanding of the domain of geometry. Intelligence allows for combining knowledge from other domains, e.g. non-geometric, technical, organizational. He asserts that various problem solving strategies use different types of reasoning which make different use of object representations. Hierarchies are thus relevant, and so are the kinds of relations between objects and classes implied by such hierarchies.

During the Q&A period, Koegel asked if Bijl's system can deal with things other than architectural objects. Bijl said that his example of the wall junctions unfortunately made his talk sound too specific to architecture, while this is really not the case. Mac an Airchinnigh commented that Bijl's lecture seemed to be focussed on drawing. Bijl's reply was that CAD systems originally were drawing systems only, neglecting the problem of semantics of the drawings in terms of materials, physical properties, and costs. His system has been used for ten years without modifications, thus forcing designers to think in terms of the system.

Hoeltzel told Arbab that it is a better idea to accept that functionality precedes geometry. For example, in mechanical applications functionality comes first. In

---

† Centre for Mathematics and Computer Science, Kruislaan 413, 1098 SJ Amsterdam, The Netherlands.

motion conversion, one can clearly separate functionality from geometry. Arbab agreed but also noted that we cannot yet in a nice way derive functionality from things other than geometry.

Genin asked Arbab about his views on inheritance. According to Arbab, the main disadvantage of inheritance is its rigid structure. One cannot manipulate classes. (He is talking about Smalltalk's inheritance.) An alternative mechanism is delegation. Delegation is more flexible, and helps soften the sharp distinction between classes and objects.

Session 2

3. Intelligent Systems Interconnection: What Should Come After Open Systems Interconnection?

P. Bernus and Z. Létray†

Centre for Mathematics and Computer Science
Kruislaan 413, 1098 SJ Amsterdam, THE NETHERLANDS

Abstract: *The battle for the best knowledge representation language seems terminated, not because anyone had found one but because fairly obviously there is none. We propose to turn our attention to general architectures of knowledge representation. If theories to be represented become specified, we can use such general architectures in the development of individual ones. Theories are generally used by some agent for specific purposes. Agents need to communicate with each other in order to make use of their theories. We claim that a theory of agenthood should describe communication and conversation among agents and the way they interact with the environment. A functional architecture for intelligent systems interconnection (ISI) is proposed. We identify functional and hierarchical layers of representations and theories. We attempt to show the road to bring together represented conscious and unconscious as well as not-represented inherent knowledge. We do it in order to combine intelligence with effectiveness.*

Keywords: representation languages, agents, CAD, conversation theory, open systems interconnection, intelligent systems interconnection.

1. INTRODUCTION

Many attempts and proposals aimed at finding a universal, "best" knowledge representation language. The scope is very broad and encompasses several dimensions, depending for example on the choices between procedural and declarative, between intensional and extensional representations (Tomiyama and ten Hagen 1987a), on the choice of blending them into so-called multiparadigm systems (Bobrow 1985), or on the choice of the right expressive power. There have been tasks for which a given representation language was inappropriate, either too poor or too powerful.

---

† Computer and Automation Institute, Hungarian Academy of Sciences, Kende u. 13-17, H-1111 Budapest, HUNGARY

We need to recognise that intelligent systems need more than one representation scheme. Many examples suggest (Doyle 1983; Minsky 1980) that intelligent behaviour may not be achieved by single physical entities acting in one scene and having one single goal only. Intelligent behaviour supposes, instead, *controlled*, *self-conscious actions* with appropriately concentrated or diversified motivations. The actions are carried out in environments possibly populated with other similar agents and are caused by internal actions on varying higher levels. Represented (conceptualised) conscious and unconscious as well as not-represented but implemented knowledge must be brought together in order to give an account on *effective intelligent behaviour*.

We claim that the integrated effect of mental and physical actions is decomposable into a cooperative function of communicating agents. Computational limitations and efficiency requirements do not allow one single theory to deal with real world problems in an intelligent manner. Intelligence can only be achieved if different theories about the same subject compete. Agents on one hand have to implement a theory of communication and conversation in order to confront the theories they represent. On the other hand an agent is not merely a part of an environment; it must be treated as a possible set of cooperating subagents.

The interconnection of agents together with the representational task is functionally abstracted into a multilayered architecture. We show how representational layers are formed in computer based implementations and how it is possible to build strata of theories on their top. In Section 3.1. we identify four representational layers:

1. Senses and actions layer.
2. Computer representation layer.
3. Logic representation layer.
4. Abstractions layer.

With the help of these, the theory of agents can be built up. Theories of agents can possibly be clustered to form the *theory of communicating agents*. Section 3.2 presents the stratification of such a theory, giving an account on:

5. Model management layer.
6. Situation and discourse modelling layer.
7. Motivations and agenthood layer.

These seven layers form the *Intelligent Systems Interconnection* (ISI) *proposal*. ISI comprises independent layers of functional subsystems. Functions defined on the highest level will always be boiled down to functions on the lowest level. The generality and the functional nature of the ISI architecture allows for many different implementations in the similar way as the ISO-OSI architecture does (ISO 1983; Zimmermann 1980). The difference between ISI and OSI is that they are concerned with communication functions on different levels of abstraction. An OSI implementation for instance could be utilised as a medium by an ISI implementation. Brachman similarly stratified the possible meanings of constituents in a knowledge representation tool (Brachman 1979) in investigating semantic nets. This technique was successfully used for representing a theory of activity and time in (Sathi, Fox, and Greenberg 1985).

As mentioned, there are no single prescribed or specific choices made in the general architecture. This is necessary because practically the functionality of every layer can be made specific depending on the actual set of agents to communicate through that layer. In the discourse modelling layer, for instance, different languages can be used (e.g. special purpose interface languages, graphic languages, natural language, etc.).

In the layer of logic representations the chosen logic cannot be frozen once for all. Although agents should not carry the overhead of implementations of obscure logics if they do not need them, the trickiest modal logic should also be introducible into this layer in case applications, i.e. agents, need it.

The architecture presented in this article serves two purposes. One is the analysis of the requirements which an intelligent CAD system's central modelling language should meet. Here the essential intellectual activity based on the motivations of the involved agents is design in a broader sense. Therefore one of the agents or a group of agents represent a design theory.

The second aim is to investigate what are the *useful abstractions* to help us describe theories of intelligent agents. In this sense the present general model is an attempt. We hope to stimulate readers to make criticisms and try to contribute to the formulation of a more widely acceptable, well specified architecture and conceptual framework. This is essential for achieving commensurable research results.

2. FUNCTIONAL ARCHITECTURE FOR KNOWLEDGE REPRESENTATION

2.1. Implementations, Representations, and Theories

A CAD system must operate on the principles of a theory which gives sufficient account on design. An implemented CAD system is at the ultimate a model of that theory. A theory needs a language for its representation. In a sense, we could say that there is no theory without a representation language. The two, however, are different, for apparently a theory can be represented in several *different* languages. Two theories are considered the same, if they have the same set of possible models in the sense of model-theoretic semantics (Hayes 1985).

The concepts appearing in design theories (such as models, theory of mechanical machines, strategies, paradigms, goals, criteria, etc.) should be representable in such a language. The distance between the available computer languages (for possible implementation of a computer model of the design theory) and the above concepts is fairly big. Therefore the expressive power of the available representation languages must be extended with great care. There might be several reasons to do so, be it for extending the possible set of representable theories, or for reducing the complexity of the representation. A good overview of possible meanings of expressive power can be found in (Israel and Brachman 1984). We shall collectively call them useful *enrichments*. The result of enrichment is a language of which the concepts subsume the concepts of the simpler language.

In practice the search for an adequate representation language can also proceed in the opposite direction. Given all the concepts in a theory we may wish to reduce the set of these concepts to a smaller set of more general concepts. Assume that we are able to describe the way the subsumed concepts can be used to create new concepts. There are various kinds of *abstractions* to do so. Two important ones are specialisation and aggregation (see (Fox 1979) for more).

Suppose we use a logic language for representation (i.e. there are axioms which assign truth values to well-formed formulae of the language). In this case the theory in question (here theory of design) should be represented in forms of axioms as well. For a language with first-order predicate logic, see for instance (Copi 1979). In case we already have a logic language with insufficient expressive power, we can do the following:

a) Extend the syntax of the language (that will also require part of b below).

b) Extend the logic by:

- inventing new truth values.

- modifying/extending the set of independent postulates which pin down those well formed formulae of the language which invariably have the same truth value.

- modifying/extending the rules of inference.

c) Add symbols to enrich the language, so that one can express the same thing in several different ways.

Given an enriched language it will be possible to express our theory in it. It is, however, natural to seek for some structures in the concepts of our theory and (as mentioned above) for layers of concepts. The language is used for creating the representation of a "basic theory." The concepts of such a basic theory can be utilised to *define* concepts of a theory based (layered) on it. A theory of mechanisms for example can be represented in terms of concepts and their epistemological relations. A theory of concepts can be represented in terms of a language with an appropriate logic (see (Brachman 1979)). Of course the actual forms of definition (various kinds of abstraction) should be expressible in the basic representation language as operations, thus determining the largest set of real world models our theory can possibly have no matter what layering strategy was used to describe the theory. It must be noted that concept definition is sometimes very demanding, concerning the expressive power of the representation language.

The layering of theories is useful, and in our view unavoidable, because of the complexity and unstructuredness of the representation that result otherwise. From the viewpoint of higher level (layered) theories the more basic theory's concepts can be considered as enrichments of the representation language, or alternatively, as just another representation language provided the underlying theory carries along all the representational features of the "real" representation language. The layering upwards into strata of theories has a counterpart downwards, in that several changes of representation language take place before the representation is actually tied to a physical implementation. This especially holds for computer models.

By extending the expressive power of the representation language, the completeness status of the language also changes. The gain of being able to say something that was impossible to state before should be weighted against the added complexity of theorem proving needed when the represented theory is set to work. One attempt to trim this trade-off is discussed in (Brachman and Levesque 1983).

We think that the decision on what order of logic and what kind of modalities to include depends on the complexity of conceptual relations within a problem area. The abstraction techniques for defining new concepts based on known ones have an important effect. Care must also be taken, because the decision has consequences to the applicable inference control strategies. In one case one might wish to get guaranteed response within predictable time for specific problems. In other situations, like searching for an adequate assembly which satisfies given functional constraints, we need to contend with a relaxation strategy or appropriate heuristics. The complexity of this second task is a result of functional abstraction in the reasoning process.

Consequently, representation languages of different agents are different. Applicable inference control strategies are also problem- or situation-dependent and thus not common to all agents.

The functional architecture of agents in general must be layered for the reasons of:

- gaining the appropriate expressive power and system qualities,
- mastering the complexity of the agent's theory.

From the above discussion it seems natural to admit the multiple agent assumption (i.e. agents should consist of multiple subagents). Only competing theories of the real world can resolve the contradicting nature of two requirements: solving complex problems on one hand and acting within limited time on the other.

Let us call the users of the theories in question *agents*. The agents, having to compete with each other, need to implement a theory about themselves (i.e. a psychology). This is what enables them to reason about activities and knowledge of their own. (Accepting agents as implemented psychological theories follows (Doyle 1983).) Competition involves at least two parties, and the reasoning has to cover activities and knowledge of the opposite parties as well. Therefore, the multiple agent assumption must be accompanied by a theory of the communication of such agents. Information exchange with the rest of the environment through perception and manipulation should also be accounted for. For purposes of design theory (see (Tomiyama and Yoshikawa 1987)) only a limited ability of communication is needed, if compared with conversational situations such as learning, teaching, and negotiating.

It is instructive to look at the representational problem from the point of view of conversation theory (Gergely and Szots 1982; Pask 1975, 1980, 1982) where the representation of an agent's theory is not only related to the communication channel through which conversation with other agents can take place but also to the sensory apparatus through which the agent can sense and the motory system through which the agent can manipulate its environment or moves. We think that the much argued question of whether intelligent beings "really" use representations or representations are only external observers' fabrications may become a non-issue in this light. The word "really" becomes meaningless in this question, since the use of representations is optional for a large part of the theories. If an interrelated set of attributes of biological or artificial structures *can* be accepted as a model for a theory, then tokens of that theory do not *need* intermediate representations. Electric impulses in a computer just happen to persist without being represented in the computer itself, but that does not exclude the possibility of representing impulses as concepts in a computer simulation language. On the other hand, the representation of the impulses in a computer becomes a self-knowledge only if the mapping from the theory of electric impulses to the real impulses can directly or indirectly be established. The presented arguments can be applied to the mapping function as well: it should be an attribute of the agent — not necessarily represented but by all means exhibited. The mapping function can be implemented directly through identity relations or indirectly through sensory means. In the light of this discussion we see that the need for a particular agent to have representations of unconscious concepts depends on the physical quality of the agent (flesh and blood, wires and N-P transitions, etc.). An agent can act without representing the act itself but hardly can reason about the act without having represented it first (be the "act" physical or mental).

Since senses and manipulative capabilities of present day computers are desperately limited (see efforts for getting off the ground in (Negroponte 1970)), the reinforcement between representations and real world models in computers lies quite ahead in time.

It would be interesting to find real world models where the representation of percepts and of motory controls were identical to the percepts and controls. An example is a case where an established connection causing a motory action is at the same time the implementation of the action's conceptual representation.

The lowest level of an agent's functional architecture is thus the layer of senses and actions (interpreted in CAD as the implementing computer together with its I/O system). Through this layer denotations finally can get connected to the real world and not to artifact models.

The whole theory of design (just as our human knowledge about the surrounding world) will not be able to rely entirely on such first hand denotations but on knowledge engineered, human made, built-in representations or to some extent on "second-hand" denotations acquired through communication with other agents (including humans). The semantic integrity of the CAD system will, on the ground of the previous reasons, rely on the users and maintainers of the system.

2.2. Choices in the General Architecture

The presented views allow for a multitude of various types of knowledge representation systems, hence theories represented in those systems. One aspect of the permissiveness lies in the allowable physical architectures. If we consider agents as autonomous systems, this autonomy is always relative to the environment, i.e. the only thing common to two agents is their environment. Exactly in computer based implementations the definition of what should be considered as the environment is left to us (the external observers).

The trivial implementation is to use separate computers for separate agents and connect those machines. It is also possible to implement agents in a multiprocess environment and provide interprocess communication. Even one language with coroutining possibility can be considered a candidate. The need for parallel processing depends on the discourse situations the agents are expected to enter. Many discourse situations can essentially be described without using parallelism. We think that future

CAD systems will need a little more than that. A sequentially described discourse between agents may do, but in-between agents should be capable of processing in parallel (e.g. a human user plus the interface to the human user as one agent and other agents of the CAD system).

The choice of representation language remains with the agent itself and is basically dependent on the agent's theory. Even on the level of equivalent expressive power there are basic choices which influence the effectiveness and the maintainability of the agent's representation of its theory. In general we expect an agent not only to represent its theory in a static way but also its functions: e.g. a simulator program should represent both the target of simulation (say a conveyor) as well as the functions of the simulator (say being able to gather statistics). If so, the control of the theorem proving task (or equivalent procedure) must also be representable in general. Various kinds of conceptual dependencies, like abstractions, can have an important role in reasoning. Inheritance is one special kind of inference that may need control.

For the design of a representation language, the language concepts should form an orthogonal system. The same holds for independent theory layers. Orthogonality assures locality of effects caused by modification and brings about longer life expectations and less vulnerability for a system in general. Orthogonality is defined here as the following properties:

- meaningful transformations on the data in that layer should be expressible using combinations of the functions in that layer (completeness criterion of the functions).

- combinations of functions within one layer must not be restricted but by the meaning of the composite transformation that the combination is intended to perform: e.g. no restriction from lower level layers (of implementation or theory) is permitted to penetrate the layer in question, unless this restriction is functional (independency criterion).

3. A LAYERED KNOWLEDGE REPRESENTATION ARCHITECTURE

3.1. Intra-Agent Representation of Knowledge

We identified layers 1-4 (Section 1) in the functional architecture of an agent.

The *Senses and Actions layer* is responsible in the end for the computer interpretation of the agent's theory. Any theory in the layer's language will be related to the computer's pieces, states, and relations over pieces and states, including the tokens of information passing through the I/O system of the computer.

In computer systems this heterogeneous representational vehicle is a sort of assembly language which directly relates to the physical parts of the machine: I/O registers, memory, interrupts, graphic I/O, etc. In order to hide all this from the higher level representational tasks, these things are abstracted into a general purpose representation language (high level programming language).

The *Layer of Computer Representations* is built on top of the previous layer and features everything that is necessary from the view point of software engineering. It supports good system design and maintenance practices and makes it possible for computer based systems to survive their original implementation environment. The change to this higher level of representation layer improves system qualities and adds at the same time expressive power (e.g. emulation of parallel computation on sequential machines).

While the concern for introducing a new layer on top of layer 1 was "good engineering," tuning the knowledge representation language to the needs of an agent may require the introduction of the third layer with adequate expressive power and computational efficiency.

We call this the *Layer of Logic Representations* because it is straightforward to compare knowledge representation languages on this level in terms of their logics. This is not to say that the representation language of every agent must by all means be a logic language. Especially in this layer it is necessary to maintain extensibility of the language. The completeness criterion of orthogonality will take care of not building theories into the representation language, while the independency criterion separates representational issues from inference control. The representation should not know about the inference control and the inference control mechanisms must not know anything specific about the representations (except what the possible inference operations are). Heuristics should be separated from general inference control functions.

In addition, in multiparadigm systems, although the automatic propagation of facts among representational parts goes along with the limitations in expressive power, the behaviour of the system is predictable. If we tolerate the inconsistency between parts, we can circumvent these limitations, but only at the cost of unpredictability in general (task-specific debugging of representations).

Augmenting the richness of the representation language by abstractions is a common way and has much been discussed by different authors (Brachman 1979; Fox 1979). The *Layer of Abstractions* enriches the language by defining conceptual dependencies (such as aggregation, generalisation, etc.) needed to economically express theories. The expressive power of the language is not effected, for the definition of these conceptual dependencies would otherwise have not been possible.

3.2. Intelligent Systems Interconnection (Inter-Agent Communication)

Layering can be continued in many different directions, depending on the actual theory to be built up. Additional system layers are called layers of theories. After realising that we need multiple agents in order to implement intelligent systems which cope with real world problems (e.g. intelligent CAD systems), we must consider how they are interconnected. Any agent taking part in the communication should have a sub-theory of communication which is valid for the given "society" of agents. This sub-theory is a possibly multilayered architecture of theories, generally accepted by all the participants (general architecture). The value of general architectures (such as the famous ISO-OSI model (Zimmermann 1980)) convinces us of the utility of putting considerable efforts into the elaboration of such framework for the interconnection of intelligent agents.

Stated in a simplified way, the intelligent systems interconnection (ISI) architecture should be a framework for the description of how intelligent agents talk to each other. The approach of the architecture is functional, allowing space for different specific cases as well as for variations in the actual implementations. For practical purposes we need to limit the generality of the architecture to the needs of interconnecting agents within a CAD system (including users as agents). It should be noted, however, that not every aspect of information exchange can be accounted for by communicating via language discourses. Sometimes a common referential model is also needed to which both agents can point.

If an agent's theory is complete, it has not too much to communicate with others. This is not the case especially in CAD. Here, incomplete theories ("models of the solution") must be dealt with. The *Layer of Model Management* could be the basis for dealing with incomplete information, building alternative approaches and representing assumptions. In a given moment several, maybé contradictory, states of affairs can be accepted as attainable by augmenting the agent's knowledge. Information exchange among agents needs to be accompanied by the agent's ability to reason about the effects of receiving new information and reasons about what actions can be taken to get some new information.

Nonmonotonic assumptions can help the agent in that reasoning. In the end, these assumptions should be checked by directly consulting the real world through manipulations and percepts — e.g. a CAD system for program design could run a profiling program and measure its performance — or asking other agents about the status of a given assumption. It is also possible that the underlying theory is not complete. Exchanging theories among agents is a most interesting phenomenon (learning, unlearning, reinterpretation, etc.). A part of this can be described by conversation theory (Pask 1975, 1980, 1982).

We can use the Model Management Layer's theory of multiple models to build a theory about situations, especially, discourse situations. The result forms the *Situation and Discourse Modelling Layer*. We think that conversation theory provides substantial basic material for working out this layer's exact specifications.

The present layer introduces the notion of situation. It should provide concepts which describe, how situations are effected by actions and communication. Conversation, data capturing and interpretation, message generation, and other actions all influence:

• the models of the situations,
• the models underlying the situations.

These concepts can be used to talk about the roles, goals, and motivations of an agent. So again we can work out a layered theory which sits on top of another theory. We call this new layer *Layer of Motivations and Agenthood*. Primitives of self-knowledge (Doyle 1983) could be defined for self-reference and reference to other agents implemented as psychological theories. The agent can take actions rooted, via conceptual or direct links, in the senses and actions layer. Whenever the theory of agenthood proves an action, this action takes place.

The multiple models of the model management layer make it possible to divide an agent into sub-agents and let them have their own theories about the topics. The theory of agenthood can use the situation modelling layer's theory to model the fact that not a single agent is acting on his behalf.

Language understanding and generation is one of the complex actions that an agent can indulge in. Potentially every layer of theories as well as representations can take part in it. Understanding and generation is not a function defined in one of the layers. It is rather a part of the total architecture. The roots are in the senses and actions layer for perception and articulation and the results are used by higher level actions. Such actions can be, for example, morphological or syntactic transformations. The agent is allowed to use representations as in an analysis using symbolic computations. In other cases it may not make use of them, as in construction based on memory.

Higher level semantic functions carried out on concepts have a place in the basic theory layer (model management). Some semantic functions require the use of situations for understanding and selecting adequate responses (thus should be defined in that layer). Interactions of discourses and motivations, on the other hand, go into the layer above (motivations and agenthood).

Since different theories call for different representations, there should not be a single language for communication, either. Humans, beside using natural language, also make use of several different forms of languages.

Some interaction between agents does not need the highest level functions of these, and even mere context-free question-answering can take us far enough for exchanging information between compatible but incomplete representations of the agent's theories. In the other extreme we take into account all the aspects of the communication. These are the syntactic construct, the functional structure (how it conveys the semantic information) and the interstructural aspect (how it functions in the given communication situation) (Hankiss 1985).

In practical CAD projects (until appropriate theories are worked out) substantial simplification is necessary. The communication between agents (especially those which are built from extant tools) can be managed by a "design manager" which has a "general model" as in the General Modeling System in (Bernus and Hatvany 1979) or a "metamodel" as in the Integrated Data Description Schema of the IIICAD system (Tomiyama and ten Hagen 1987b). The design manager, implemented as an agent, represents all the relevant information about the object of design and works basically in query answering mode with other agents (or slightly generalised, it acts as a moderator-mediator between agents).

To simplify matters it is also possible to accept the representation language of the design manager as a language for communication. In that case the specific ISI implementation will have much in common with the representational architecture of the "design manager." While agents in general represent their theories about the target of design, the design manager will represent a theory of design (for details see (Tomiyama and Yoshikawa 1987)).

The layers to build on top of the representational layers (1-4) must therefore contain a theory of design evolution (Tomiyama and ten Hagen 1987b). One of its sub-theories should describe the communication between the agents or at least between any agent and the design manager.

4. CONCLUSION

a) We argued that intelligent agents should be decomposable into interconnected agents. Each of them should have its own theory, representation language, and specific implementational details.

b) We proposed a functional layering approach for knowledge representation with four layers of representation and three for one particular theory, that of communication. Layering the knowledge representation architecture is not a goal *per se*. Changing representation languages once or more could be replaced if appropriate technology was available for different implementations. Layering of theories is much more inherent in the knowledge to be captured because of the basic complexity of conceptual structures.

c) We also proposed that conversation theory could be incorporated into this multilayered architecture and provide artificial agents with the possibility of directly relating their theories to the real world (not only to theories of other agents).

d) We conclude that intelligent CAD systems should be integrated in the following way.

  d.1) Create computer based agents which contain different CAD tools.

  d.2) Represent a theory of the tool incorporated by the agent.

  d.3) In every agent implement the same *theory of communicating agents*.

  d.4) Implement a design manager agent.

ACKNOWLEDGEMENTS

The idea of ISI has first been laid down in February 1986 at a preliminary oral presentation at the Winter School on Conceptual Modeling in Visegrád, Hungary, under the title, "The Idea of a FRPRSM System for Advanced System Analysis and CAD," by the present authors. At that time FR stood for Frames, PR for production rules, and SM for Smalltalk (referring to object modelling). The present article is a refinement of these ideas reflecting valuable discussions with members of the IIICAD group at Centre for Mathematics and Computer Science, Amsterdam. We would also like to thank the editors for their useful comments.

REFERENCES

Bernus, P. and Hatvany, J. (1979): "Computer aids to the design of integrated manufacturing systems," *Computers in Industry*, **1**(1), pp. 11-19.

Bobrow, D. G. (November 1985): "If Prolog is the answer, what is the question? or What it takes to support AI programming paradigms," *IEEE Transaction on Software Engineering*, **SE-11**(11), pp. 1401-1408.

Brachman, R. J. (1979): "On the epistemological status of semantic networks," in *Associative Networks: Representation and Use of Knowledge by Computers*, Findler, N. V. (ed.), Academic Press, New York, pp. 3-50.

Brachman, R. J. and Levesque, H. J. (October 1983): "Krypton: A functional approach to knowledge representation," *IEEE Computer*, **16**(10), pp. 67-73.

Copi, I. M. (1979): *Symbolic Logic*, Macmillan Publ. Co., New York.

Doyle, J. (1983): "A Society of Mind," Technical Report No. CMU-CS-83-127, Carnegie-Mellon University, Department of Computer Science, Pittsburg, PA.

Fox, M. S. (1979): "On inheritance in knowledge representation," in *Proceedings of the Seventh International Joint Conference on Artificial Intelligence* **2**, Tokyo, pp. 282-284.

Gergely, T. and Szots, M. (1982): "About representation of semantics," in *Progress in Cybernetics and Systems Research, Proceedings of the Symposium of Austrian Society of Cybernetic Studies* **11**, Trappl, R., Findler, N. V., and Horn, W. (eds.), Hemisphere Publ. Co., Washington D.C., pp. 227-234.

Hankiss, E. (1985): *The Literary Work as a Complex Model*, Magveto, Budapest. (In Hungarian).

Hayes, P. J. (1985): "The second naive physics manifesto," in *Formal Theories of the Commonsense World*, Hobbs, J. R. and Moore, R. C. (eds.), Ablex, Norwood, NJ, USA, pp. 1-36.

ISO (May 1983): "Open Systems Interconnection, Basic Reference Model," ISO International Standard 7498.

Israel, D. J. and Brachman, R. J. (1984): "Some remarks on the semantics of representation languages," in *On Conceptual Modeling: Perspectives from Artificial Intelligence, Databases and Programming Languages*, Brodie, M. L., Mylopoulos, J., and Schmidt, J. W. (eds.), Springer-Verlag, Berlin, Heidelberg, New York, Tokyo, pp. 119-146.

Minsky, M. (1980): "K-lines: A theory of memory," *Cognitive Science*, **4** , pp. 117-133.

Negroponte, N. (1970): *The Architecture Machine*, MIT Press, Cambridge, MA, USA.

Pask, G. (1975): *Conversation, Cognition and Learning*, Elsevier, Amsterdam.

Pask, G. (November 1980): "Developments in conversation theory," *International Journal of Man Machine Studies*, **13**(4), pp. 357-411.

Pask, G. (1982): "Concepts, coherence and language," in *Progress in Cybernetics and Systems Research, Proceedings of the Symposium of Austrian Society of Cybernetic Studies* **11**, Trappl, R., Findler, N. V., and Horn, W. (eds.), Hemisphere, Washington D.C., pp. 421-427.

Sathi, A., Fox, M. S., and Greenberg, M. (September 1985): "Representation of activity knowledge for project management," *IEEE Transaction on Pattern Analysis and Machine Intelligence*, **PAMI**-7(5), pp. 531-552.

Tomiyama, T. and Yoshikawa, H. (1987): "Extended general design theory," in *Design Theory for CAD, Proceedings of the IFIP Working Group 5.2 Working Conference 1985 (Tokyo)*, Yoshikawa, H. and Warman, E. A. (eds.), North-Holland, Amsterdam, pp. 95-130.

Tomiyama, T. and ten Hagen, P. J. W. (June 1987): "Representing Knowledge in Two Distinct Descriptions: Extensional vs. Intensional," CWI Report No. CS-R8728, Centre for Mathematics and Computer Science, Amsterdam.

Tomiyama, T. and ten Hagen, P. J. W. (1987): "Organization of design knowledge in an intelligent CAD environment," in *Expert Systems in Computer-Aided Design, Proceedings of the IFIP W.G. 5.2 Working Conference 1987 (Sydney)*, Gero, J. S. (ed.), North-Holland, Amsterdam, pp. 119-147.

Zimmermann, H. (April 1980): "OSI reference model — The ISO model of architecture for open systems interconnection," *IEEE Transaction on Communication*, **COM-28**(4), pp. 425-432.

4. Multi-Expert Systems for CAD

B.T. DAVID

Ecole Centrale de Lyon, Département Mathématiques-Informatique-Systèmes, B.P.163, 69131 Ecully Cédex, FRANCE

Abstract: *Expert systems are more and more commonly used in CAD. Their performances are better if the problems are limited. However in CAD several designers cooperate for the design of a sophisticated product. In the expert system approach a super-expert system which is able to deal with all aspects of design seems unrealistic. Thus a multi-expert system, which is a federation of smaller expert systems seems more appropriate. In this paper we study this multi-expert system approach mainly from the view point of strategies for cooperation, which are related to the design process model. We also present a knowledge management system which assures information management for the whole system and supports the exchange of information between different tools including expert and classical systems.*

Keywords: CAD, multi-expert system, knowledge management system, design process model.

1. INTRODUCTION

Research in expert systems is becoming very important. Studies have been undertaken in knowledge modelling and representation, in inference engine design, etc. From the user's point of view, the expert system approach is very useful, allowing an incremental collection of knowledge. It is however a black box approach — the inference engine working with its strategy. It is, of course, able to explain the reasoning used but this is not sufficient.

In large domains usually several experts work together to produce an object. That is the case in *computer aided design* in which several designers cooperate in the design of a sophisticated product (Gero 1985). In the expert system approach two methods can be expected:

- construction of a *super-expert system* which is able to take into account all aspects of the design.

- construction of a *multi-expert system* which is a federation of smaller expert systems each corresponding to a special domain of expertise.

Our project is of the second type. Our research is concerned with *different strategies of cooperation* between the expert systems using the same or different inference engines and/or the same or different knowledge modelling. We are also working on a specific problem concerning the use of *multi-expert systems in design*, mainly in design process modelling permitting the use of multi-expert system approach. This paper briefly presents our project.

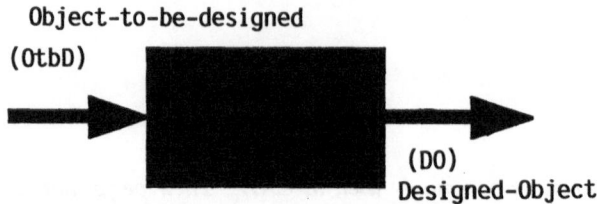

Fig. 4.1. Implicit approach — black box

2. DESIGN PROCESS MODELLING

To be able to support a design activity by a computer-aided process it appears indispensable to have a design process model. For CAD several design process models have been developed. Each captivates a particular characteristic of the design activity. We can present the most current models:

- *Implicit approach* (*black box*: The design process occurs without apparent method and justification, based mainly on experience and intuition (Fig. 4.1). This approach is useful for individual work but completely inappropriate for collective work.

- *Explicit approach* (*glass box*: The design process is organized in steps and phases with pre-defined relationships (Fig. 4.2). This approach proposes a more organized way to design and permits the collaboration of several partners (designers or design tools).

- *Sequence of design phases*: From a macroscopic point of view the design process, or more generally *design-development-manufacturing process* (DDM), is organized in large phases each of which has a particular objective that can be characterized by the result (conceptual object, virtual object, and prototype). This approach defines a global structure and organization of the DDM process (Fig. 4.3).

- *Design loop*: From a microscopic point of view each step of design is characterized by an *analysis-synthesis-evaluation-decision sequence* with several possibilities for

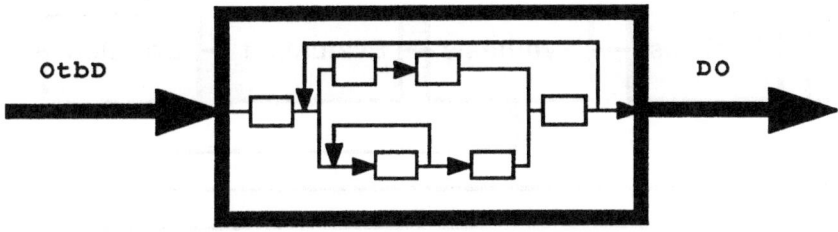

Fig. 4.2. Explicit approach — glass box

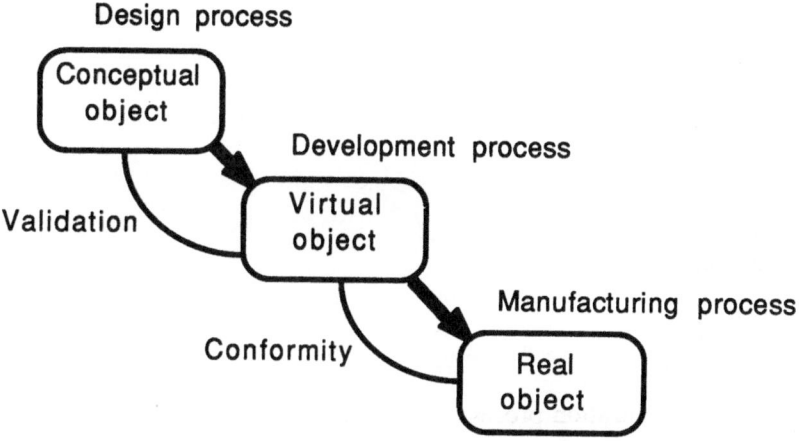

Fig. 4.3. Sequence of DDM phases

loops according to the evaluation of the results. This approach defines a micro structure of each design step (Fig. 4.4).

- *Design as optimization*: The design process can be also viewed as an optimization process with variables, parameters, constraints, and an objective-function. This approach is not so useful because the mathematical formulation of a design process is difficult (Fig. 4.5).

- *Well-defined and ill-defined problems*: Usually a design process works on a problem for which a trivial solution does not exist. For well-defined problems in which all aspects or parameters have been identified, the design is a parameterized design. For ill-defined problems the design process works not only on the designed-object but also on the object-to-be-designed which is elaborated during different phases of the design process (Fig. 4.6). This approach points out to the necessity to manage

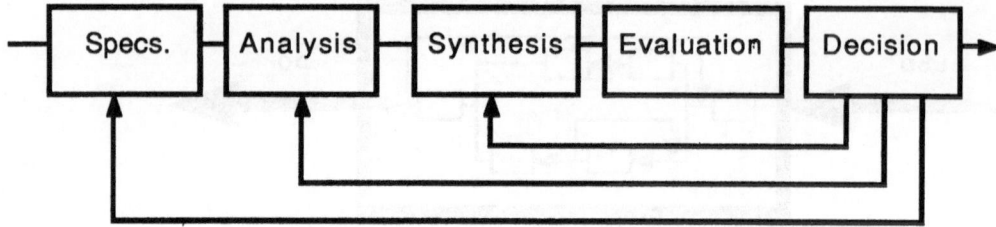

Fig. 4.4. Design loop

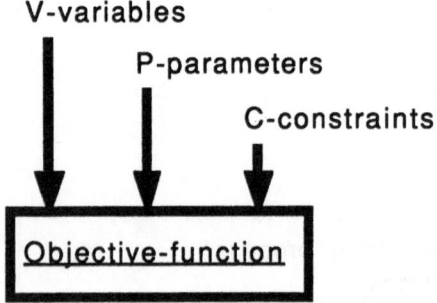

Fig. 4.5. Design process as optimization process

not only the result of the design process (i.e. the designed-object) but also the description of the specification of this future object (i.e. the object-to-be-designed).

- *Design process* considered as activities based on *induction, deduction, intuition, experience, and creativity* (Fig. 4.7): This approach points out to several activities on which the design process is based and which are difficult to formalize. Until recently, all these activities were necessarily under the responsibility of the designer. With the expert system approach it is possible to transfer progressively the experience, the deduction, and the induction to an artificial expert. Intuition and creativity are still in the field of competence of a human expert, but for how long?

In this expert system based approach we need a unique *comprehensive design process model* which is able to take into account all the aspects of the design process presented above. We now present our comprehensive design process model which was elaborated long ago (Rivero 1977) but which has become particularly important for this new approach.

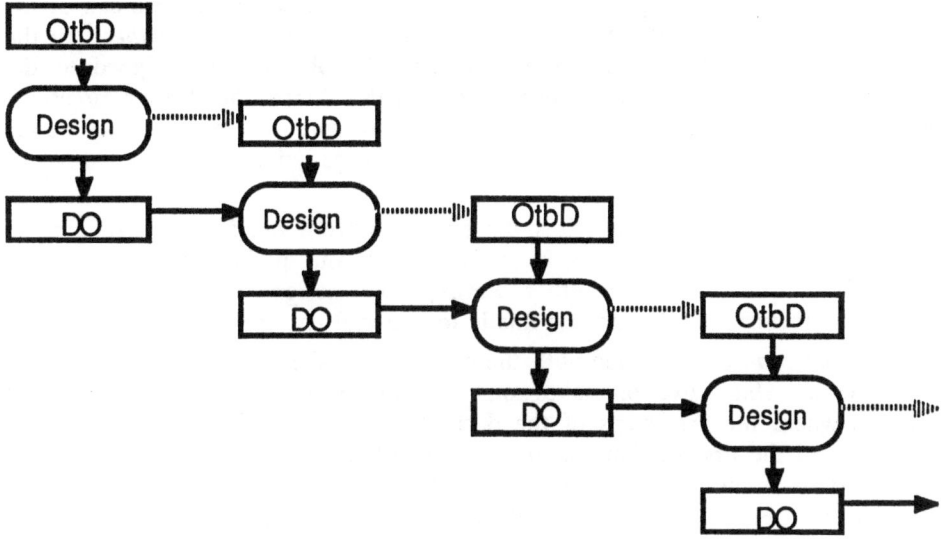

Fig. 4.6. Ill-defined problem

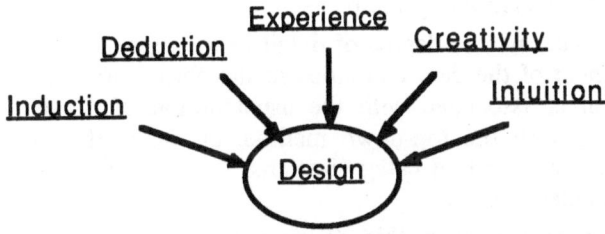

Fig. 4.7. Design activities

3. COMPREHENSIVE DESIGN PROCESS MODEL

The comprehensive design process model (CDPM) has as objective to be able to define all authorized design processes and to support and manage the applied design process. Then CDPM must take into account the evolution of the *object-to-be-designed*, which is a specification of the product and the *designed-object*, which is a possible answer (i.e. a proposal for the designed product). The CDPM must also manage the design activities by using artificial or human processes based on induction, deduction, experience, creativity, and intuition.

We can consider that design takes place within a *design field* which we can divide into two sub-fields: the *field of the objects* and the *field of the design activities*. In the "object field" we describe the transformation from the object-to-be-designed to the designed-object as a result of a design procedure. In this "object field" four directions corresponding to different types of operations were brought in to light. They define different semantic operations either on the object-to-be-designed or on the designed-object or both.

- *Morphological direction* characterizes the evolution from the function toward the form, toward the structure, that is, from the abstract and qualitative information to concrete and quantitative (in other words, from design notions to relational notions, to dimensional notions and finally to material and technological notions).

- *Definition direction* is concerned with the evolution of the specifications (object-to-be-designed). This direction takes on importance in the case of ill-defined problems where the first steps of the design process contribute not only to the elaboration of the first versions of the designed-object but also and mainly to the understanding of the object specification and the progressive elaboration of these specifications. For well-defined problems, displacement in this direction rarely happens.

- *Structural direction* is related to the method for problem solving to organize the structure of the design process. For example, top-down approaches, bottom up approaches, or the mixture of these two. Other related concepts include objects, sub-objects, and the articulation of sub-objects. We can also count up composition and decomposition for constructing the problem space.

- *Resolution direction*: we call "resolution" the degree of detail in the representation of an object. This is the refinement of the data contained in the object model (in-depth definition). This direction is associated with the usual notion of "design phase," which itself is associated with the top-down method, or the method of successive refinement. In this context we find categorizations such as preliminary design, schematic design, and detailed design.

A design process is a particular way to evolve in this *design space* (Fig. 4.8). The classical design process model corresponds to the diagonal run of the design space.

These four directions are independent, i.e. displacement in each direction is possible independently from displacement in the other directions. Moreover displacement in each direction is characterized by the nature of participation of the object-to-be-designed and the designed-object in this operation (Fig. 4.9). In this way each operation is characterized with relation to the design space movement.

The "field of design activities" is characterized by the above mentioned characteristics, i.e. experience, deduction, induction, intuition, and creativity. For each particular activity its action on the design data with respect to the design space displacement must be defined. As for the relationship which exists between these two sub-fields (the design activities field and the object field), it is the design process as a whole which is composed of data and activities.

Two other notions are important in this CDPM:

- *Weighted objectives and constraints* are used to describe in a more precise manner the nature of information which appears in the specifications (object-to-be-

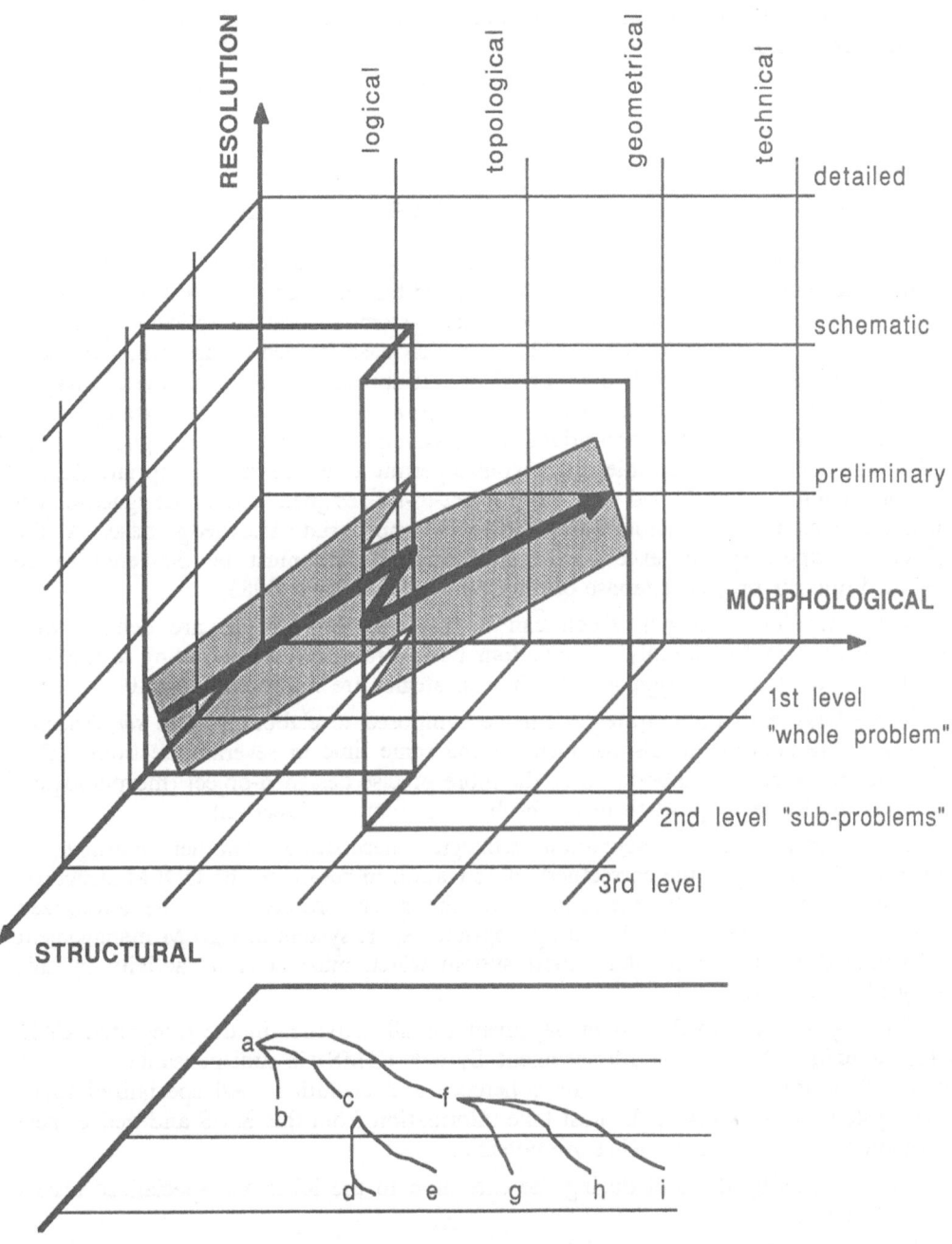

Fig. 4.8. Design space

designed), which include objectives with the requirements to be met and the constraints representing the design decision that have also been anticipated in these requirements.

- *Traceability of the design process* is the ability to explain all design decisions and also to be able to go back to the previous design steps.

4. FROM EXPERT SYSTEMS TO MULTI-EXPERT SYSTEMS

Expert systems are usually able to solve complex but particular problems. Their performances are better if the problems are limited but detailed. In CAD we find these kinds of problems, but in an integrated CAD system, several complicated problems occur often in completely different fields. In this case it seems impossible to use a single expert system to solve all these problems. It appears more reasonable to imagine a set of expert systems and to define a *strategy of cooperation*. This strategy must permit the activation of the appropriate expert systems each time the designer (human or artificial) considers it desirable. It must also permit access to the appropriate data: a selection of information issued from the object-to-be-designed and/or designed-object which will constitute the factual base, and to the appropriate knowledge base. At the end of an expert system execution the elaborated data must be delivered to be integrated into the project database of the CAD system (David 1983).

The activation strategies which can be used in the practice are defined with relation to the CDPM. Mainly we have four basic strategies corresponding to the four directions of the CDPM design space. These strategies are shown in Fig. 4.10.

These basic activation strategies can be composed to elaborate more *sophisticated strategies* corresponding to displacement at the same time in several directions. This means, for instance, the elaboration of the more precise designed-object (morphological direction) and its decomposition into sub-objects (structural direction).

This organization of activation strategies necessitates another *manager of activations* which verifies the correctness of activation in regard to the CDPM structure, i.e. with the design space organization for the given problem and the authorized relations. This can be realized by an appropriate expert system in CDPM management or distributed to each particular expert system which must start its activity by this management sequence.

To support the information management for all activities in the integrated CAD system a unique Knowledge Management System (KMS) is indispensable. It must receive all information and assure its coherence and evolution. All specialized tools, expert systems, or classical tools must take information from this KMS and deliver new information to it. Two approaches are possible:

(1) *Direct access* by the tool during the execution to the KMS via specialized access procedures;

(2) Isolation of the tool from the KMS, i.e. *initial acquisition* of information at the beginning of the execution and delivery of new or modified information at the end.

The first approach is more cooperative; the second one permits the choice of appropriate knowledge representation. This KMS is presented in the next section.

Direction	modified	read only
Definition	object-to-be-designed	designed-object
Morphological	designed-object	object-to-be-designed
Resolution	designed-object	object-to-be-designed
Structural	object-to-be-designed designed-object	

Fig. 4.9. Directions and operations

Direction	input information	results
Definition	object-to-be-designed/i designed-object/j	object-to-be-designed/$i+1$
Morphological	designed-object/j object-to-be-designed/i	designed-object/$j+1$
Resolution	designed-object/i object-to-be-designed/j	designed-object/i' object-to-be-designed/j'
Structural	object-to-be-designed/i designed-object/j	sub-objects-to-be-designed sub-designed-objects

Fig. 4.10. Elementary strategies for cooperation

5. KNOWLEDGE MANAGEMENT SYSTEM

The Knowledge Management System is basically a *frame-like system*. It allows the separate definition of the structure and the semantics attached to this structure.

The *structure* is expressed by the *entities* and their *hierarchical relationships*. The *semantics* is expressed by the *properties* attached to the *entities* and *semantical relationships*. The *type* of property, entity and relationship must be indicated for a particular occurrence before it is used. This type defines not only the type of value but also the authorized operations, persons, and tools allowed to use this information and the *design process steps* in which this information is available. In this way the KMS is able to verify the *coherence* each time that a manipulation occurs. A *Frame Prototype Editor* (FPE) is used for the creation and updating of the types.

To facilitate the elaboration of information, several levels of entities are possible:

- *Generic entity* which can receive a specialization from the FPE.
- *Type entity* which is produced by the FPE.
- *Partially valued entity*: first step of occurrence production.
- *Completely valued entity*: occurrence of entity.

The coherence of five natures is assured (Fig. 4.11):

- *Dependence* which expresses the relationship between properties of a same entity;
- *Correspondence* which expresses the relationship between the same property of different entities located in different levels of the structural hierarchy (synthesize, inherit, or other);
- *Transformation* which expresses a particular transformation during a displacement in the design space;
- *Conceptual accessibility* which expresses access restriction related to the design space organization;
- *Organizational accessibility* which expresses access restriction related to the design team organization.

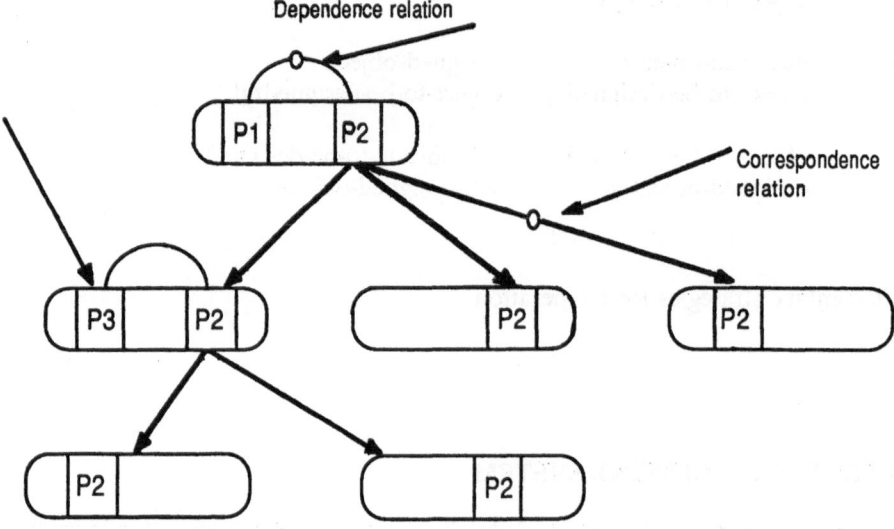

Fig. 4.11. Coherence relations

Extendibility and evolutivity are possible because access to FPE is possible each time. In this way a particular tool or designer can add a new property or elaborate it from an existing one.

The operations which are used in the KMS are semantically coherent, working on the structure (and elaborated once and for all) or on the semantics and extendible by the activation of the operations associated with the properties (and created if necessary at this time). This organization of the KMS allows it to be extendible and assure the coherence of managed information.

6. CONCLUSION

In this paper we have presented an approach which allows for the generalization of the use of experts systems by introducing the concept of multi-expert systems. This approach preserves the usual performance of the expert systems and increases their use in a complex design environment. The cooperation of different expert systems is managed by the activation strategies related to the design process model. The information used by these expert systems is managed by a unique knowledge management system. We presented a comprehensive design process model as a framework for a multi-expert system for CAD and the basic strategies for cooperation between the expert systems related to this model. We also outlined the main principles of our knowledge management system. We are experimenting with this approach in two different fields: architectural programming (Berger, David, and Bernanose 1986) and mechanical design for electronics. Two prototypes are under development and the first results are promising.

REFERENCES

Berger, D., David, B. T., and Bernanose, C. (June 1986): "Le système MARS: Un système de modélisation architecturale à référence spatiale," *Proceedings of International Joint Conference on CAD and Robotics in Architecture and Construction*, Marseille, France.

David, B. T. (1983): "Conceptual framework for CAD systems construction," in *CAD Systems Framework, Proceedings of the IFIP W.G 5.2 Working Conference 1982 (Rolos, Norway)*, Bø, K. and Lillehagen, F. M. (eds.), North-Holland, Amsterdam.

Gero, J. S. (ed.) (1985): *Knowledge Engineering in Computer-Aided Design, Proceedings of the IFIP WG 5.2 Working Conference 1984 (Budapest)*, North-Holland, Amsterdam.

Rivero, V. (June 1977): "Une Contribution à la Conception Architecturale Assistée par Ordinateur (A Contribution to Computer Aided Architectural Conception)," Thèse de Docteur-Ingénieur (Doctorial thesis), Université Scientifique et Médicale de Grenoble, Institut National Polytechnique de Grenoble, Grenoble, France.

Report on Session 2

Chair: T. Takala †
Cochair: P. Veerkamp ‡
Edited by: V. Akman ‡

We can pose three questions and try to see what each paper has to say about them.

1. *What is design?*

    Bernus has not touched the question of designing too much in his presentation, but rather discussed intelligence in general. To him, design is a cooperative process of many concurrent active agents, some of which are human and the others artificial.

    David defines design in his comprehensive design process model as a transformation process from the object-to-be-designed to the designed-object. In other words, design is a mapping from the design specifications to the design solution. It is an evolutionary process in four orthogonal directions (morphology, structure, definition, and resolution) and can be depicted as a trace in the 4-D Cartesian space.

2. *What is the role of CAD systems now and in the future?*

    A simplistic answer would be to define the role of CAD as supporting the design activity. We can summarise both presentations by stating that the main role of a CAD system is to manage all the information related to the design and production processes, i.e. to bring the right information to the right places when it is needed by different concurrent agents.

3. *What/Where is intelligence?*

    Bernus gives a particular but important view of intelligence: Intelligence is an agent's ability to communicate with other agents, using concepts on multiple levels of abstraction. This requires transformation and representation of the intended ideas "up" (making abstractions of ideas) or "down" (making models of theories) to a level where communication is physically possible.

---

† Laboratory of Information Processing Science, Helsinki University of Technology, Otakaari 1, 02150 Espoo, Finland.
‡ Centre for Mathematics and Computer Science, Kruislaan 413, 1098 SJ Amsterdam, The Netherlands.

For David, intelligence appears in different forms in different places. A user interface, a problem-solving expert, a process-planning expert, etc., all have their own specialized intelligence. An intelligent design process combines many different ingredients like induction/deduction, experience, creativity, and intuition.

During the Q&A period, Bijl told Bernus that belief and proof cannot be exercised by formal systems in a manner meaningful to people and that we may not be successful in reconciling formal proofs with the intentions of a designer. Nadin asked about the representational aspects of semantic integrity. Bernus replied that an agent refers to some implementation of a theory. Since there is a mapping from the agent to the implementation, one needs a model to ensure the integrity of the mapping.

Schramel advised David that he should start from the beginning in order to deal with integration. The beginning consists of the needs of the user. Nadin asked about the relevance of circumscription, a theory developed by John McCarthy. David told him that he uses two paradigms: generalization and specialization. Akman raised his doubts about the usefulness of expert systems in general, for they have barely a superficial appreciation of their expertise domain. David thinks that his problem is to integrate several expert systems. Hoeltzel observed that design optimization is powerful and natural despite David's claims. David said that a user cannot hope to discover good objective functions for complex design optimization problems.

Session 3

5. An Adaptive, Generic Planning Model for Large Scale Integrated Engineering Design

D.A Hoeltzel and W.H. Chieng

Laboratory for Intelligent Design, Department of Mechanical Engineering
Columbia University, New York, NY 10027, USA

Abstract: *Intelligent computer-aided design (CAD) emulates the human activity of design so that production planning, decision making, and inventive design can be performed by computers. Based on the history of human experience in engineering design, a formalized approach to design methodology should include procedures from (1) conceptual design, (2) layout design, and (3) numerical optimization design. The highest design level in such a system should be responsible for generating skeleton structures of entities within the design process which are eventually to be specified uniquely, and to be optimized. Planning plays a key role in such a system. Planning has been utilized as a tool for process organization within the knowledge domains of chemical engineering, electrical engineering, and manufacturing, as well as for general problem formulation and solution. State estimation, subtask scheduling, and constraint propagation are factors of prime importance in this type of problem. A methodology for planning and the problems associated with it within large scale design are discussed. The generality of our hypothesis for the design optimization process is examined within the context of a prototypical mechanical design model. An example which demonstrates the applicability of this approach to mechanical power transmission design and which is representative of a large scale design problem, is provided.*

Keywords: hybrid design optimization, mechanical design, artificial intelligence in CAD, planning models, scheduling, constraint propagation.

1. INTRODUCTION

In engineering design, the main task is to apply scientific knowledge to the solution of technical problems and then to optimize the solutions within given material, technological, and economic constraints. Pahl and Beitz (1984) show that the main phases of the design process can be separated into four: (1) Clarification of the task and specification of the design requirements, (2) Conceptual Design, which establishes the solution principles and concepts, and estimates a rough design, (3) Embodiment Design, which determines the design layout and form and develops a technical and economical evaluation, and (4) Detailed Design, which optimizes the design principle, layout, and form. The concept of of this arrangement as it might exist in the form of a design expert system is depicted in Fig. 5.1.

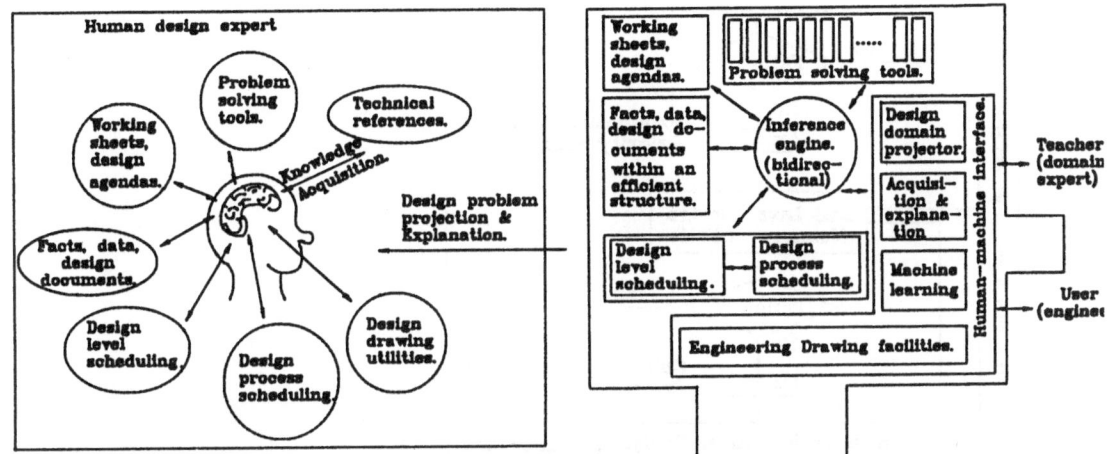

Fig. 5.1. A design expert system to emulate human design expertise.

Within OPTDEX (Optimal Design Expert) (Chieng and Hoeltzel 1987), a cognitive CAD system (Fig. 5.2), concentration is focused on design optimization. This prototypical system contains levels corresponding to those described above including conceptual (top) level design, layout (cell) level design, and numerical (optimization) level design. In particular, the Top Level Design Manager (TLDM) within the OPTDEX system generates skeleton structures (interconnected "generic" power transmission components, i.e., gears, shafts, pulleys, etc. without predetermined specifications, Fig. 5.3) for the optimum design of mechanical power transmissions and is considered as a planner. Planning has been used as a tool for process organization within the knowledge domains of chemical engineering (MOLGEN (Stefik 1981)), electrical engineering (Gongaware et al. 1983), manufacturing (GARI (Descotte and Latombe 1985)), as well as for general problem formulation and solution (NOAH (Sacerdoti 1977) and SIPE (Wilkins 1984)).

State estimation, subtask scheduling, and constraint propagation are factors of prime importance in this type of problem and are currently undergoing investigation. Here, a methodology for planning and the problems associated with it within large scale design are discussed. The generality of our hypothesis for the design optimization process is examined within the context of a prototypical mechanical design model, which we refer to as the Top Level Design manager (TLDM) within OPTDEX.

Based on the design procedures included in OPTDEX, our hypothesis for a systematic design optimization methodology is postulated as follows:

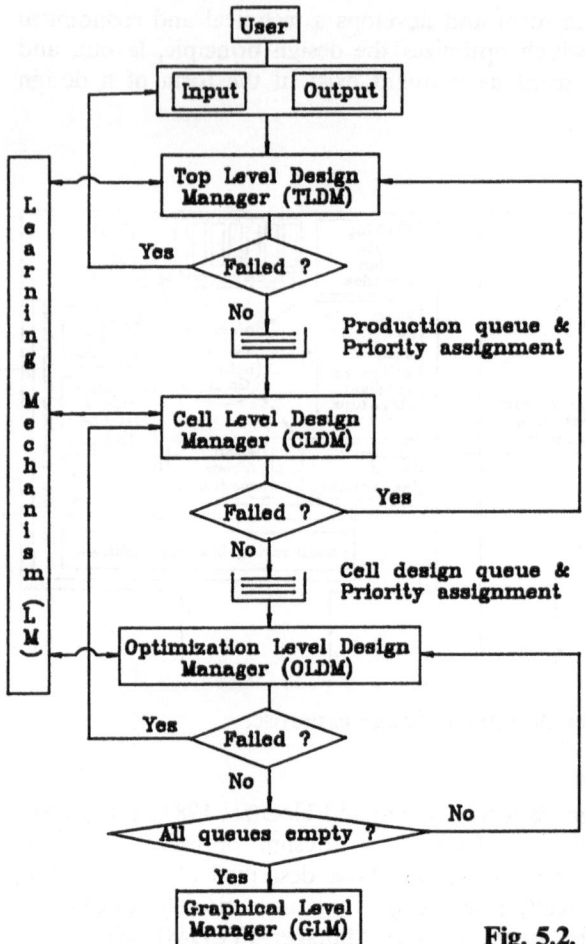

Fig. 5.2. Flow control in the OPTDEX system.

1. Design procedures are propagated gradually from a qualitative domain to a quantitative domain, from synthesis to analysis, from estimation to evaluation. This hypothesis is based on observations from real world systematic engineering design (Pahl and Beitz 1984), which imply that design procedures are generally hierarchical in nature.

2. Based on the framework of a planning model, the design process can be separated into a generic portion and a domain specific portion. These, in turn, may be further subdivided into a creative (or inventive) design portion and a routine design portion. This hypothesis further stresses the fact that design knowledge is generally hierarchical.

3. An abstract design optimization process which is based on a hierarchical data structure and monotonic reasoning is guaranteed to converge during the search for the optimum solution.

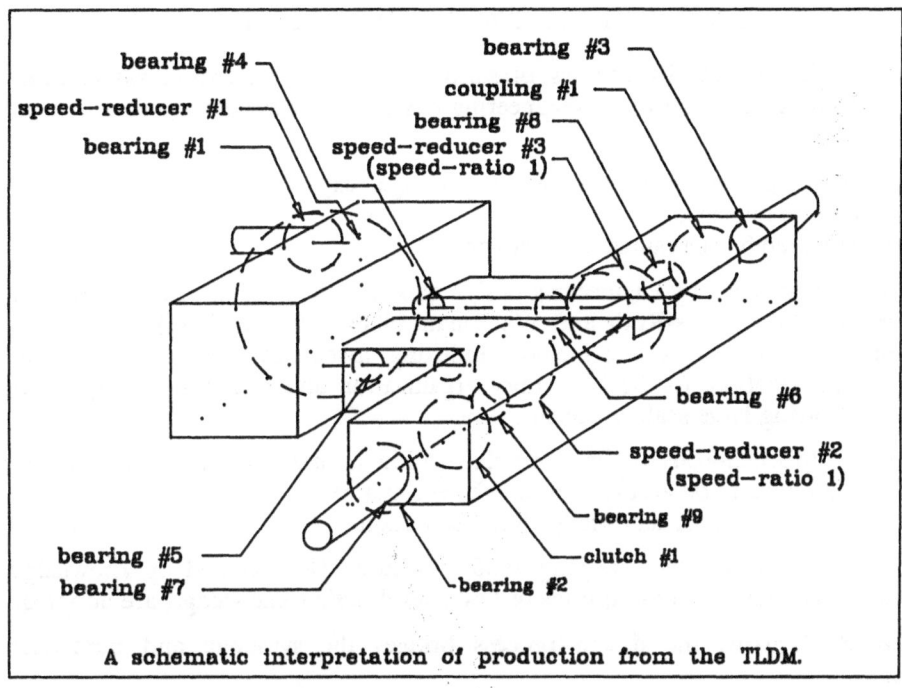

bearing #4

bearing #3

speed-reducer #1

coupling #1

bearing #1

bearing #8

speed-reducer #3
(speed-ratio 1)

bearing #6

speed-reducer #2
(speed-ratio 1)

bearing #9

bearing #5

clutch #1

bearing #7

bearing #2

A schematic interpretation of production from the TLDM.

Fig. 5.3. Generation of generic power transmission components.

Yoshikawa's paradigm model (Yoshikawa 1981), coupled with the concept of monotonic reasoning (Choy and Agogino 1986; Israel 1983), serves as a proof of the third hypothesis. This proof is based on a hierarchical data (design knowledge and design procedures) structure and may be stated as follows:

For a given total design goal, T, composed of n subgoals,

$$T = T_1 \cap T_2 \cap T_3 \cap \cdots \cap T_n,$$

there are two intermediate designs S_1 and S_2. The first design, S_1, exists and satisfies design subgoals T_1 through T_k. The second design, S_2, which does not presently exist, i.e. a proposed design, satisfies the combination of subgoals $S_1 \cap T_{k+1}$.

(1) The "policies" (Yoshikawa 1981), which transform design S_1 into S_2, can be produced by retrieving data from the data hierarchy and comparing their relative performance in parallel.

(2) Assuming that the similarity between the design subgoals, T_i, and the design attributes increase monotonically within the data hierarchy, the functional and attributive design insufficiencies, propagated from higher entity levels in the data structure, can always be resolved from a lower level in the data structure.

Based on (1) and (2) above, the optimization process is guaranteed to converge.

2. DEVELOPING AN ENGINEERING DESIGN EXPERT SYSTEM

In accordance with the above hypotheses, prescriptions for the problems encountered in the design and implementation of an engineering design expert system are summarized in the following list:

Problem 1: Large amounts of design knowledge, experience, and experimental results are to be managed.
Prescription: Memory Organization Network.

It has been recognized by engineering design expert system investigators (Charniak and McDermott 1985; Hayes-Roth and Waterman 1986; Lebowitz 1986) that frame-based system approaches, including memory indexing, associated memory networking and concept generalization, for both memory organization and rule base management are suitable for solving large scale problems.

Problem 2: Miscellaneous design formulae within a rule-based and frame-based expert system have to be accessed in an efficient manner.
Prescription: Hierarchical Knowledge Representation.

Procedural design networks (Dejong 1986; Winston 1980), guided by knowledge layer linkers, containing brief descriptions of lower level design knowledge, are adopted.

Problem 3: Usually, the design process bridges the symbolic and numerical domains.
Prescription: Hybrid (Symbolic-Numerical) System Approach.

Inference, i.e., logical deduction, operates on qualitative rules and produces numerical equations. Knowledge about numerical programming under a well understood mathematical programming package, for example within OPTDEX the ADS (Vanderplaats 1985) optimization library has been adopted, and subsequently encoded into qualitative rules. The associated decision making within the numerical execution process, based on human experience or a machine learning process, can also be qualitatively formulated.

Problem 4: Large-scale systems require overall testing and calibration.
Prescription: Knowledge Acquisition and Explanation.

It has been indicated by Quinlan (1982) that an expert system containing data explanation and knowledge acquisition facilities expedites the testing and calibration process by five fold compared to plain knowledge-based systems. In order to supervise the overall system, a centralized reasoning mechanism which surpasses the design levels and chaining method has to be used for the reasoning process.

Problem 5: Some other difficulties and their prescriptions.

There are a number of other problems which typically occur in large-scale design. For example, "where to begin" and "how to continue?" are questions frequently asked. To remove these obstacles from computer-aided design, the DOMINIC-I system (Dixon et al. 1986), for example, presented an analytical design prototype containing heuristic decision making. Also, regarding the representation of common sense design knowledge, Zadeh (1983) states that, *"in the design of expert systems, $\cdots$ conventional knowledge representation techniques based on predicate calculus and related methods are not well suited for representing common sense knowledge."*

Experience-based common sense design knowledge includes design procedures for scratching, prescription assigning, manufacturing process configuring, and prediction of design difficulty.

3. PLANNING IN DESIGN OPTIMIZATION

The generic design optimization planning model incorporated within the TLDM of OPTDEX is constructed from among (1) design requirement realization, (2) problem reduction, (3) design objective compromising, (4) scheduling, and (5) redesign. The design optimization planning process which we have formulated based on these phases of design is depicted in Fig. 5.4. Each of these five areas will be described in detail.

3.1. Design Requirement Realization

Transcending traditional CAD tools, an intelligent CAD system should employ a reasoning capability that includes design requirements and design specification assignment. These capabilities include natural language processing, projecting (projection of common design specifications onto specific design domain tasks), and refinement acquisition (querying the user about the degree of design vagueness). It has been pointed out by Ullman and Dietterich (1986) that the capability of specification development and assessment are also important within design requirement realization.

According to these requirements, there are two tasks for specific design expert system development:

(1) Abstract user design specifications which emulate human designer response for a specific design domain. These are usually specified via a video taping session between the customer and the human designer during a design session. Combining machine learning techniques based on the concept generalization process, from among the real world designer's responses to the user's requirements, a protocol is constructed which simulates the human interface, thereby removing the bias knowledge (from the knowledge engineer) within the expert system. From the machine learning process a generic protocol model for a specific design can be constructed.

(2) Projecting and refining real world design requirements onto specific design domain knowledge. From the standpoint of design efficiency and memory organization, the machine learning and emulation procedures which have been described above decrease the modularity of an expert system. Projecting and refining processes which are based on machine internal design problem solvability, reasoning, and understanding can be utilized to compensate for that disadvantage.

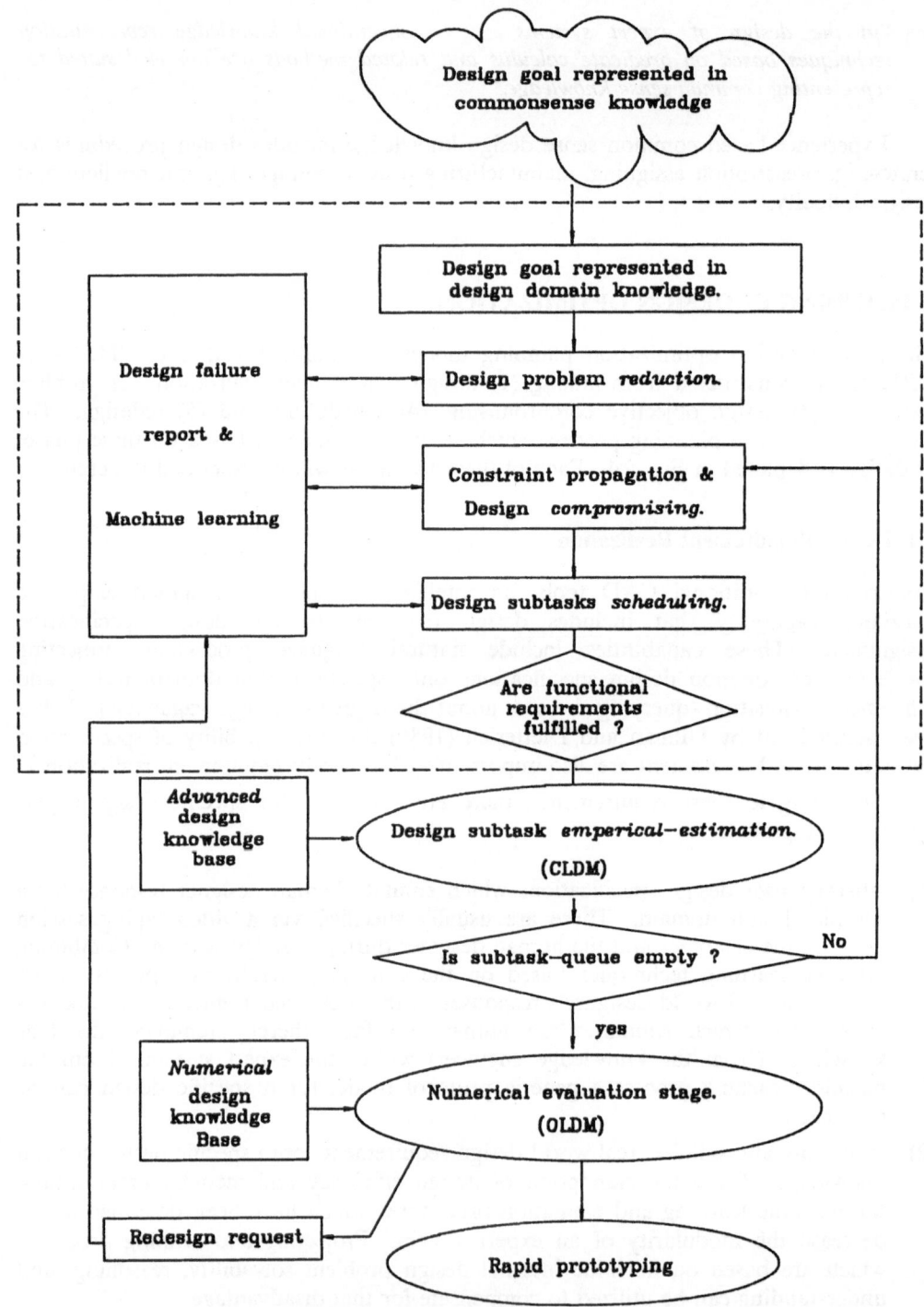

Fig. 5.4. The optimization planning process is enclosed within the dotted lines.

3.2. Design Problem Reduction

Problem reduction represents an approach to reasoning backward from the problem, generating subproblems until the original design objectives can be fulfilled. Some basic design problem reduction terminology follows:

1. Deterministic design problems using logical AND/OR graphs — some of the design objectives can be broken down into alternative sets of successor problems using a tree structure (Love 1980). For example, in the design of a power transmission, to confine the range of allowable transmitted torque values, bounds are established. The lower bound, *?lb*, is specified by the engine's power at idle while the upper bound, *?ub*, is specified by the engine's maximum torque and is maintained by a power clutch. This knowledge has been represented in the following rule:

   > IF   *(Torque-bounds (ranges: from ?lb: N-m to: ?ub: M-m))*
   > THEN
   > *(Engine-power-at idle: ?lb: N-m)*
   > *(Clutch maximum-allowable-torque: ?ub: N-m)*

2. Means-ends analysis (Israel 1983) — a general and heuristic problem reduction methodology whereby the problem reduction process is driven by the elimination of differences between the current state and the goal state.

3. Nondeterministic design problems using AND/OR graphs — the majority of design and manufacturing processes reside in the nondeterministic domain where the design variables are tightly coupled. A special strategy is usually used where nondeterministic problems are linearized at specific states (Quinlan 1982), so that the nondeterministic design knowledge can still be. represented in AND/OR graphs at those given states. As an example, in the design of a power transmission, if the transmitted power (transmitted torque, *?torque*, times the rotary speed, *?speed*) is higher than 10,000 watts, then cooling is required. If the transmission shaft contact stress is higher than 100,000 Pa, then forced lubrication is required. This knowledge has been represented in the following rules:

   > IF   *(?shaft operating_speed: ?speed: rpm load: ?torque: N-m)*
   > *(function (>(* ?speed ?torque) 10,000))*
   > THEN
   > *(cooling ?shaft)*
   > IF   *(?bearing thrust area: ?area: m-m load; ?side_thrust: N)*
   > *(function (> (/ ?side_thrust ?area)) 100,000)*
   > THEN
   > *(forced_lubrication ?shaft)*

As depicted above, deterministic problems can be solved by using a divide and conquer (AND/OR graph) approach. Nondeterministic problems can be solved by using means-ends analysis aided by constraint propagation. For a general purpose intelligent CAD system both types of problem reduction have to be implemented.

3.3. Design Compromising

Since optimization processes are dependent on boundary conditions (design requirements), there is no general methodology that will guarantee global design optimization. While "global optimization" represents a confusing concept in both numerical and symbolic search domains, the concept of "design compromising" can be used more effectively since it concentrates more on overall design goal achievement. Through design compromising, the global design constraints integrate with local design constraints and are propagated throughout the overall design. The temporary design objective for the current subproblem is pinned down according to the design constraints. To construct the procedures for pinning down the subgoals, Love (1980) states that,

> " · · · Many design projects that missed their targets for performance, cost, and time have shown that there was a lack of trade-off control, · · · ".

It has been emphasized by Dixon *et al.* (1986) that without trade-off reasoning, the optimization process always lacks the ability to track the true optimum.

Trade-off control is an art in engineering design which defies generic representation, but always has an explicit representation for a specific design problem. To simplify the model for design compromising such that the design system can easily follow it, a logical resolution theory may be applied. Based on a hierarchical problem solving structure, a temporary subgoal from the antecedent is transmitted to the children and the constraints among the children are propagated to each other. In this way the logical resolution theory is applied to the constraints and subgoals, so that temporary subgoals, corresponding to individual subproblems, are generated.

3.4. Subproblem Scheduling

As each stage of problem reduction decomposes the original design objective(s) into sets of preliminary subproblems recursively, systematic design planning continues to decompose them until primitive designs for detailed (cell level design, Fig. 5.2) design are found. The conflict-resolution phenomena among these subproblems induces a scheduling problem: "Which one should first be expanded?".

Heuristic scheduling (Nilsson 1971) introduces human design expertise into cost evaluation of design-scheduling and attempts to solve large scale design problems efficiently. Design scheduling is a queuing problem. Control of scheduling differs from one design domain to another. Thus, explicit knowledge representation and metaknowledge representation (Buchanan and Shortliffe 1984) are employed for generic design purposes.

3.5. Redesign

Redesign can be subdivided into two categories (Pahl and Beitz 1984). One is failure-driven redesign, the other is new-design-concept-driven redesign. Failure-driven redesign is invoked by design failures (design constraint violations). The main difficulties encountered in redesign for large scale hybrid optimization include:

1. The lack of an absolute design goal: Choosing a temporary design goal is an additional optimization process. Design failures corresponding to temporary design goals are difficult to define and detect.

2. The lack of obvious design constraints: The majority of design constraints are dynamic and temporary design goal independent.

Design violations also involve trade-offs. However, as was discussed in the previous section, the design architecture of symbolic optimization is hierarchical and the design knowledge base (design levels, rules and facts) is also organized in a hierarchical sense. Based on this systematic structure, the design failures can be easily exposed to the design hierarchy until the failure can be totally resolved by the antecedent node (higher design level). Within TLDM, redesign is localized using a least commitment judgement, which allows failure driven redesign to be achieved in an efficient manner.

4. PROTOTYPE PLANNING MODULE WITHIN TLDM

The schematic design planning module incorporated within TLDM of OPTDEX is depicted in Fig. 5.5. Planning in TLDM has been implemented using a central

Link from higher level design plans.

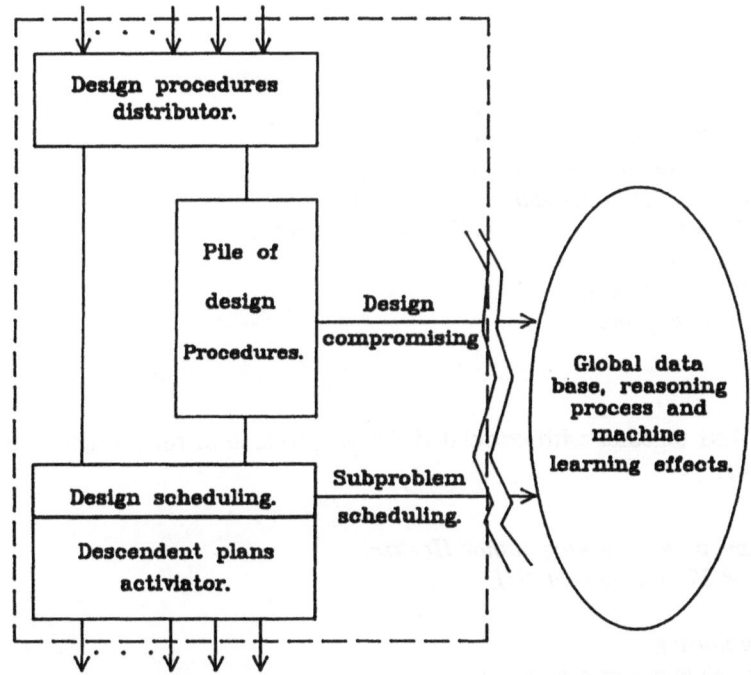

Link to lower level design plans.

Fig. 5.5. Planning module (enclosed within dotted lines) within TLDM.

81

reasoning process which includes global control for design compromising, redesign, and design scheduling. A strong reason for a centralized control configuration is that design compromising, redesign and scheduling are not necessarily hierarchical.

A planning module example which demonstrates the concepts described in the preceding sections follows. A partial planning structure within TLDM is shown in Fig. 5.6.

```
(PLAN:name speed-reducer:author Hoeltzel
     :date  06-10-86:comment NIL
     :link-from
     (:name  power-trans-problem-reduction
      :activate-procedures T)
      :name  request-speed-reducer
      :activate-procedures NIL)

     :procedures
     (:1 rotary-to-rotary-speed-reduction-formating
      :2 rotary-to-reciprocate-speed-reduction-formating..)
     :link-to
     (:scheduling
      :test default
      :priority NIL

     {Machine learning process creates learned scheduling, e.g.
     :test (speed-reduction-ratio (?Sr > 30))
     :priority (first rotary-rotary-speed-reducer)
     may be created in a context-free format})
     (:name rotary-to-reciprocate-speed-reducer
      :test (difference ?name1 ?name2 (?speed-difference <> *1.))
     (difference ?name1 ?name2
     (?motion-type-difference
     *rotary-to-reciprocate))
      · · · ))
```

Procedures are chunked together with correlated design production rules which can be represented as follows:

```
(PROC:name power-checking:author Hoeltzel
      :date 08-27-86:comment NIL

      :compromising
      (:info (input-power-actuator
      (actuator > (step-motor hand))
       :effect-on power-violation
       :test (input-power-sum ?input-power)
```

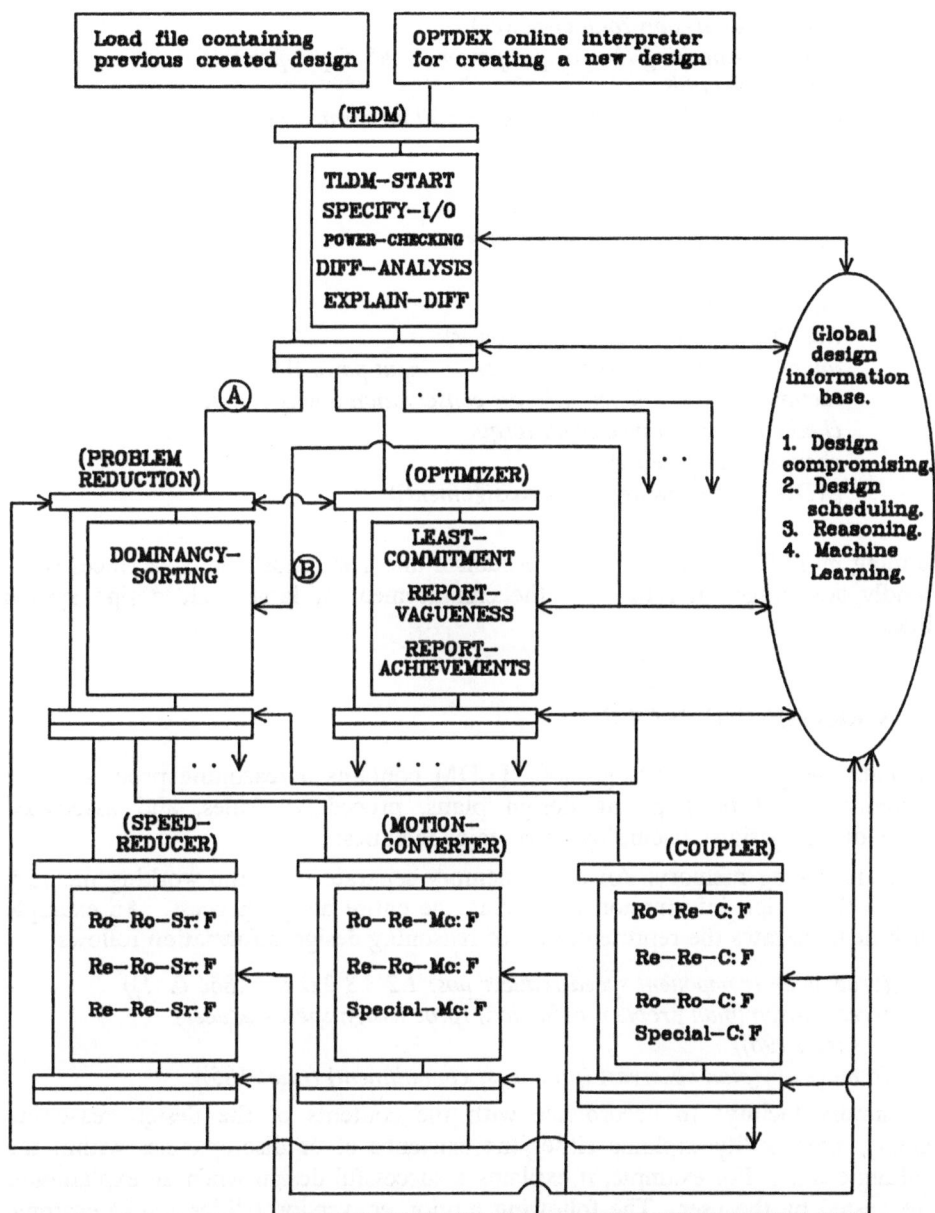

Where, R_o: Rotary, R_e: Reciprocating, S_r: Speed reducer, M_c: Motion converter, C: Coupler, F: Formulation.

A: Inter module constraint propagation between two different levels.

B: Inter module constraint propagation within the same level.

↓: Transfer of control to next lowest design level module.

Fig. 5.6. Design planning hierarchy for a power transmission design problem within TLDM.

```
(output-power-sum ?output-power)
(function-checking (> ?output-power (* .85 ?input-power)))
 :action remains
{Other sets of design compromises will be created by
machine learning.})
 :rules
 (:name input-power-summation
:test  <test10>:action <act10>
  . . .

:name power-violation
:test (input-power-sum ?input-power)
(output-power-sum (?output-power > input-power))
:action (BIND ?choice (Power conservation law is violated,
choose action (change-input-torque
change-input-speed?))
(ADD (direct-invoke I/O-reassignment))
 · · · ))
```

Planning in TLDM employs more formalism and restriction. This sacrifices some user friendly advantages in return for the enhancement of large scale design system capabilities.

5. DESIGN REASONING PROCESS

For automatic design failure recognition, TLDM contains a reasoning process which assigns credits to all participating design plans, procedures, rules, and associated productions during design. It employs two major facilities:

1. Design reasoning memory: An extra memory separate from the working memory to store the design information by tracing the entire design process. An example which demonstrates the representation of reasoning design information follows.

    ((production (component speed-reducer pos: 1.2 3.3 2.0 vec: .866 0. .5))
    (created-from(plan problem-reduction) (proc assign-speed-reducer)
    (rule r96))
    (removed-by(plan observer) (proc least-commitment) (rule r185))

2. Explanatory facility: In accordance with the contents of the design reasoning memory, this facility explains either the existence of or discrepancies within the resulting design. For example, it explains a successful design when an explanation is requested by the user. The following motion conversion (slider-crank) example illustrates this. (Figure 7 provides a schematic representation of this).

    Reasons for the existence of: slider-crank mechanism SC-1.
    Input 1, rotary motion & 100 rpm & pos 4 2 3 & vec 1 0 1 · · · ,
    [i/o-specification], connects Output3, reciprocate motion and 200 rpm & pos
    [create-new-pos]
    Input-1 and Newpoint-4 a speed reducer, SR-2, shaft 1 0 1 & · · · ,
    and [least-commitment] [motion-converter] by add-in-between Newpoint-4 and
    Output-3 a slider-crank mechanism, SC-1, crankshaft 1 0 1.

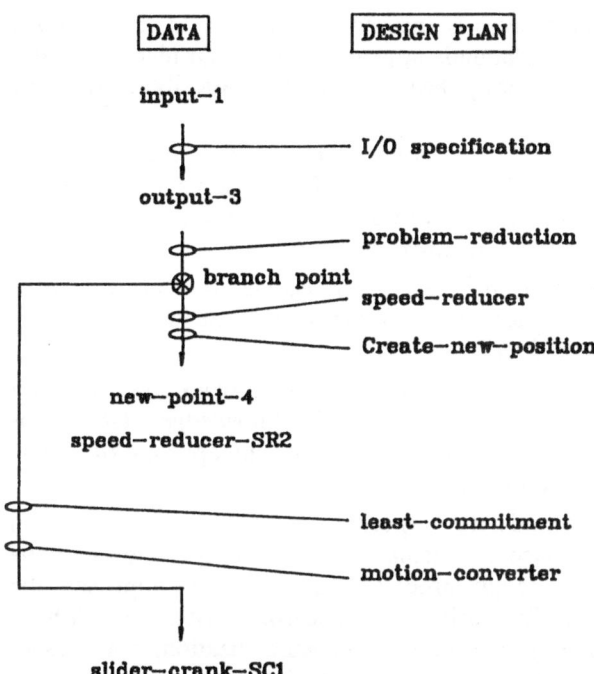

Fig. 5.7. A schematic representation of the explanatory reasoning process.

In order to account for a new design concept assigned by the user (i.e., a design failure), TLDM simply traces the design reasoning memory and attempts to locate the break point between the design failure and the existing successful plans. For example,

··· *[You say that slider-crank mechanism, SC1, should not exist, and you gave the reasons you gave the reasons*

    (and (component slider ?name1 ?p1 ?p2 ?xvec ?yvec ?zvec)
        (component crank ?name2 ?p3 ?p4 ?xv ?yv ?zv)
        (not (function (almost-equal (/ ?xv ?xvec)
        (/ yv ?yvec) (/ zv ?zvec)))))

    I found that, this reason contradicts
        [plan problem-reduction proce assignmotion-converter] and
        [plan observer proce least-commitment]

    How can I do this?
        modify rule, procedure or plan ?

6. ALTERNATIVE DESIGN, INVENTIONS, AND MACHINE LEARNING

Is has been indicated by Middendorf (1986) that creative capability and experience are used to create successful alternative designs or inventions. Our third design hypothesis, that the symbolic design optimization process is insured to converge in the planning model. is based on the assumption that no bounds appear in the design hierarchy. If a subtask cannot be resolved within a lower design level, it will automatically move to the next level resolved.

It has been stated by Brown andd Chandrasekaran (1986) that existing expert systems are first generation expert systems. Winston (1982) suggests that machine learning and design analogy capabilities may be useful in overcoming the difficulties which have been encountered in first generation expert systems:

" · · · , practice with specific situations in one domain enables the invention of a specific law. In the other direction, once several forms of the same sort of law are known, comparison enables the generalization · · · ".

Expert systems must be organized in a structured fashion to avoid confusion and to achieve efficiency. Furthermore, in order to attain extensible knowledge relationships (i.e., relationships which can be utilized in multiple domains) an appropriate frame structure for design knowledge representation must be chosen.

The adaptability of this planning model to a specific design domain is based on the use of knowledge acquisition and machine learning. These concepts represent the capability to transcend the classical redesign process and to further approach the activities of human invention in design. The basic machine learning criteria which have been used in TLDM are design analogy, design method generalization, new design concept learning, design compromise learning, and design schedule learning.

7. CONCLUSION

This research has concentrated on a systematic methodology for design optimization within an overall engineering design expert system, and has stressed design planning for design optimization. Based on the hypotheses that (1) design problems can be separated into a generic portion and a specific portion, and that (2) design problems can be further subdivided into a creative design portion and a routine design portion, a procedural design strategy has been applied to a prototypical design planning model.

This generic design model represents a shell for developing an engineering design expert system. Combined with knowledge acquisition facilities (for example, video taping sessions which directly feed real world design knowledge into the design system), the dominancy of a hierarchical expert system structure, and machine learning, this planning model is adaptable to specific engineering design domains.

ACKNOWLEDGEMENTS

This research was supported by grants from the IBM and NCR Corporations and by the U.S. Army Research Office/DoD Instrumentation Program.

APPENDIX: DESCRIPTION OF TERMINOLOGY

Acquisition facility: This symbolic programming language facility is equivalent to an interpreter within a traditional programming language.

Concept generalization: The design knowledge can be generalized in accordance with design knowledge interdependency.

Constraint propagation: In accordance with the correlations between new and existing design subtasks, constraints are detected and used to eliminate design contradictions.

Coupled design variables: Design decision making depends on the effect of interrelated design variables.

Design difficulty prediction: According to the prescription assignment, the design constraints which compromise the design difficulties are formulated from design estimation.

Explanatory facility: This symbolic programming language facility is equivalent to a debugger within a traditional programming language.

Knowledge layer linker: The interrelations between design procedures are specified within design plans. These interrelations are referred to as layer linkers.

Least commitment judgement: using a minimal cost path to cover a least commitment failure.

Logical resolution theory: Resolving the relations between conjuncted and disjuncted knowledge in order to generalize the design goal into a canonical form.

Memory indexing: The majority of basic knowledge in a large scale design system is parsed within the design knowledge base. Memory indexing records a keyword from each design knowledge cluster and creates an artificial hierarchical path.

Memory networking: According to the existing design hierarchy, the indexed design knowledge is accessed as a network.

Prescription assignment: A temporary design objective which solves the problems found in the rough design is referred to as a prescription assignment.

Scratching: The fast creation of a rough design.

State estimation: Numerical evaluations are the most time consuming in the design optimization process. State estimation is based on design domain experience and is utilized in large scale design to reduce the number of alternatives in the design process.

Subtask scheduling: At each stage of the design process, many parallel design proposals (subtasks) are queued to wait for further processing. A policy, referred to as scheduling, assigns priorities to those subtasks to increase the efficiency of the overall design process.

Trade-off reasoning: To escape from a locally optimal design, which is not the global optimum, it is necessary to trade-off portions of a design achievement in order to gain freedom in the design variables for the investigation of further design alternatives.

REFERENCES

Brown, D. C. and Chandrasekaran, B. (July 1986): "Knowledge and control for a mechanical design expert system," *IEEE Computer (Special Issue on Expert Systems in Engineering)*, **19**(7), pp. 92-100.

Buchanan, B. G. and Shortliffe, E. H. (1984): *Rule-Based Expert Systems: The MYCIN Experiments of the Stanford Heuristic Programming Project*, Addison-Wesley, Reading, MA, USA.

Charniak, E. and McDermott, D. (1985): *Introduction to Artificial Intelligence*, Addison-Wesley, Reading, MA, USA.

Chieng, W. H. and Hoeltzel, D. A. (in press, 1987): "An interactive hybrid (symbolic-numeric) system approach to near optimal design of mechanical components," *Engineering Optimization*, **11**(4), Gordon & Breach Science Pub.

Choy, J. K. and Agogino, A. M. (1986): "SYMON: Automated symbolic monotonicity analysis system for qualitative design optimization," *Proceedings of ASME International Computers in Engineering Conference*, Chicago, IL, USA, pp. 207-212.

Dejong, G. (1986): "An approach to learning from observation," in *Machine Learning: An Artificial Intelligence Approach* **II**, Morgan Kaufmann, Palo Alto, CA, USA, pp. 571-590.

Descotte, Y. and Latombe, J. C. (1985): "Making compromises among antagonist constraints in a planner," *Artificial Intelligence*, **27** , pp. 183-217.

Dixon, J. R., Howe, A., Cohen, P. R., and Simmons, M. K. (1986): "DOMINIC I: Progress towards independence in design by iterative redesign," *Proceedings of ASME International Computer in Engineering Conference*, Chicago, IL, USA, pp. 199-206.

Gongaware, T., Hseih, L., McGuire, P., and Volgel, S. (1983): "Application of automated design process planning to electronics manufacturing," *CAM-I's 12th Annual Meeting & Technical Conference*, pp. 5.

Hayes-Roth, F. and Waterman, D. A. (1986): *Building Expert Systems*, Addison-Wesley, Reading, MA, USA.

Israel, D. J. (October 1983): "The role of logic in knowledge representation," *IEEE Computer*, **16**(10), pp. 37-41.

Lebowitz, M. (1986): "Concept learning in a rich input domain: Generalization-based memory," in *Machine Learning: An Artificial Intelligence Approach* **II**, Morgan Kaufmann, Palo Alto, CA, USA, pp. 193-214.

Love, S. F. (1980): *Planning and Creating Successful Engineering Designs*, Van Nostrand Reinhold, New York.

Middendorf, W. H. (1986): *Design of Devices and Systems*, Marcel Dekker Pub.

Nilsson, N. (1971): *Problem Solving Methods in Artificial Intelligence*, McGraw-Hill, New York.

Pahl, G. and Beitz, W. (1984): *Engineering Design*, Springer-Verlag, Berlin, Heidelberg, New York, Tokyo.

Quinlan, J. R. (1982): "Semi-autonomous acquisition of pattern-based knowledge," in *Introductory Reading in Expert Systems*, Michie, D. (ed.), pp. 192-207.

Sacerdoti, E. (1977): *A Structure for Plans and Behavior*, North-Holland, Amsterdam.

Stefik, M. (1981): "Planning and meta-planning (MOLGEN: Part 1 and 2)," *Artificial Intelligence*, **19** , pp. 111-139 and 141-169.

Ullman, D. G. and Dietterich, T. A. (1986): "Mechanical design methodology: Implementation on future developments of computer-aided design and knowledge-based systems," *Proceedings of ASME International Computers in Engineering Conference*, **1** , pp. 173-180.

Vanderplaats, G. N. (June 1985): *COPES/ADS A Fortran Control Program For Engineering Synthesis Using The ADS Optimization Program*, Engineering Design Optimization (EDO) Inc.

Wilkins, D. E. (1984): "Domain-independent planning: Representation and plan generation," *Artificial Intelligence*, **22** , pp. 269-301.

Winston, P. H. (1980): "Learning and reasoning by analogy," *Communication of ACM*, **23** , pp. 689-702.

Winston, P. H. (1982): "Learning new principles from precedents and exercises," *Artificial Intelligence*, **19** , pp. 321-350.

Yoshikawa, H. (1981): "General design theory and a CAD system," in *Man-Machine Communication in CAD/CAM, Proceedings of the IFIP Working Group 5.2 Working Conference 1980 (Tokyo)*, Sata, T. and Warman, E. A. (eds.), North-Holland, Amsterdam, pp. 35-58.

Zadeh, L. A. (October 1983): "Commonsense knowledge representation based on fuzzy logic," *IEEE Computer*, **16**(10), pp. 61-65.

6. A CAD System with Declarative Specification of Shape

G. Sunde

Center for Industrial Research
Box 350 Blindern, 0314 Oslo, NORWAY

Abstract: *This article describes a CAD system that allows for* declarative specification *of drawings, in contrast to the* construction procedure *that has to be carried out in conventional CAD. As a smaller part this fits into a picture of design as a series of refinements from a "functional" description at the beginning to a description sufficiently detailed for manufacturing. Declarative, but non-complete, specifications are also convenient for representing and recognizing standard form-features of geometric products. In future the level of this kind of descriptions has to be increased and formalized in the direction of functionality. For performing efficient geometric reasoning about spatial relationships between elementary geometric entities, e.g. line segments and arcs, we propose a system architecture combining a* production *mechanism and a* verification *mechanism, both based on rules. Central in this architecture are new terms defined for an intermediate representation of how entities are related.*

Keywords: CAD, declarative specification, design, interference, geometric reasoning.

1. INTRODUCTION

The use of CAD systems has shown good results both in terms of time and cost. But for the creative design phase itself, CAD systems turn out to be of less use. In traditional CAD systems users make their drawings by carrying out a construction procedure, and their description of the drawing is therefore *procedural* .

This article describes a system that carries out this construction procedure automatically. The user only specifies the shape and size of the figure by declarative statements and the system takes care of making the drawing in accordance to the specification. The user specifies *what* to draw, not *how* to draw it.

The declarative specification of shape and size is done by dimensions and other geometric constraints like parallelism and tangency. The system is not intended to have any knowledge of what is drawn or designed in the system.

The particular system and system architecture described in this article is not yet implemented, but some of the ideas are included in another system we have designed and implemented. Unlike the proposed system, this system requests the drawing to be completely determined, and produces no explanations when it is over- or underdetermined. Its number of available geometric entities and constraints are larger, including line segments, circular arcs, circles and symmetry lines as entities and a long list of different constraints, tangency, symmetry, etc. The language used is Fortran, and the method used is similar to a production system. See (Sunde 1986) for a description.

2. THE SYSTEM BEHAVIOUR

The users build their drawings by successively adding geometric entities and geometric constraints. For each geometric constraint added, the system will detect if it will lead to some inconsistency and if not, do the correct updates of the drawing to meet both this new constraint and the already existing ones. For each geometric entity added the drawing is updated to include this entity. The system specified is so far limited to line segments as the geometric entity available and to geometric constraints for distance values, angle values and common endpoints of line segments.

An example of an interaction sequence. Before discussing system design a sample description will be given (Fig. 6.1).

Explanation when overdetermined. An important feature of such a system is its ability to explain to the user why and how drawings are overdetermined. In Fig. 6.2 if the distance value of 30 (circled) is added, the response may be:

This new constraint is in conflict with earlier specifications. This distance is already implicitly constrained since (...explanation...). If any one of the constraints in the explanation is removed, the new constraint may be added without a conflict.

Suggestions between alternatives when underdetermined. An important feature is to deal with underdetermined drawings. As long as a drawing is not completely determined, there will exist several different solutions that all meet the specified constraints. The system is to produce some likely alternatives that the user may choose among, possibly in a priority sequence.

Figure 6.3 shows some of the possible solutions if the user specifies a constraint on the angle between the lines *a* and *b* to be 90 degrees (and no other value constraints).

Impossible solutions. The system will also recognize specifications that are impossible to accomplish, e.g. a triangle with one side longer than the sum of the two others.

Variant design by parameterization. Included in this method are the possibilities of varying distance and angle values. (See example in Fig. 6.4.) When the user changes any angle or distance value, the system automatically recomputes and redraws according to the new values.

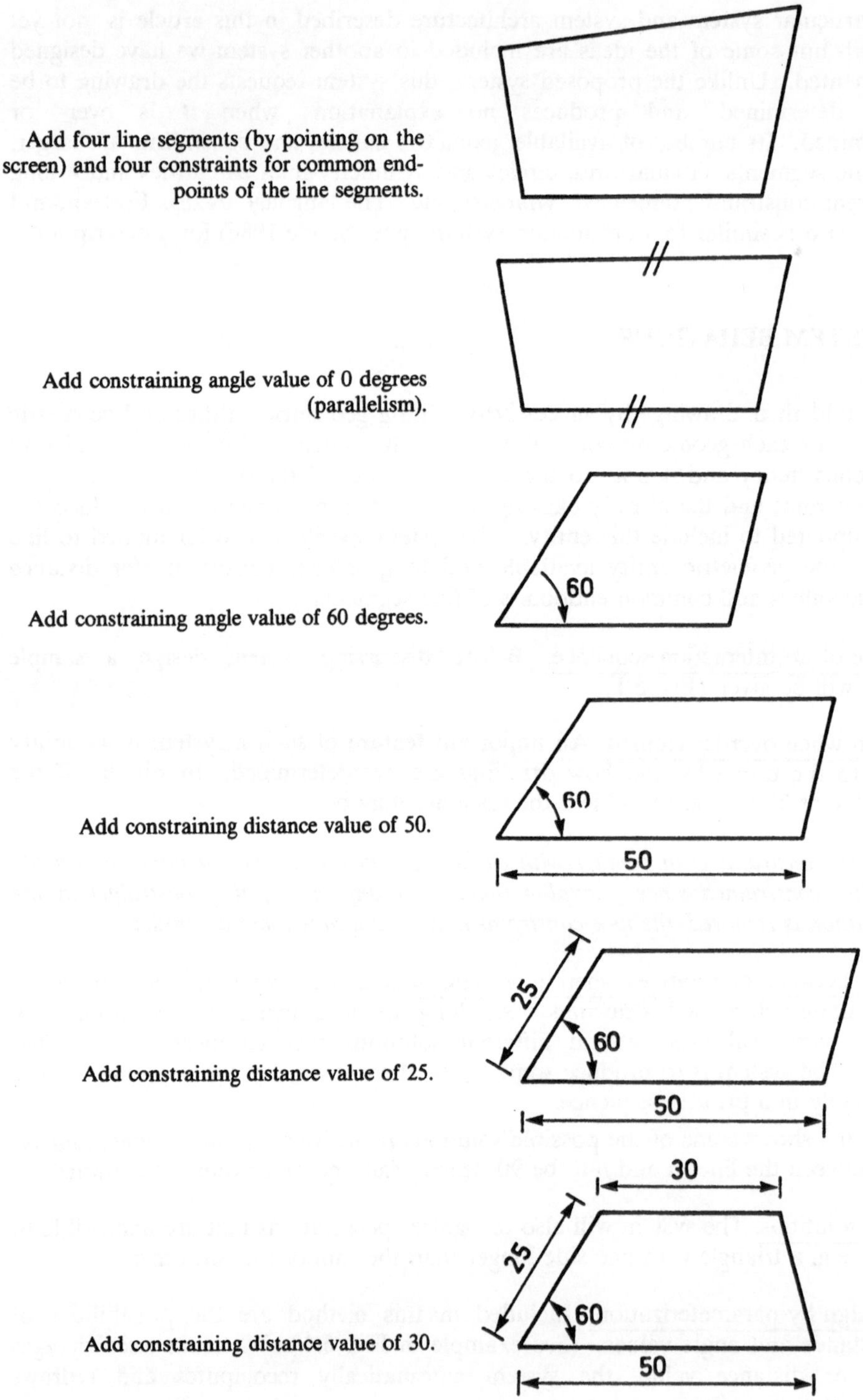

Add four line segments (by pointing on the screen) and four constraints for common end-points of the line segments.

Add constraining angle value of 0 degrees (parallelism).

Add constraining angle value of 60 degrees.

Add constraining distance value of 50.

Add constraining distance value of 25.

Add constraining distance value of 30.

Fig. 6.1. An interaction sequence

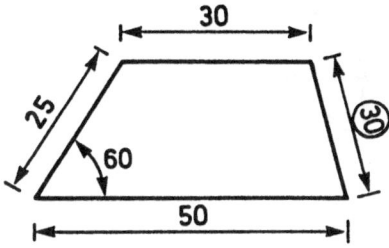

Fig. 6.2. Adding a conflicting constraint (overdetermined case)

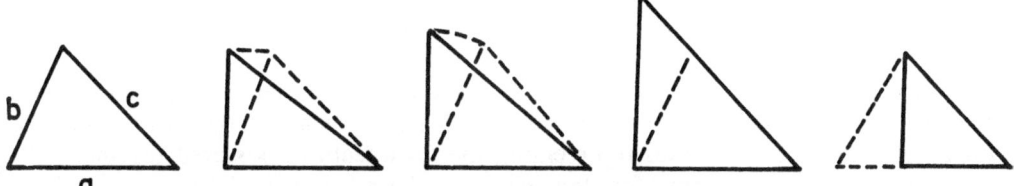

Fig. 6.3. Example of possible solutions (underdetermined case)

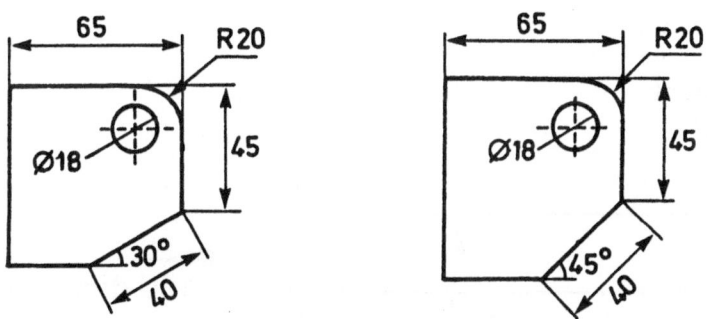

Fig. 6.4. Example of parametric design

<u>Alteration of the specification.</u> An even more important advantage is the possibility to modify the specification itself. In Fig. 6.5, this could be done by removing a specified angle and adding a distance. After this modification the system automatically recomputes and redraws the drawing. This is not possible in conventional CAD systems in which a new construction procedure has to be formed and carried out.

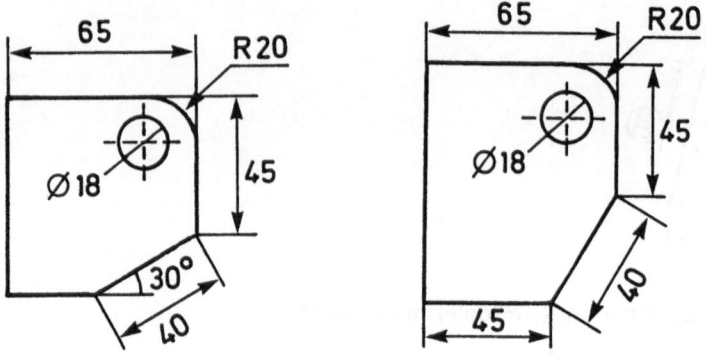

Fig. 6.5. Example of alteration of the specification

Representation of standard form-features. A non-complete declarative specification is convenient for representation of standard form-features such as holes, slots, threads, grooves, etc. Geometric entities representing the form-features is stored together with constraints that are common for this class of form-features. The unconstrained relationship that may vary between different instances of the form-features remain so. In Fig. 6.6 is shown a drawing of a groove. Common properties of grooves are the two sides, their parallelism, the two half cylindric ends and that these are tangential.

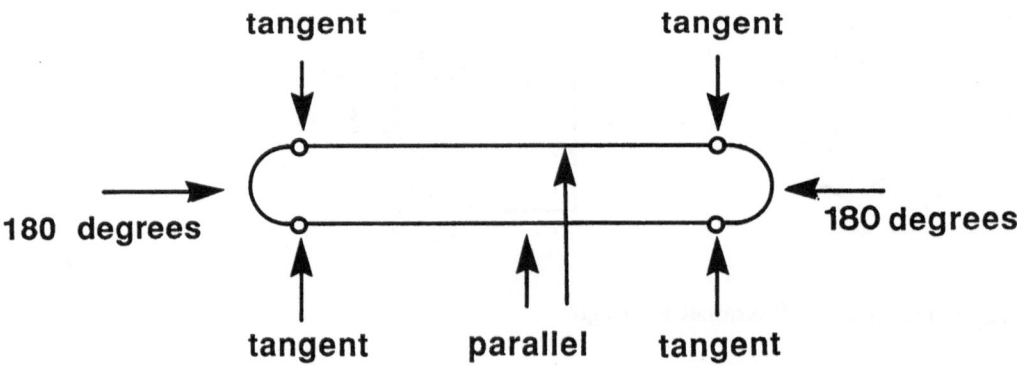

Fig. 6.6. Common relationships of a groove

Using common constraints as common properties of shapes is therefore one way of classifying shapes.

3. SYSTEM DESIGN

3.1. The Tasks to be Performed

The tasks (reasoning) to be performed in order to attain the above system description are as follows:

- to detect whether a specific distance or angle is 1) explicitly constrained by the user, 2) implicitly constrained, or 3) unconstrained.
- to produce explanations of overspecifications.
- to produce alternative drawings that meet the specified constraints.

Whether or not these tasks serve the characteristic of geometric reasoning may be left as an open question.

3.2. The Basic Geometric Terms

The basic geometric terminology and facts specified by the user consist of "the existence of":

Geometric entities

- line segments, including endpoints.

Geometric constraints

- distance values, referring to endpoints.
- angle values, referring to endpoints.
- common endpoint locations of line segments.

An example of the common geometric knowledge available to us, based on this basic terminology is:

If all three side lengths of a triangle have constrained distances, then the angles are all implicitly constrained.

To perform reasoning about properties of simple geometric shapes like triangles and squares, based on this terminology and simple knowledge, is fairly easy.

For arbitrary complex shapes we have no such knowledge. In this case the approach is to combine the simple knowledge to make conclusions about the complex geometry. Different lines of arguments about different subparts of a drawing must be combined to perform reasoning about larger portions of the drawing.

For this purpose and as a higher level of abstraction, we define two new terms.

3.3. The New Geometric Terms Defined

The new terms defined are represented by the two types of facts:

- PPWMCA-set (or CA-set for short) is a set of Pair of Points With Mutually Constrained Angles.
- PWMCD-set (or CD-set for short) is a set of Points With Substitute all over Mutually Constrained Distances.

A *Point* is any endpoint of a line segment, and a *Pair of Points* is any pair of Points (existing or virtual).

In Fig. 6.7, the Pair of Points PP1, PP3 and PP4 has mutually constrained angles. This means that their orientation relative each other is constrained — the angle between any two Pair of Points in a CA-set has a constrained angle value. The orientation of the Pair of Points PP2 is free to change relative to the others. Whenever two Pair of Points belongs to different CA-sets the angle value is unconstrained.

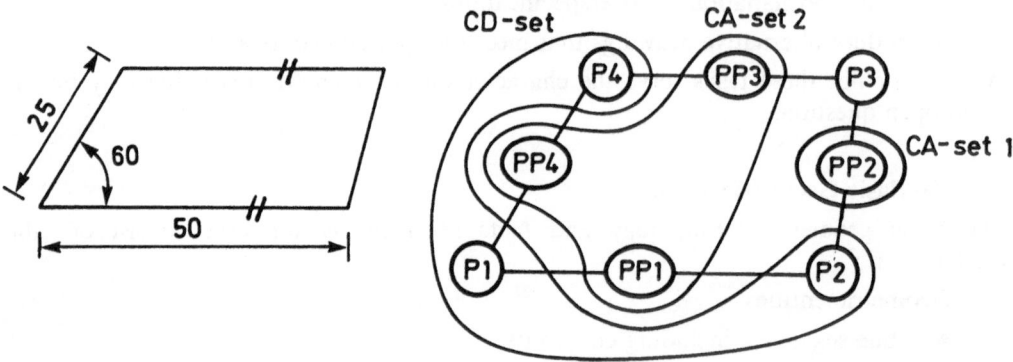

Fig. 6.7. CA- and CD-sets

In the same figure, the points P1, P2 and P4 have mutually constrained distances. This means that their position relative to each other is constrained. The distance between point P3 and any other point is free to change, but since the point is a part of the Pair of Points PP2 and PP3 there is a dependency among its position and their orientations.

These terms, and the knowledge represented by them, is inferred and stored explicitly in the database of the system. In some sense, this representation serves the purpose of "keeping in mind" or "keeping in memory" which parts of the drawing are already completely constrained, implicit or explicit, and how these (completely constrained) subparts are loosely constrained and related to each other.

3.4. The System Architecture

The system architecture consists both of a production mechanism and a verification mechanism. The geometric knowledge, represented as rules, is divided into a set of production rules and a set of verification rules. (See Fig. 6.8).

The production mechanism. The purpose of the production mechanism is to successively produce CA-sets and CD-sets, as the user builds the drawing. We call these sets the intermediate higher level facts in Fig. 6.8.

The production mechanism makes use of 1) the facts specified by the user, 2) the production rules, and 3) the verification mechanism.

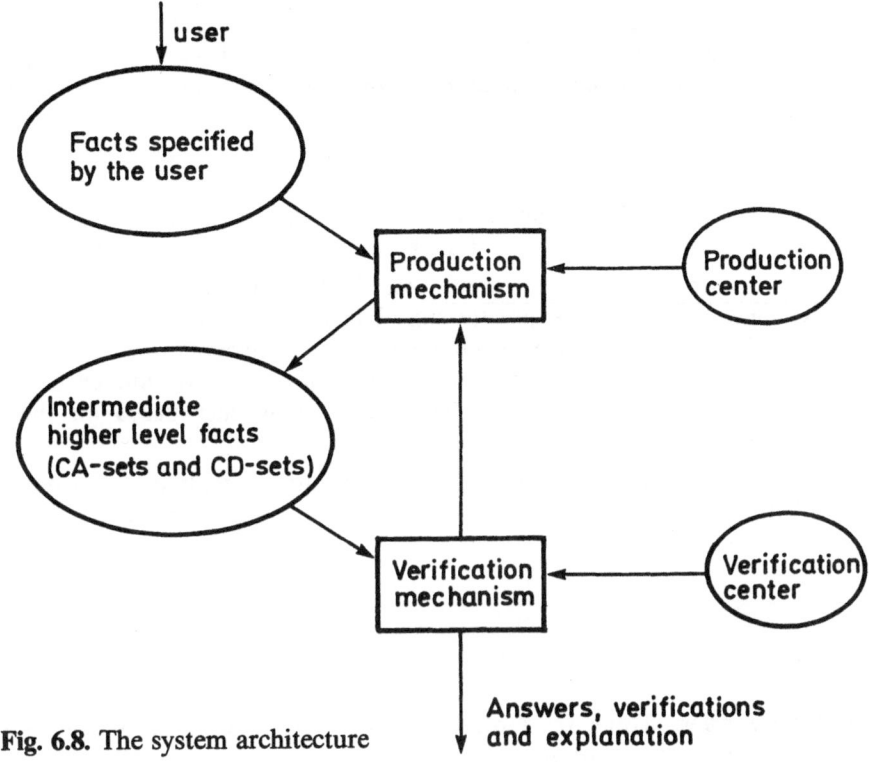

user

Facts specified
by the user

Production
mechanism

Production
center

Intermediate
higher level facts
(CA-sets and CD-sets)

Verification
mechanism

Verification
center

Answers, verifications
and explanation

Fig. 6.8. The system architecture

1) The facts specified by the user are the existence of the geometric entities and geometric constraints.

2) The production rules are "looking for" situations where the conditions for combining two or more CA- or CD-sets into larger CA- or CD-sets are fulfilled. Some examples are shown in Section, 3.5.

3) Some elements of these conditions are not stored explicitly in the database. The production mechanism uses the verification mechanism for testing the truth of fact elements.

The reason for this is to avoid the storage of a large number of less important facts that are easily verified whenever needed. On the other hand, this leads to the problem that rules may not be triggered when they should be, since the triggering fact is not produced. An example of such a goal/fact is whether two CA-sets have a common Pair of Points. (See later example of production rules in Section 3.5.1.) The production is done by inference and the mechanism is triggered whenever a new fact is created by the user or produced by the production mechanism. It is therefore a forward chaining mechanism.

Overdetermined drawings. Whenever a new constraint is added the system verifies whether this distance or angle value is already constrained. If it is constrained the new constraint is rejected. In this way the specification stored in the database will never cause an overdetermined drawing.

In the cases where the specified distance (or angle) are already constrained, the system will produce the explanation of why or how this particular distance (or angle) is constrained. If any constraint in this explanation is removed, the user may insert this new constraint without a conflict.

The verification mechanism. It is the verification mechanism that is doing the actual "high level" reasoning, providing answers to the above mentioned questions. Typically the verification mechanism verifies whether a particular distance (or angle) has a constrained value or not. This verification is done by a backward chaining search.

The definitions of terms are not explicitly stored. An important point is that the definition of these new terms is not explicitly stored in any database in the system. Neither are the definitions of the basic terminology of line segments, distances, etc. stored. It is important to realize, that these definitions strongly influence the production rules and the verification rules. The rules are in fact based on these definitions.

The problem we have touched on here is the difficulty of making a reasoning system that deals with information on different levels of abstraction at the same time.

3.5. The Rules

3.5.1. The production rules: Three (currently selected from nine) typical rules for building this intermediate representation is:

Rule A: *When a distance is added, a CD-set containing the two referred points is created.*

Initially each Pair of Points is regarded as a CA-set of its own.

Rule B: *When an angle is added, the two CA-sets of the referred two Pair of Points (line segments) are combined into one CA-set.*

The first two rule examples are triggered by new facts from the user. The next rule example is triggered by the CA- and CD- set facts produced by the production mechanism.

Rule C: *If two CD-sets contain a common Point and the angle between them is constrained then the two CD-sets are combined into a single CD-set.*

The correctness of this is shown by using a triangle between the common point and two arbitrary points from the two CD-sets. The length of the side between the two arbitrary points is constrained since the other two lengths are constrained (common CD-sets) and the angle between them is constrained (common CA-set).

3.5.2. The verification rules:

Verifying constraints. To verify whether a particular angle value is constrained or not is done by checking whether the two Pair of Points are members of the same CA-set. If the two Pair of Points referred are members of the same CA-set, then the angle is constrained. This condition is accomplished by a rule. The same applies to distances.

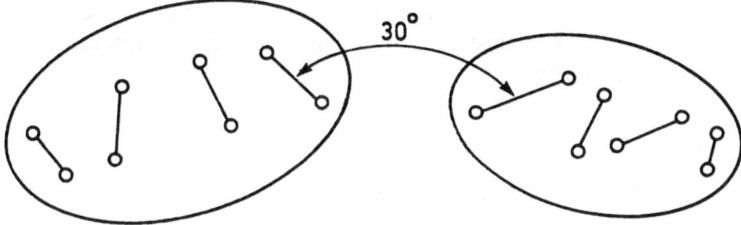

Fig. 6.9. When adding the specified angle, the angle between any two Pair of Points from the two CA-sets is constrained (for geometrical reasons), and the two CA-sets are therefore combined into one single CA-set.

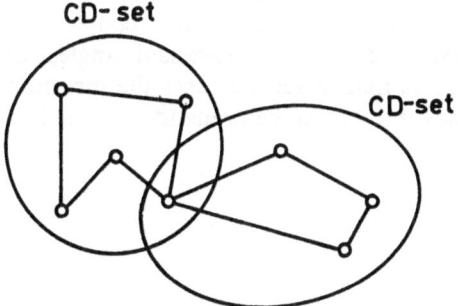

Fig. 6.10. If the angle between the two CD-sets is constrained (common CA-set of Pair of Points), then the distance between any two Points from the two CD-sets is constrained, and the two CD-sets are therefore combined into one CD-set.

Producing explanations. The effect of the production rules is that Points are collected into CD-sets and Pair of Points are collected into CA-sets, but mostly that CA-sets and CD-sets are collected into new, more comprehensive CA- and CD-sets. None of these sets are removed, so there exists in the database a structure of sub-CA-sets and sub-CD-sets that the topmost set consist of. This structure is to serve as the basis for producing explanations of why particular distances and angles are constrained. For a particular distance, the minimum CD-set covering the two points will contain the explicit constraints needed to explain why this distance is implicitly constrained.

The verification mechanism is used by the production mechanism. As mentioned above, and as seen in the examples of the production rules, the production mechanism makes use of the verification mechanism. An example would be to verify that two CA-sets contain a common Pair of Points. This is done by performing union and intersection. These operations are accomplished by rules.

3.6. Why This Architecture?

The main reasons to use this system architecture instead of using a pure production or a pure verification mechanism are:

- The large number of facts that would have to be produced by a pure production system.
- The large search space that a pure verification system would have.
- To avoid generating a lot of different proofs for the same fact as would be the case in a pure production system

Large number of possible facts. In case of a pure production system every possible constrained distance (or angle) value has to be deduced and stored as a fact in the database. The number of such facts is large. For a medium size drawing with 100 points and 100 line segments there will be approximately 5 000 possible constrained distances and approximately 12 000 000 possible different constrained angles to produce. The last number is large, but the reason is that in several cases the direction of "virtual" line segments is needed. It is not possible to say beforehand when a fact will be needed.

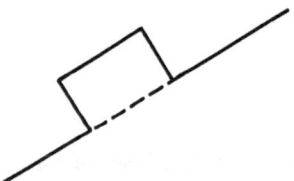

Fig. 6.11. Even the direction of "virtual" line segments may be needed.

Large search space of a pure verification mechanism. The geometric knowledge may be stated in different ways, but one alternative (later to be referred) is:

> In a polygon, consisting of both real line segments and "virtual" line segments, if a specific set of the distances of the line segments and the angles between the line segments are constrained ($2n - 3$ constraints for a n-gon) then the residing distances and angles (three constraints) are also constrained.

Using the above geometric knowledge and a backward chaining verification approach, the search space will be large and the search difficult to control. A distance value may be proved to be implicitly constrained by using a polygon, including this distance, where the other (except for 2) distances and angles are constrained. But the number of such polygons is large, and for each condition of the polygon that is not fulfilled explicitly, the mechanism has to move down a level to see whether this condition could be verified. This in turn implies a large number of new alternative polygons to investigate, and so on. In addition there is a problem of handling and controlling the search, in particular to avoid using the same polygons over and over.

Different proofs exist for the same fact. There is a problem in that there exist several different explanations and proofs for why a specific distance or angle is constrained. An example demonstrates this.

The question in Fig. 6.12 is whether the distance marked by a question mark is constrained (implicitly) or not.

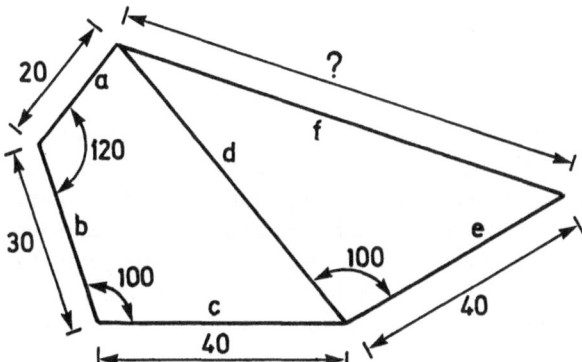

Fig. 6.12. Verify marked distance as being constrained.

In some way it is obvious that this distance is constrained, but even for a human being it is somewhat difficult to produce an explanation. Two alternative explanations are:

1) That the distance of the line segment *d* is implicitly constrained because of the explicitly constrained distances of, and angles between, the line segments *a*, *b* and *c*, and the wanted distance of line segment *f* is implicitly constrained because of the constrained distances of, and angles between, the line segments *d* and *e*.

2) That the angle between the line segment *c* and *d* is implicitly constrained because of the constrained distances of and angles between the line segments *a*, *b* and *c*, and the angle between the line segments *c* and *e* is implicitly constrained because of the constrained angles between the line segments *c* and *d* and *d* and *e*, and the wanted distance is constrained because of the constrained distances of, and angles between, the line segments *a*, *b*, *c* and *e*.

For more complex drawings the number of different explanations increases as the number of different polygons and combinations of polygons increases. In a pure production system all of these different explanations would be produced, unnecessarily. In the proposed system this problem is handled by the new terms defined.

4. FUTURE

One picture of design is as a series of refinements from a "functional" description of the product to a description satisfactorily detailed for manufacturing. One future goal is therefore to clarify the correspondence between functional descriptions of products down to their exact geometry. We have specified a system dealing with a part of this process. From a possible non-complete declarative specification of a shape the system can deduce unspecified properties in terms of implicit constrained relationship, and the system transforms this description into exact coordinates so the drawing may be produced on a terminal. We intend to increase this level of abstraction by introducing recognized form-features, relationship between form-feature and between different shapes, and moving towards a functional description of the product.

5. CONCLUSIONS

A CAD system has been proposed that allows for declarative specification of drawings in contrast to the construction procedure that has to be carried out in conventional CAD systems. Efficient performance for the geometric reasoning tasks needed in order to attain this entails an intermediate higher level representation of the spatial relationships in the drawing. A separation of the geometric reasoning mechanism into a production mechanism and a verification mechanism is also needed. The geometric knowledge, represented as rules is separated into production rules (producing the intermediate representation) and verification rules (examining the intermediate representation).

ACKNOWLEDGEMENTS

This work has been sponsored by the Joint Norwegian German APS project. I would like to thank my colleagues Ø. Aronsveen, V. Kallevik and S. Schjerve for valuable discussions.

BIBLIOGRAPHY

Cugini, U., Devoti, C., and Galli, P. (1985): "System for parametric definition of engineering drawings," *Proceedings of MICAD'85*, pp. 50-64.

Dixon, J. R. and Simmons, M. K. (November 1983): "Computers that design: Expert systems for mechanical engineers," *Computers in Mechanical Engineering*, pp. 10-18.

Eastman, C. M. and Preiss, K. (March 1984): "A review of solid shape modelling based on integrity verification," *Computer-Aided Design*, **16**(2), pp. 66-60.

Gero, J. S. and Coyne, R. (1984): "The place of expert systems in architecture," in *Proceedings of CAD84*, Wexler, J. (ed.), Butterworth, pp. 529-546.

Gossard, D. C. and Lin, V. (1983): "Representation of part-families through variational geometry," in *Advances in CAD/CAM*, Ellis, T. M. R. and Semenkov, I. I. (eds.), North-Holland, Amsterdam, pp. 47-53.

Hillyard, R. C. and Braid, I. C. (May 1978): "Analysis of dimensions and tolerances in computer-aided mechanical design," *Computer-Aided Design*, **10**(3), pp. 161-166.

Kawagoe, K. and Managaki, M. (January 1984): "Parametric Object Model: A Geometric Data Model for Computer-Aided Engineering," NEC Research & Development No. 72, Nippon Electric Co., Tokyo.

Kimura, F., Suzuki, H., and Wingard, L. (1986): "A uniform approach to dimensioning and tolerancing in product modelling," in *Proceedings of CAPE '86, Second International Conference on Computer Applications in Production and Engineering*, Bo, K., Estensen, L., and Warman, E. A. (eds.), Copenhagen, pp. 165-171.

Light, R. A. and Gossard, D. C. (1982): "Modification of geometric models through variational geometry," *Computer Aided Design*, **14**(4), pp. 209-214.

Lin, V. C., Gossard, D. C., and Light, R. A. (1981): "Variational geometry in computer-aided design," *Computer Graphics*, **15**(3), pp. 171-178.

Parden, G. and Newell, R. G. (April 1984): "A dimension-based parametric design system," *Proceedings of CAD84*, Brighton, UK, pp. 252-259.

Preiss, K. and Kaplansky, E. (September 1983): "Solving CAD/CAM problems by heuristic programming," *Computers in Mechanical Engineering*, pp. 56-60.

Requicha, A. A. G. (May 1977): "Part and assembly description languages: dimensioning and tolerancing," Technical Memo No. 19, Production Automation Project, University of Rochester, Rochester, NY, USA.

Robin, H. (1978): "Dimensions and tolerances in shape design," Technical Report No. 8, University of Cambridge Computer Laboratory, Cambridge, UK.

Shinohara, K., Hanakawa, R., Kawagoe, K., and Managaki, M. (1985): "Parametric object modeler: A conceptual object modeler for computer aided engineering," in *Design and Synthesis, Proceedings of the International Symposium on Design and Synthesis, Tokyo, Japan, July 11-13, 1984*, Yoshikawa, H. (ed.), North-Holland, Amsterdam, pp. 477-482.

Stiny, G., "Introduction to shape and shape grammars," *Environment and Planning*, **7**, pp. 343-351.

Sunde, G. (May 1986): "Specification of shape by dimensions and other geometric constraints," *Proceedings of the IFIP WG 5.2 Workshop on Geometric Modeling*, Rensselaerville, NY.

Sutherland, I. E. (May 1965): "Sketchpad: A man-machine graphical communication system," MIT Lincoln Laboratory Report No. 296, Cambridge, MA, USA.

Suzuki, H., Kimura, F., and Sata, T. (1985): "Treatment of dimensions on product modeling concept," in *Design and Synthesis, Proceedings of the International Symposium on Design and Synthesis, Tokyo, Japan, July 11-13, 1984*, Yoshikawa, H. (ed.), North-Holland, Amsterdam, pp. 491-496.

Szalapaj, P. J. and Bijl, A. (1984): "Knowing where to draw the line," in *Knowledge Engineering in Computer-Aided Design, Proceedings of the IFIP WG 5.2 Working Conference 1984 (Budapest)*, Gero, J. S. and Gero, J. S. (eds.), North-Holland, Amsterdam, pp. 149-169.

Tomiyama, T. and Yoshikawa, H. (June 1985): "Knowledge engineering and CAD," *Future Generations Computer Systems*, **1**(4), pp. 237-243, North-Holland.

Yoshikawa, H. (1985): *General Design Theory: Theory and Application*, Department of Precision Machinery Engineering, The University of Tokyo, Tokyo, Japan.

7. Methodology of Intelligent CAD Systems

E.E. Berkhout

Delft University of Technology, Faculty of Building and Environmental Design, Berlageweg 1, 2628 CR Delft, THE NETHERLANDS

Abstract: *This paper deals with the characteristics of two different methodological approaches to intelligent CAD systems. The underlying paradigms are introduced and basic concepts, such as method, intelligence, and CAD systems are defined and elaborated on. One methodology is based on a decision structure oriented toward product-users. In conclusion, requirements are formulated with respect to the implementation of the two different methodologies as a language.*

Keywords: CAD, knowledge engineering, decision methodology.

1. INTRODUCTION

For a better understanding of the idea of CAD systems, we have to indicate the contents of the concept from which the discussion will be started. In the following we primarily use the concept CAD system in the sense of computerized instruments for solving problems in the field of building and environmental design and construction.

Until now there has not been much enthusiasm for the use of computers in the field of building and environmental design. Most CAD systems are considered cumbersome and of little use since only the drawing is computer aided, not the designing. Sometimes it is implicitly assumed that the use of CAD guarantees a better quality of design. Sometimes CAD demonstrates explicitly that the components are mixed up and are insolvable by the use of CAD systems.

These characteristics of computerized design processes are confusing (Boden 1981), not only for novices but also for financial directors and experts who want to be efficient. Investments that will not pay off are discouraging as are computers that only demand more labour. Moreover, architects have great difficulty accepting new instruments. At the time when other disciplines were discussing computers, architects were discussing adjustable drawing boards. They want things to be easy, very cheap, and comprehensive. They also want things justified in the sense that the instrument produces results that are acknowledged and approved as designs by architects themselves. Designs have to advertise the personal qualities of the designer.

Obviously all these deficiencies force the elaboration of a methodology.

2. PARADIGMS

At the outset of applied computer science, some people assumed that the availability of the correct data would automatically lead to their meaningful use. Conversely, they assumed that if meaningful and valid theories were available, all that was required was the addition of the correct data.

This brings us to the following paradigm:

I: *Scientists formulate theories.* Theories have to be validated by data. Data are produced by experiments or taken from real life. Validated theories control the concepts for designing and are suitable as such for implementation as a design that solves the needs and demands in question.

Several large scale environmental design projects were based on this paradigm. After several years, at the end of the 1970s, the people involved had to state publicly that "the study did not match the aims of today." Since then we have not heard much of paradigm I. (See also (Lee Jr. 1973).)

To designers it is obvious that a theory is not a design.

This unsuccessful paradigm has been replaced by another with less deductive pretensions. Nowadays, attention is directed mainly to the appropriate managing and selection of data. Selection criteria are derived from concepts which are produced by the designers.

This brings us to the following methodological paradigm:

II: *Concepts are produced by experts.* Sets of data are used in transforming concepts into designs. Data are produced in previous designs and in real life. Which data are needed at what stage has to be decided by the designer. Instruments have to offer a quick survey of supposedly relevant data. They also have to offer the facility of easily selecting even a huge amount of data. Finally they have to offer the ability of a sound evaluation of all different combinations of data (clusters, layers, etc.).

At the moment, most of the instruments are available only in a rudimentary form. A substantial effort has to be made in order to produce mature tools corresponding to the program implied by the paradigm (Benneth 1983). Several pitfalls have to be avoided. Some of these will be mentioned and discussed further.

In environmental design, there is another problem to be solved. The real decision making 'designers' are never behind the screen, but somewhere in the bargaining field of interest groups and politics. This situation may lead us to an interface for 'remote designing' by non-designers.

There are a number of fundamental points that are evaded by the latter paradigm:

● The concepts of designers are not always congruent with the needs or demands of principals nor are the resulting designs congruent with the original concepts.

- For principals and similar other people, there are hardly any possibilities to correct either the concepts or the resulting design. Losing lots of time or paying damages are of course no alternative for giving in to an unsuitable design.
- No designer ever guarantees that a second attempt will be better.

Recently we have seen that several of the most successful industrial organisations are, in a way, able to solve these problems either by internal selection or by customized production.

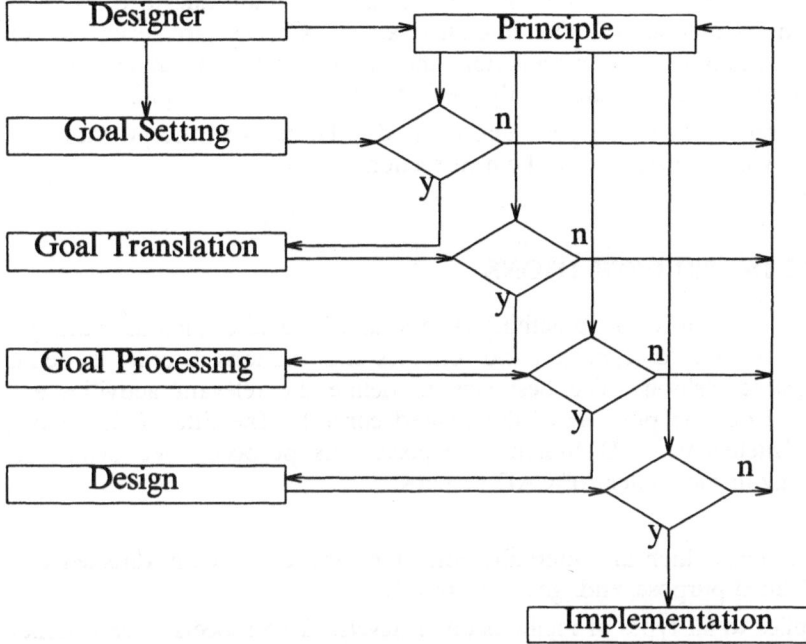

Fig. 7.1. A new methodological paradigm

A new methodological paradigm is fundamentally oriented toward product-users (Berkhout, Micheels, and van Loon 1982), has some characteristics of customized behaviour and may provide an answer to some of the problems (Fig. 7.1):

III: *Principals formulate their aims, goals/objectives and needs in a goal-setting language.* This language is translated into expressions that are processed by formal goal-processing methods. Formal goal-processing methods reliably produce goal-optimized designs. Since no deformations originate from a sound formal method itself, unsuitable designs are due to formulation, interpretation and translation errors. This type of communication errors is detected in feedback loops, based on any representation of the (provisional) designs. The list of goals and the corresponding translations are not only to be corrected for errors, but also for reconsideration by principals. Adding and deleting from the list of goals must be easy to carry out. Feedback is repeated until the last error has been detected and the final design has been approved by the representatives of the demand.

This paradigm is based on the major theoretical elements of cybernetics, self-organizing principles, and communication theory. Self-organizing systems observe innate behaviour and external inputs and react accordingly. In our paradigm the list of goals condenses the observations and builds the starting point of adaptation, this being the major principle of self-organizing. Goal-processing realizes information processing without deformations. This methodology means that all errors are due to translation, formulation, or interpretation of the intended aims. Checks are executed on incompleteness and redundancy, contradiction and discrepancies, misunderstandings and errors by communicative feedbacks to the decision level of the principal.

This paradigm relies heavily on goals and methods. These two concepts, however, are not unrelated as will be discussed in the next section.

3. BASIC CONCEPTS AND DEFINITIONS

Whenever we aim at the control of something chaotic and indefinite, such as creating a method, we have to be able to separate basic activities and relations in order to define them in a manageable fashion. The best way to define the relevant activities and relations is to start from the purpose of the desired control. The title of this paper, "Methodology of Intelligent CAD Systems," exposes this purpose. We define the following terminology in the context of CAD systems:

Method:

> a set of actions which are internally consistent and consistently directed to a clearly defined purpose, end, goal, or objective.

This definition applies to all types of methods on all levels. Two aspects of consistency are introduced at the end of this section. However, all starts with purpose. The definition is as follows:

Purpose, end, goal, or objective:

> the intended future state of a system, degrees of freedom implied, and means provided.

Degrees of freedom are necessary to create variation, flexibility and the opportunity to select. Through means we are able to realize our intentions. We accordingly define:

Means:

> materials and concrete actions for realizing a desired state of a system.

Of course methodology refers to a set of methods.

Methodology:

> characteristics and properties of a set of related methods.

Now that we have explained the first key word of the title we shall proceed with the second, i.e. intelligence. Initially psychologists and AI researchers placed emphasis on learning faculties as an important part of intelligence. Recently, flexibility and creativity have been discovered.

Intelligence:

a creative and flexible understanding of context and change.

Creativity:

the faculty of being able to imagine and conceive a new and realistic future state of a system.

The realism (constructiveness) is to be based on expertise and a profound understanding of the relationship between goals and means.

Mainly instrumental issues are involved with the last key word:

CAD System:

a set of computerized and related instruments directed to the design of objects.

Instrument:

a set of well-arranged and clearly defined (design) techniques dedicated to the execution of well-defined tasks.

While we show in Section 7 that techniques essentially are empty tasks, we define tasks as:

Task:

a set of actions realizing the requested future state of a system; no degrees of freedom are allowed.

A design condenses all method and creativity into materialized entities. We define:

Design:

a fully defined set of means, allocated to functions and activities of the system.

Object:

materialized system usable for certain classes of related actions.

Scanning these eleven basic concepts we identify some items that are closely related and repeated several times. The items concern variations of a state of a system which is not yet available such as 'future,' 'intended,' 'desired,' and 'requested.' Five concepts are directly related to 'a future state' and moreover, four other concepts are indirectly related to the same concept. The fact that nine of the eleven concepts are related to the same key item, justifies that the definitions are chosen sufficiently consistently in accordance with the prerequisite, i.e the consistence required by the concept method. The eleventh concept (object) may occasionally be deleted since no method is involved.

Another aspect of consistency is shown by a relational graph (Fig. 7.2). All defined concepts are chained easily and in an orderly fashion. No confusion should arise since the relationships between the items are straightforward.

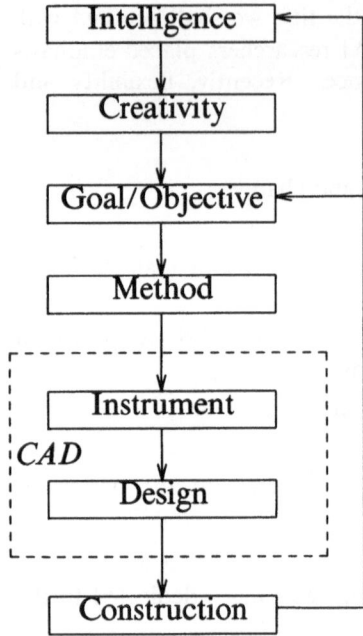

Fig. 7.2. Related concepts

The key item is fully represented by the definition of a goal/objective. Accordingly we state that the methodology to be developed, must be oriented toward the realisation of goals/objectives. Moreover, main concepts such as methods, creative intelligence and design are also closely related to goals/objectives. We may therefore expect that only methods based on a goal/objective language will be effective.

4. GOALS/OBJECTIVES

As we try to state something more definite about the methodology to be developed and since a set of methods is valid only if it is an operational, we have to investigate several basic concepts with respect to operationality. We start with the key item, goal/objective, which intends 'a desired future,' and with the concept method, which relates actions to 'a desired state.' The concepts purpose, end, and aim which are related to goal/objective are concepts for the long-term and less operational.

One of the major points in operationality is whether we are able to identify clearly enough the expressions which satisfy the definition of a goal. Ten years of 'goal-watching' has brought the following, somewhat surprising, observation. In practice there are only two basic different expressions which represent all goals that are produced in (observed) situations. Both basic goal-expressions have two variations and hybrids exist.

There are goals expressed as entities to be optimized, e.g.:

- "We want the greatest, highest, largest" or otherwise maximized entities that are possible in the actual circumstances (Fig. 7.3a).

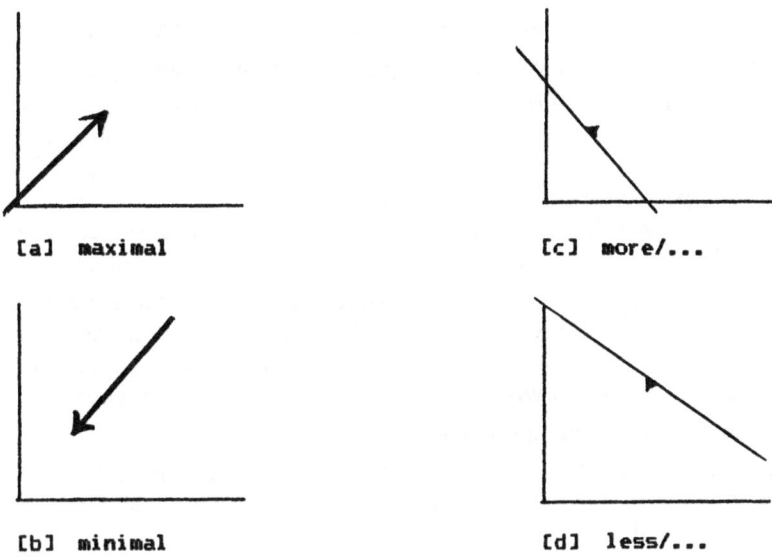

[a] maximal	[c] more/...
[b] minimal	[d] less/...

Fig. 7.3. Goals/Objectives: Collection/Translation

- "We want the smallest, lowest, tiniest" or otherwise minimized entities that are possible in the actual circumstances (Fig. 7.3b).

Other goals are expressed as levels or restrictions, e.g.:

- "The situation allows only for a limited space, limited time, and limited money," or other means restricted by upper levels (Fig. 7.3c).
- "The (intended) object requires a basic amount of material, sufficient quality, at least fifty centimetres width," or other entities stimulated by lower levels (Fig. 7.3d).

Hybrids are often found in expressions such as "this activity has to be balanced with respect to the demand." The demand may be represented as a level/restriction and the 'balancing' is represented as minimizing the actual gap between activity and level.

All other expressions appeared to be expressions that have no intentional meaning, but claim truth in the form of 'the only solution' or 'facts' or 'tasks which have to be executed at any rate.' These last expressions in particular do not solve a problem in a more general way, but claim means for experts' own interests in or view to the problem.

5. METHODS

As stated before, methods are directed to a clearly defined goal, but essentially they are a set of actions. This set has to correspond with the type of the goal. Two main classes of goals differentiate between the corresponding sets of actions:

a: goals that express intentions to modify context and objects, by construction.

b: goals that express intentions to increase knowledge, possibly by combining it.

Sets of actions that aim at the first class of goals — design and construction methods — cannot be used for creating knowledge, with the exception of the type of knowledge that is known as skill. On the other hand, methods directed to the creation of knowledge are not equipped to create designs or constructions. This characteristic of methods (they only can be used for the purpose for which they are designed) also applies to the more detailed differences between methods belonging to the same class.

Methods therefore offer a specific connection between actions and the state of a system. This notion begs the questions: What method is the best way to realize a certain state of the system? Will there be any guarantee that the selected solution realizes the truly desired state and not a wrong interpretation of it by the designer?

Generally one of the two following approaches is chosen (Berkhout 1986):

● Trying to derive the best solution by searching a hierarchically or relationally organised set of data or tasks, solving problems by means of several kinds of database management, combining, and calculation.

This approach generates selected sets of tasks.

● Trying to construct the best possible subset of means by simultaneously considering all relevant goals according to certain criteria and within restrictions and solving problems by means of (intelligent) communication and mathematics.

This approach generates selected sets of requirements, which are to be converted into sets of tasks.

A design method has to provide the opportunity to specify one's goals and the faculty to realize the inherently best subset of means for the desired state of the system. In keeping with the two approaches above, designing today uses one of the following methodologies, corresponding with paradigm II and paradigm III, respectively.

I: Evaluating a full enumeration of all relevant combinations of data, possibly integrated with the 'if-then-else' clauses of branch and bound methods. Evaluating a full enumeration guarantees the best subset, but is very laborious. Branch and bound methods reduce evaluating labour to a greater or lesser degree but are accordingly either more or less reliable in finding the best subset (Fig. 7.4). The subset of selected data marks the solution as a group of predefined tasks; within the data or tasks there is no degree of freedom. Solving may be described as searching the best combination in a discretely valued space, going from point to point. Points are represented by sets of meaningfully related data (Fig. 7.5a) (Berkhout, Dekker, and van Loon 1979).

II: Using multicriteria design methods which operate on goals and therefore have been built in evaluation (cf. goal class 1) of the best subset of means. The solution is created by optimization algorithms and chosen out of many degrees of freedom and continuous alternatives. The design or 'solution' is defined as a set of requirements, that may be realized like an ordinary set of tasks. Solving may be described as optimal selection in a continuous realm. The realm is restricted by the restrictive goals and optimal selection is performed by the optimizing goals (Fig. 7.5b) (Berkhout 1971).

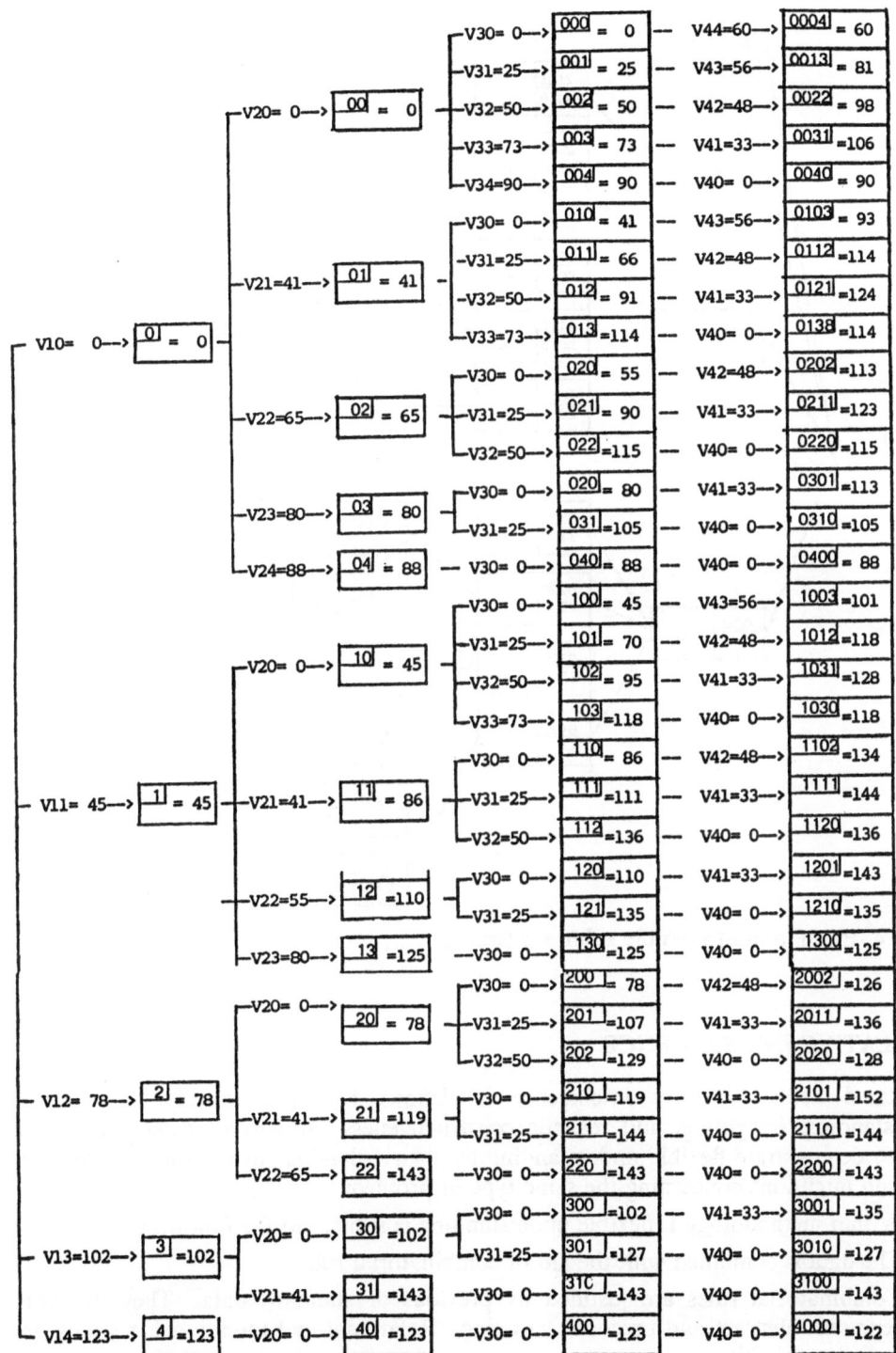

Fig. 7.4. Evaluation of possible solutions

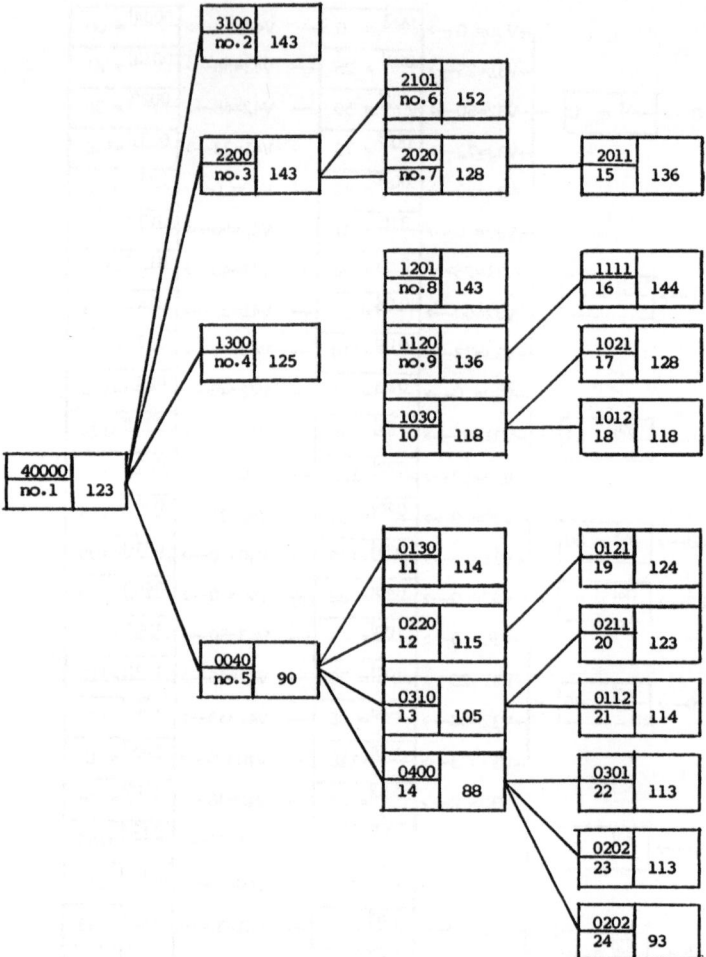

Fig. 7.4. Evaluation of possible solutions (cont.)

6. CREATIVE INTELLIGENCE

As stated before, creative intelligence is defined by three criteria, i.e. flexible understanding, imagining, and realistic constructiveness. Intelligent devices therefore have to demonstrate flexible understanding by their behaviour in response to different data and intentions concerning the same type of problem.

Within methodology I, flexible understanding is subject to the following:

- The data is combined with the aid of combinatorial rules.
- Combinatorial rules are justified by previous or standard data. They therefore represent abstract and precoded meaning. Relational and intentional meaning are indistinguishable.

114

Solving by data-processing   Solving by goal-processing

DISCRETE   **CONTINUOUS**

solving space   solution area

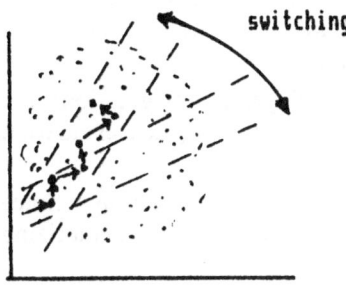
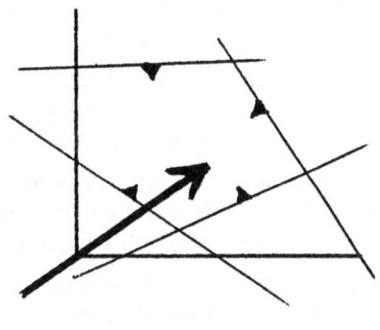

[a]   heuristic combination   [b]   optimal solution

Fig. 7.5. Two ways for solving problems

- The precoded meaning is applicable only to the same type of problems. Data may vary within the allowed variance area.
- Variances beyond the boundaries of the variance area are reported to and examined by the designer, without producing combinatorial results.
- The concept of 'switching' sometimes deals with those variances and provides an adapted answer to different data, but most probably not to different intentions (cf. Fig. 7.5a).
- Other impossibility handling routines are necessary in order to detect oversight and other inconsistencies.

In summary, methodology I (in fact a knowledge processing methodology) processes precoded meaning, but flexible understanding of intelligence has to be supplied by the designer. Only the concept of 'switching' allows for a restricted type of precoded intelligence.

Within methodology II, flexible understanding is characterised as follows:

- The data are supplied by the goals.
- Goals represent intentional meaning. Processing by mathematical matrix methods provides for relational meaning (cf. Fig. 7.7).
- Goals and inherent data are easily added or deleted from the set that defines the problem, if an appropriate interface is available.

- Evaluation and combinatorial selection are implied by the mathematical algorithms, as well as by the detection of impossibilities.

- Changing goals and data affects the combined intentional meaning as well as the implied relational meaning. The evaluation and selection is affected accordingly.

In summary, methodology II (a meaning processing methodology) processes meaning and data flexibly. Several aspects of flexible understanding of intelligence are supplied by the methodology. Procedural requirements provide for the recovery of impossibilities, misunderstandings, and other interpretational issues (see Section 5).

With respect to the second criterion of creative intelligence (imagination) the imagining explicitly is a task of the designer in both methodologies. Quick evaluation surveys are to be provided by the CAD instruments.

Unlike production, intentions are seldom or never the same in designing. Since the imagining implies intentional change at will, it is doubtful whether the memorizing of arbitrariness by learning capabilities would be useful.

With respect to the third criterion of the definition of creative intelligence (realistic constructiveness) there are also differences present between the two methodologies.

Within methodology I, realistic constructiveness is subjected to:

- Reliance on the quality of the experts and the quality of their knowledge. The former quality may differ from the latter, depending on the quality of the questionnaire and how it is understood.

- Designer dependent selection of contradictory knowledge, derived from experts within the same field of expertise. Justification of the selection is difficult to obtain.

- Designer dependent selection of conflicting knowledge, derived from experts in different fields of expertise. The intentions of the designer imply the selective priorities.

- Simulation on the effects of different priorities may provide information about which priorities are preferable.

In summary, methodology I has to rely on the quality of the consulted expertise. Continuous simulation may inform the designer about the agreeability of the combinatorial results concerning both the parts and the whole of the design.

Within methodology II, realistic constructiveness has the following characteristics:

- Reliance on the quality of experts and their knowledge in an interactive mode. Quality is improved in feedback loops and judged by both experts and principals from the provisional designs (cf. Fig. 7.7).

- Contradictory and conflicting knowledge are translated into goal-expressions, which display overlap or concurrency. Selection is conducted by the priorities of the optimizing goals. Indications are automatically produced which knowledge was the most appropriate to a solution. Corrections are easily fixed in feedback.

- From the start of a design, full design cycles are to be performed in order to facilitate the judgement of the realistic constructiveness from the earliest stage of the design.

- In most cases it appears to be possible to formulate specific goals with respect to realistic constructiveness. These goals are processed together with all other goals, enlarging the set of intentional and related meaning.

In summary, methodology II has considerable checking routines with respect to the quality and the realism of the data used and the intentional meaning. Provisional designs accomplish the continuous simulation.

The type of feedback called 'learning,' is not essentially a characteristic of creative intelligence. Imagination cannot be learned from data. Essentially the key data of the next design are the same only by accident. Otherwise we are probably making variations to a standard design. Learning methods will pay off in the event of setting norms and executing standard tasks or statistical processes. Generally opposite view is adopted (Feigenbaum and Barr 1986).

7. TASK, INSTRUMENT, AND DESIGN

As stated before, a design system consists of instruments consisting of tasks.

Tasks are essentially a set of actions, realizing a future state without any degree of freedom in their execution. This means that all parameters and requirements are known as data with respect to materials, budget, places, time, etc. The composition of the whole and its parts is also fully known. Each amount, place, and characteristic are set. But parts have to be put together, computations have to be made, and objects have to be placed in space. These activities that are applied to data, knowledge, or other representations of means are techniques. Techniques are essentially tasks without parameters, i.e. empty tasks.

Instruments are simply characterised as a set of well-arranged and clearly defined design techniques. The set is an implementation of a method and therefore only the intended types of results can be obtained. For instance, data management, computing, drawing, and designing all call for different basic instruments.

Since a method is directed to a certain purpose, i.e. obtaining specific results, we may speak of instruments as being directed to a specific methodology. Nevertheless we sometimes find common parts in different instruments. This does not mean that these instruments are directed to the same methodology. They only have — depending on the project — some strong points for coupling.

Designs essentially are programs for allocating means to the parts and relations of a system, in order to construct the object system. The means are allocated to functions of the object system, including internal functions, such as maintaining internal distances, resisting transport forces, or putting up with net weight. The completeness of (the description of) the means will be dependent on the design level. Sub-designs are generally supposed to be standard details or supposed to be easily generated without repercussions with respect to the surroundings of the detail.

8. LANGUAGES FOR INTELLIGENCE

We tried to trace the essentials of intelligent design systems. In summary:
- instrumentation of the flexibility in conceiving means.
- instrumentation of the evaluation and selection of solutions.

Intelligent design methodology has to be an easily accessible and decisive methodology. Since creative intelligence is the integrated combination of flexibility, imagination and decisive selection, the concept does not create a separate stage within the design process, but offers the ability to decide while conceiving.

The requirements and characteristics of design methods that may be called intelligent were gradually introduced. With respect to methodology I, we have to consider it mainly combinatorial. After producing numerous combinations we evaluate and select an alternative. Purposeful strategies are almost impossible (for a different view, see (Sowa 1983)). We may summarize the characteristics as follows:

I-a. The methodology is based on the identification of combinatorial design rules and their application on expert design knowledge (Fig. 7.6).

I-b. Combinatorial rules represent precoded meaning which has to be derived from previous designs. At present there are no combinatorial rules for imagination.

I-c. If intelligence requires flexible understanding, this requirement contradicts the intelligence of precoded meaning and oppositely requires the generation of uncoded meaning.

I-d. Flexible understanding may be simulated by a real time rule editor. The problem of justification of the 'instant rules' has to be solved.

I-e. Justification by empirical theory is the usual approach, but apparently does not work. The problem remains unsolved at present.

I-f. As a design is defined as a intended state of a system and as there is no theory of intentions or desires, no justifiable combinatorial rules can be derived that way. Moreover, as intentions and desires essentially are inductive there is little probability that such a theory will ever be developed.

I-g. Alternatively we may try to detect intentional knowledge within the derived expert design knowledge. It is doubtful whether intentions are knowledgeable. Essentially they may be the way we treat knowledge.

I-h. When we derive knowledge from experts, pointing out to contradictions is needed. Contradictions may occur when knowledge is derived from designers in the same expert field.

I-i. The detection of impossible solutions is needed. Impossibilities may occur when we combine knowledge from designers in different expert fields. Impossibilities are solved by applying priority scheduling. As a result several aspects may seem to be 'forgotten.' Essentially they are only deprived of the means for serving their interest.

I-j. The designer always has to evaluate the resulting design as a whole and will detect that the 'forgotten' aspects have to be reintroduced.

```
:-
orient(north),
orient(south),
    ...
opposite(north, south).
opposite(south, north).
    ...
main(R1) :-
    read R2,
    R2 nextto R1,
    main(R1).
R2 nextto R1 :-
    orient(Or1), orient(Or2),
    opposite(Or1, Or2),
    location(R2, Or2, X1, Y1, Z1).
location(R, Or, X, Y, Z).
    Or = south,
    Y2 = Y + 1.
location(R, Or, X, Y, Z).
    Or = north,
    Y2 = Y - 1.
    ...
```

In first instance
resulting in:

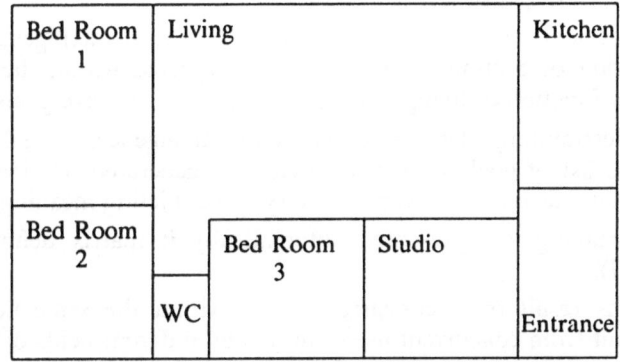

Floor 1

Other approaches tried on the same design activity:

- graphic/numerical definition (creation) of chains of related activity-spaces.
- manual computer aided allocation of activity-spaces (using layers, clusters, etc.).

Fig. 7.6. Prolog approach to housing design (relational operators)

I-k. Intelligence is assigned for the greatest part to the evaluation by the designer. The intelligence of realistic constructiveness is implied by the collected knowledge.

I-l. Recently another type of design concept has been in demand. Ordinary discrete and determined concepts ought to be replaced by 'trajectorial' concepts (cf. (Faludi 1973)), i.e. concepts that are valid for a certain range of values.

I-m. In architectural design the computer implementation of methodology I is presently used as a copying device. Proliferation of the same design appears to be the most efficient use of CAD.

With respect to methodology II, which produces designs by processing goals, we summarize the requirements and characteristics as follows:

II-a. This methodology is basically oriented toward principals and product users. Most effectively it is to be based on the communication between design and principal.(cf. Fig. 7.1).

II-b. Goals represent intentional meaning. Consistently specified methods process the goals. Several methods already exist; more are to be developed.

II-c. Concepts from the product users are be interpreted as sets of goals. Expert concepts may be interpreted as knowledge with respect to realistic constructiveness.

II-d. Goal-setting, goal-translation, and additional reconsideration are concept-specifying routines. Accordingly, adequate user interfaces are to perform an easy update of the set of goals.

II-e. Conceiving devices, as an implementation of artificial intelligence according the second criterion of Section 6, are not to be expected within the near future. It seems to be a function of living intelligence solely to conceive goals.

II-f. Flexible understanding, the first criterion of intelligence, is realised by an easy update of the list of goals by the automatically generated relational meaning and by the automatic dealing with contradictory or conflicting meanings.

II-g. Relational meaning is represented automatically if matrix defined methods are used (Fig. 7.7).

II-h. Contradictions result from concurrent views within the same field of expertise. Conflicts result from concurrent use of means by different fields of expertise.

II-i. Contradictions are signalised by mathematical algorithms. Conflicts are solved by means of priority optimizing. Some conflicts resemble contradictions and are processed accordingly. The recovery has to be performed by the designer. The result has to be approved by the principal.

II-j. Degrees of freedom, i.e. alternative choice, and the continuous solving area create trajectories for concepts. These trajectories create also flexible answers to changes in conditions.

II-k. The third criterion of intelligence, realistic constructiveness, is represented either by the data resulting from expert knowledge or by goals that represent this aspect purposefully.

II-l. Learning capabilities are only useful in the case of varying standard designs. Since intentions vary fundamentally in different projects, memorizing of intentions appears not to be of much use.

Creating a matrix of goals and selectable means:

GOALS	MEANS				
	room1	room1a	room1b	room2	room2a
area1		X		X	X
area2	X	X		X	
demand1	X	X	X		
demand2	X	X			X
priority	X		X		X

In first instance resulting in:

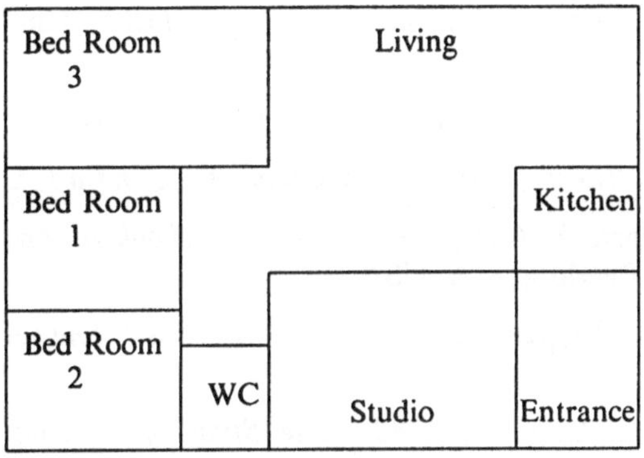

Floor 2.

Based on goals with respect to
● *costs*
● *space differentiations*
● *priority for costs over space.*
Result: *demanded space* with *25% less costs*.

Fig. 7.7. Housing design by goal-processing

We may summarize the characteristics of both methodologies as follows.

Knowledge oriented methods are effective in manufacturing. They are only intelligent with respect to one of the three criteria. Furthermore the elimination of conflicts and contradictions is cumbersome.

Goal-processing methods are effective in designing. They are intelligent with respect to two of the three criteria. Conflicts and contradiction are easily remedied. However, more goal-processing methods are to be developed.

REFERENCES

Benneth, J. L. (1983): *Building Decision Support Systems*, Wiley, New York.

Berkhout, E. E. (1971): "Optimizing in Spatial Design," Technical Report, OPM Group, Faculty of Building, Technical University of Delft, Delft, The Netherlands. In Dutch.

Berkhout, E. E., Dekker, A. P. M., and van Loon, P. P. (1979): "Methods, Techniques, Models," Lecture Notes, Faculty of Building, Technical University of Delft, Delft, The Netherlands. In Dutch.

Berkhout, E. E., Micheels, S., and van Loon, P. P. (1982): *Design and Planning Methodology*, Delft University Press, Delft, The Netherlands. In Dutch.

Berkhout, E. E. (1986): "Methodological Survey," Research Report, OPM Group, Faculty of Building, Technical University of Delft, Delft, The Netherlands. In Dutch.

Boden, M. (1981): *Artificial Intelligence and Natural Man*, Basic Books, New York.

Faludi, A. (1973): *Planning Theory*, Oxford University Press, Oxford, UK.

Feigenbaum, E. A. and Barr, A. (eds.) (1986): *The Handbook of Artificial Intelligence III*, Addison-Wesley, Reading, MA, USA.

Lee Jr., D. B. (May 1973): "Requiem for large scale models," *Journal of the American Institute of Planners*.

Machlup, F. and Mansfield, U. (eds.) (1983): *The Study of Information*, Wiley, New York.

Shannon, C. E. and Weaver, W. (1949): *The Mathematical Theory of Communication*, Urbana, IL, USA.

Sowa, J. F. (1983): *Conceptual Structures*, Wiley, New York.

Weizenbaum, J. (1976): *Computer Power and Human Reason*, Freeman, San Francisco, CA, USA.

Wiener, N. (1948): *Cybernetics*, Wiley, New York.

Yovits, M. C. and Cameron, S. (eds.) (1960): *Self Organizing Systems*, Pergamon Press, London.

Report on Session 3

Chair: M. Mac an Airchinnigh †
Cochair: J. Rogier ‡
Edited by: V. Akman ‡

At first glance, it might appear that the papers in this session bear no relation to each other. A framework into which the papers might be placed has been chosen by the session chairman for the purpose of developing a coherent summary. Viewing the session as a small world which mirrors the entire workshop, one might order the papers in the sequence Berkhout, Hoeltzel, Sunde.

A setting for the conceptual basis of an intelligent CAD system methodology is attempted by Berkhout. Concepts and definitions are elaborated. From the discussion which followed the paper, it is clear that the author is presenting his ideas to computer scientists for the first time. Therefore, some of his terminology appears to be non-standard to the latter group.

The design methodology of Hoeltzel is characterised by a concern for design optimization. His approach is inspired by the discipline of planning which provides concepts and techniques upon which he constructed his methods. He validated his approach in so far that a prototype was constructed for designing power transmission components — a very specialised problem domain. In particular, Hoeltzel would like to see a separation of the overall design process into a generic and a domain-specific portion. A significant aspect of his work is that he focuses on the design of industrial components. From his experience, conclusions about the nature of intelligent CAD systems for specific engineering design problems may be drawn.

A very specific design technique used in the construction of 2-D drawings of geometric objects is proposed by Sunde. He is aware of the fact that he offers no design methodology. Rather, he is trying to address the issue in a narrow sense. To design an object, it is important that the basic tools are available. One such tool, as described in the paper, allows the designer to construct a product (as represented by a drawing) by providing declarative specifications based on constraints. This approach recalls a pioneering work by Ivan Sutherland, the SKETCHPAD system.

† Department of Computer Science, Trinity College Dublin, Dublin 2, Ireland.
‡ Centre for Mathematics and Computer Science, Kruislaan 413, 1098 SJ Amsterdam, The Netherlands.

Having taken the papers in the order suggested, one returns to that of Berkhout and asks how the other two fit into his conceptual framework. Of course, this session is just a part of the entire workshop and its relevance may only be gauged by viewing it in the wider context. To facilitate and enable this shift of view, the authors were asked to make brief statements on their understanding of (1) design, (2) future role of CAD, and (3) intelligent CAD. Their views are given below:

Berkhout

Design: Design is to be seen as a communication between designer and consumer. A design communication language has to be based on goal-processing capabilities.

Role of CAD: A CAD system is a set of computerized and interrelated instruments directed towards the design of objects. An instrument is a set of well-arranged and consistently elaborated design techniques dedicated to the execution of well-defined tasks.

Intelligent CAD: The author is primarily concerned with the user's view of the user interface. Establishing a clear conceptual basis for the interface is a prerequisite for the development of intelligent CAD systems.

Hoeltzel

Design: The design process can be separated into a generic portion and a domain specific portion. Although this is considered very important by the author, he does not know how it can be achieved. Design process is non-deterministic. However, it must always be constrained to be deterministic if a definite engineering solution is to be realized. The planning process in design must be carefully considered in order to properly integrate (1) design requirements, (2) problem reduction, (3) design objective compromise, (4) scheduling, and (5) the redesign process.

Future role of CAD: Geometry and aesthetics are insufficient as starting points in the solution of engineering design problems. In fact, their priority is at a lower level in the design process hierarchy, namely, after basic configurations have been achieved to satisfy functional requirements.

Intelligent CAD: An intelligent CAD system should employ a reasoning capability that includes design requirements and design specification assignment. Design specification assignment is the process of coding design knowledge in a machine-understandable format such that it can be accessed and utilized to do design. Scheduling and planning strategies for propagating and satisfying design constraints are of fundamental importance in realising solutions. The concept of design optimization, traditionally considered as residing in the numerical domain, can be generalized to include symbolic optimization. This is referred to as design compromising.

Sunde

Design: Design is the description and definition of a product and the techniques by which one refines a vague description into a sufficiently detailed one for manufacturing. In addition, it encompasses the notion of invention.

Declarative specification, presented in the paper, is intended to support a part of the entire design process. We have no general design methodology but believe that design by constraints and refinement is needed. At the moment, this is done in our system for geometrical entities such as line segments, arcs, etc. The ultimate goal is to establish the relation between geometry and functionality.

Future role of CAD: One challenge of future CAD systems will be to offer better tools for the description of a product, for the incremental refinement or transformation of that description, and for evaluation. In other words, the role of CAD will continue to be the total support of the design activity.

Intelligent CAD: It is rather difficult to say when a CAD system is intelligent. It is obviously not intelligent when it behaves illogically or unnaturally. The declarative specification method presented in the paper offers a more natural way of creating drawings at a higher level of description. With respect to the design of 2-D geometric objects, constraints and reasoning facilities have been presented.

During the Q&A period, in response to Nadin's question about his future goals, Berkhout mentioned his commitment to devising a strategy to convince architects to use computer techniques. Ten Hagen asked if Berkhout has already tried to contact computer scientists in order to implement his theory in a working intelligent CAD system. The response was that trying to motivate architects to use computer techniques is so troublesome that one can hardly find the energy to convince computer scientists to implement the proposed theory.

Novak challenged Hoeltzel about his system's ability for creative design. Hoeltzel's system does not incorporate creative design processes. The reason for this is that only a part of the number of design processes can be included in the theory and creative design is thought to be one of the hardest to formulate. Ruttkay asked about the possibility of learning processes which may be incorporated to the system. Hoeltzel said that the system contains a large amount of predefined problem solving methods. These are collected in categories of methods which may be used alternatively or are selected by means of pattern recognition mechanisms. The main goal of the system is to solve design problems by using optimization techniques.

Mac an Airchinnigh posed a very original question: Can you mention some significant things to come out of your project in relation to the development of intelligent CAD? The reply was in three parts: (1) Intelligent CAD must be able to manipulate symbolic rather than numerical notations or at least bridge these notations. (2) A distinction has to be made between generic and domain-specific knowledge. Research has to be done where this distinction has to be made. (3) The technique by which design procedures are propagated is still open. Neelamkavil then posed the following question: If a problem for which the approach of optimization may not be useful is encountered, can the system still be used? The answer was two-fold. First, it is possible to use the system without making use of the tools included to optimize the result. Second, every design problem can be considered as an optimization problem. So appropriate tools are always useful.

Nadin mentioned that designers do not design without some reference to material and asked Sunde: How could you incorporate references to the used design material and how can that knowledge be used? Sunde's reply was that in his system it is possible to declare additional information linked to the objects. Such use however is somehow restricted since it is like a comment. Hofer-Alfeis wondered the number of constraints one may add to the system, and what cannot be designed with the system. The answers were that there exist no upper bounds within reasonable limits and that only line segments can be used, respectively. Tomiyama requested from Sunde a justification for his claim stating that the geometrical parts of design systems nowadays are more interesting than the functional parts. Sunde said that the geometrical part is always needed. It forms the elementary basis of a design system. Furthermore, unambiguous specification of design objects by declaring only their functionality is more or less a dream and one has to start at a more modest level. Akman commented that the constructive nature of geometry makes this a powerful system, and asked: What are the boundaries? It was then revealed that the system is limited by speed which in turn is influenced by the number of elements incorporated in an assembly.

Mac an Airchinnigh posed the following challenge to Sunde: While showing the drawing of a part with a hole in the centre, you said that the hole would be automatically translated if a length constraint was changed. But suppose instead that I want the position of the hole to remain fixed while changing the length constraint. How do I determine what item should be influenced by my next declaration? Sunde replied as follows. Declaration takes place by means of explicit mention of the items to be added. Only in the case of a conflicting declaration warnings will be provided. Unforeseen results of adding information are prevented by the fact a design object is not complete if underdetermined.

Session 4

8. Acquiring Design and Analysis Knowledge for Knowledge-Based Systems

J.H. Boose and J.M. Bradshaw

Knowledge Systems Laboratory, Boeing Advanced Technology Center
Boeing Computer Services, P.O. Box 24346, Seattle, WA 98124, USA

Abstract: *A workbench for acquiring design and analysis knowledge to build knowledge-based systems is presented. Eliciting and modeling such knowledge from a human designer is a major problem when building knowledge-based systems for CAD problems. Aquinas, an extended version of the Expertise Transfer System (ETS), combines ideas from psychology and knowledge-based systems research. Aquinas interviews design and analysis experts and helps them model, analyze, test, and refine their knowledge. Expertise from multiple designers or other knowledge sources can be represented and used separately or combined. User consultations are directed by propagating information through hierarchies. Aquinas delivers knowledge by creating knowledge bases for several different expert system shells. Help is given to the expert by a dialog manager that embodies knowledge acquisition heuristics. Aquinas contains many techniques and tools for expertise transfer; the techniques combine to make it a powerful testbed for rapidly prototyping portions of many kinds of complex knowledge-based systems.*

Keywords: expert systems, expert system tool, knowledge acquisition, knowledge-based systems, knowledge engineering, personal construct psychology.

1. INTRODUCTION: OBTAINING AND MODELING EXPERTISE

The Expertise Transfer System (ETS) has been in use at Boeing for more than three years. Hundreds of prototype knowledge-based systems have been generated by ETS. The system interviews experts to capture vocabulary, conclusions, problem-solving traits, trait structures, trait weights, and inconsistencies. It helps build prototypes very rapidly (typically in less than two hours), assists the expert in analyzing the adequacy of the knowledge for solving the problem, and creates knowledge bases for several expert system shells (S.1[1], M.1[2], OPS5, KEE[3], and so on) from its own internal representation (Boose 1985; Boose 1986).

[1, 2] M.1 is a trademark of Teknowledge, Inc.
[3] KEE is a trademark of Intellicorp.

Aquinas has been built to overcome ETS's limitations in knowledge representation and reasoning (Boose and Bradshaw 1987). Due to previous limitations, the system was usually abandoned sometime during the knowledge acquisition process. Typically it was used to explore project approaches and assess feasibility for several days or a week, and then development continued in some other expert system shell. While the use of the tool in this way saved substantial time (typically one or two months in a 12- to 24-month project), it was desirable to make the system more powerful.

2. AQUINAS WORKBENCH

Aquinas is a collection of integrated tool sets. The tool sets share a common user interface (the dialog manager) and an underlying knowledge representation and data base (Fig. 8.1). Each set of tools addresses a general knowledge acquisition task:

Dialog manager						
ETS Repertory grid tools	Hierarchical structure tools	Uncertainty tools	Internal reasoning engine	Multiple scale type tools	Induction tools	Multiple expert tools
Object-oriented DBMS						
CommonLoops / CommonLisp						

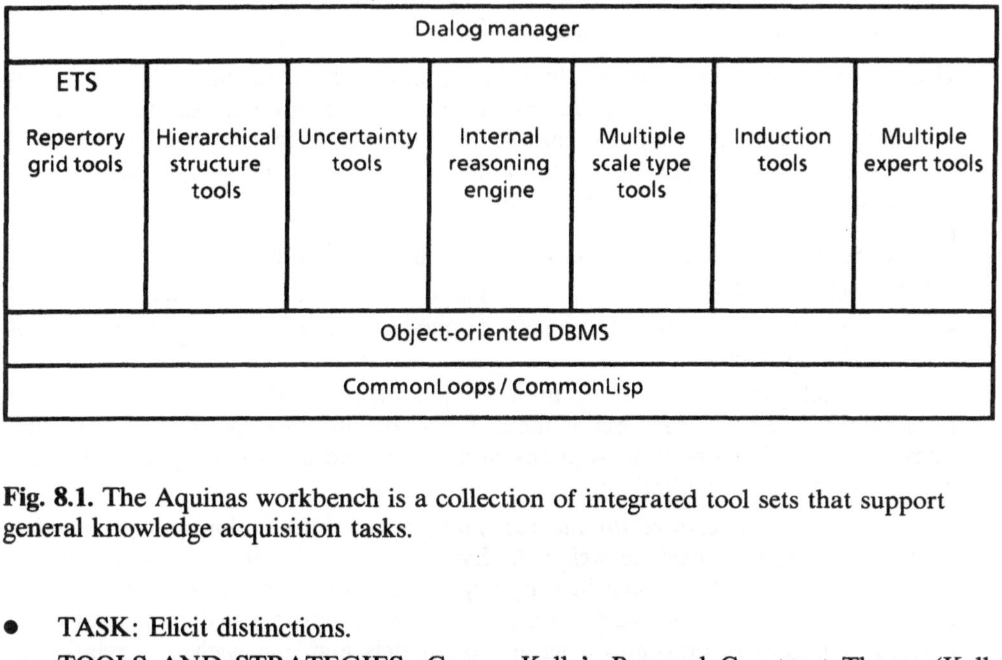

Fig. 8.1. The Aquinas workbench is a collection of integrated tool sets that support general knowledge acquisition tasks.

● TASK: Elicit distinctions.

TOOLS AND STRATEGIES: George Kelly's Personal Construct Theory (Kelly 1955) provides a rich framework for modeling the qualitative and quantitative distinctions inherent in an expert's problem-solving knowledge. Aquinas is a set of tools used by the expert to elicit, analyze, and refine knowledge as rating grids. In a rating grid, problem solutions — *elements* — are elicited and placed across the grid as column labels, and traits of these solutions — *constructs* — are listed alongside the rows of the grid. Traits are first elicited by presenting groups of solutions and asking the expert to discriminate among them. Following this, the expert gives each solution a rating showing where it falls on the trait scale.

Many of the strategies used in building a rating grid are extensions of ideas in the work of Kelly and in the PLANET system (Gaines and Shaw 1981; Shaw and Gaines 1987a; Shaw and Gaines 1987b). These strategies include triadic elicitation (combining groups of three objects together and asking for discriminating traits), corner filling (a form of N-dimensional table filling), and multiple analysis and display tools. Aquinas can analyze a rating grid in many ways to help the expert refine useful distinctions and eliminate those that are inconsequential or redundant. Distinctions captured in grids can be converted to other representations such as production rules, fuzzy sets, or networks of frames.

- TASK: Decompose problems.

  TOOLS AND STRATEGIES: In our previous work using ETS, the difficulty of representing complex problems in a single rating grid became clear. First, a single rating grid can represent only "flat" relations between single solutions and traits. No deep knowledge, causal knowledge, or relationship chains can be shown. A second limitation is that only solutions or traits at the same level of abstraction could be used comfortably in a single grid. Finally, large single grids are often difficult to manipulate and comprehend.

  Hierarchical tools help the expert build, edit, and analyze knowledge in hierarchies and lattices. Currently, hierarchies are organized around solutions, traits, knowledge sources (i.e. experts), and cases. Strategies to help the expert build hierarchies include laddering (eliciting more general or more specific objects), cluster analysis, and trait value examination.

- TASK: Specify methods for combining uncertain information.

  TOOLS AND STRATEGIES: Uncertain knowledge, preferences, and constraints may be elicited, represented, and locally applied in Aquinas using combinations of several different methods.

  Absolute (categorical) reasoning involves judgments made with no significant reservations. Experts can build these types of absolute constraints into the knowledge associated with an Aquinas rating grid, and users may specify absolute constraints during consultations.

  Many judgments involve *relativistic* reasoning, where information and preferences from several sources must be weighed. Even if criteria for the ideal decision can be agreed on, sometimes it can be only approximated by the available alternatives. In these cases, problem-solving information must be propagated in a relativistic fashion. Aquinas incorporates a variety of models and approaches to relativistic reasoning, including MYCIN-like certainty factor calculus (Adams 1985), fuzzy logic (Gaines and Shaw 1985), and the Analytic Hierarchy Process (Saaty 1980).

  In the current version of Aquinas, some limited propagation of probabilistic information is also included. Experts may make discrete distributions on rating values. Future versions of Aquinas will have more complete models.

  The availability of different inference methods within a single workbench allows users and experts flexibility in adapting Aquinas to the problem at hand. Methods are selected based on the cost of elicitation, the precision of the knowledge needed, convenience, and the expert's preference. Future research will suggest heuristics for helping experts select appropriate methods and designs for particular types of questions, e.g. (Shafer and Tversky 1985). These heuristics will be incorporated into the Aquinas dialog manager (see below).

- TASK: Test the knowledge.

  TOOLS AND STRATEGIES: A mixed initiative reasoning engine allows consultations to be run using the knowledge in Aquinas. Several inheritance, specialization, and generalization techniques are employed to propagate knowledge through hierarchies and reach conclusions. New grid ratings may be created from information derived from other grids.

  The model of problem-solving currently used in Aquinas is that of multiple knowledge sources (experts) that work together in a common problem solving context (case) by selecting the best alternatives for each of a sequential set of decisions (solutions). Alternatives at each step are selected by combining relevant information about preferences (relativistic reasoning), constraints (absolute reasoning) and evidence (probabilistic reasoning).

  For many structured selection problems, a more specialized version of this model seems adequate. After analyzing several expert systems for classification, Clancey (1986) suggested that many problems are solved by abstracting data, heuristically mapping higher level problem descriptions onto solution models, and then refining these models until specific solutions are found. In Aquinas, data abstraction is carried out within hierarchies of traits, and solutions are refined as information is propagated through solution hierarchies.

  While the current version of Aquinas works best on those problems whose solutions can be comfortably enumerated (such as those amenable to the method of heuristic classification), we are interested in generalizing Aquinas to incorporate synthetic (constructive) problem-solving methods such as those in SALT (Marcus 1987).

- TASK: Combine data types.

  TOOLS AND STRATEGIES: Aquinas expands the knowledge representation capability of rating grids from Personal Construct Theory by allowing the use of several types of rating scale types (nominal, ordinal, interval, and ratio). These trait scale types can be elicited, analyzed, and used by the reasoning engine. Embedded heuristics help the expert represent and combine symbolic and numeric rating values for varying levels of precision and convenience.

- TASK: Expand the knowledge base automatically.

  TOOLS AND STRATEGIES: Several types of tools make inductive generalizations about existing knowledge. Generalizations can be examined by the expert and used to refine the knowledge, and are used by the reasoning engine.

  Learning strategies in Aquinas include simple learning from examples (e.g., selective induction on lower level grids to derive values for higher level grids), deduction (e.g. inheritance of values from parents), analogy (e.g. derivation of values based on functional similarity of traits), and observation (e.g. constructive induction based on cluster analysis). The dialog manager (described below) also contains various learning mechanisms.

- TASK: Use multiple sources of knowledge.

  TOOLS AND STRATEGIES: Knowledge from multiple experts (or other knowledge sources) can be analyzed to find similarities and differences in knowledge, and the degree of subsumption of one expert's knowledge over another. Information from these analyses can be used to guide negotiation among experts. The reasoning engine uses knowledge from user-specified and weighted sources and gives consensus and dissenting opinions.

- TASK: Provide knowledge acquisition process guidance.

  TOOLS AND STRATEGIES: A subsystem called the *dialog manager* contains pragmatic heuristics to guide the expert through knowledge acquisition using Aquinas (Kitto and Boose 1987). Such help is important in the use of Aquinas, given the complexity of the Aquinas environment and the many elicitation and analysis methods available to the expert. The dialog manager makes decisions about general classes of actions and then recommends one or more specific actions providing comments and explanation if desired. This knowledge is contained in rules within the dialog manager. A session history is recorded so that temporal reasoning and learning may be performed.

3. AQUINAS IN USE

Aquinas is written in Interlisp[4] and runs on the Xerox[5] family of Lisp machines. Subsets of Aquinas also run in an Interlisp version on the DEC[6] VAX[7] and a C/UNIX[8]-based portable version. The Aquinas screen is divided into a typescript window, map windows showing hierarchies, rating grid windows, and analysis windows (Fig. 8.7). Experts interact with Aquinas by text entry or by mouse through pop-up menus.

Following are the steps in a Aquinas session in which an expert is building a Structural Analysis Package Advisor. Structural designers would use such a system to help select appropriate structural analysis software packages. Aquinas guides the expert in putting knowledge into Aquinas's knowledge base, and continues through the making of a knowledge base for the OPS5 expert system shell. These steps are:

3.1. Elicit Cases and the Initial Grid (Solutions, Traits, and Ratings)

The expert begins the session by entering several structural analysis packages. These will be part of the set of potential solutions. The packages are added to the solution hierarchy as children. Then Aquinas asks the expert to enter traits based on differences and similarities between packages. This is the heart of Kelly's interviewing

---

[4, 5] Xerox and Interlisp are registered trademarks of Xerox Corporation.
[6, 7] DEC and VAX are trademarks of Digital Equipment Corporation.
[8] UNIX is a trademark of AT&T Bell Laboratories.

Please enter a list of analysis-package types, one to a line. When you're done, enter STOP, or a RETURN.

AQU** <u>E3SAP</u>
AQU** <u>ANSYS</u>
AQU** <u>STARDYNE</u>
AQU** <u>GT-STRUDL</u>
AQU** <u>NASTRAN</u>
AQU** <u>ATLAS</u>
AQU** <u>MARC</u>
AQU** <u>(CR)</u>

Think of an important trait that two of E3SAP, ANSYS, and STARDYNE share, but that the other one does not. What is that trait?
AQU** <u>NON-LINEAR ANALYSIS</u>

What is that trait's opposite as it applies in this case?
AQU** <u>LINEAR-ANALYSIS ONLY</u>

What is the name of a scale or concept which describes NON-LINEAR ANALYSIS / LINEAR-ANALYSIS ONLY?
AQU** <u>LINEARITY</u>

Think of an important characteristic that two of ANSYS, STARDYNE, and GT-STRUDL share, but that the other one does not. What is that characteristic?
AQU** <u>LARGE PROBLEM CAPACITY</u>

What is that characteristic's opposite as it applies in this case?
AQU** <u>SMALL PROBLEM CAPACITY</u>

What is the name of a scale or concept which describes LARGE PROBLEM CAPACITY / SMALL PROBLEM CAPACITY?
AQU** <u>PROBLEM SIZE</u>
      :

Please rate these things on a scale of 5 to 1, where 5 means more like NON-LINEAR ANALYSIS and 1 means more like LINEAR-ANALYSIS ONLY. If neither one seems to apply, enter N(either). If both seem to apply, enter a B(oth).
   NON-LINEAR ANALYSIS(5) LINEAR-ANALYSIS ONLY(1)
      (E3SAP) ** <u>1</u>
      (ANSYS) ** <u>4</u>
      (STARDYNE) ** <u>2</u>
      (GT-STRUDL) ** <u>3</u>
      (NASTRAN) ** <u>5</u>
      (ATLAS) ** <u>3</u>
      (MARC) ** <u>5</u>

Please rate these things on a scale of 5 to 1, where 5 means more like LARGE PROBLEM CAPACITY and 1 means more like SMALL PROBLEM CAPACITY. If neither one seems to apply, enter N(either). If both seem to apply, enter a B(oth).
   LARGE PROBLEM CAPACITY(5) SMALL PROBLEM CAPACITY(1)
      (E3SAP) ** <u>1</u>
      (ANSYS) ** <u>4</u>
      (STARDYNE) ** <u>3</u>
      (GT-STRUDL) ** <u>4</u>
      (NASTRAN) ** <u>5</u>
      (ATLAS) ** <u>5</u>
      (MARC) ** <u>4</u>
      :

Please rate the relative importance of LINEARITY on a scale from 5 to 1, where 5 means more important, and 1 means less important.
AQU** <u>4</u>

Please rate the relative importance of PROBLEM-SIZE on a scale from 5 to 1, where 5 means more important, and 1 means less important.
AQU** <u>5</u>
      :

Fig. 8.2. Aquinas asks the expert for an initial set of potential solutions to the first problem case. Then, the solutions are presented in groups of three, and the expert gives discriminating traits. Ratings are entered for each solution for each trait.

methodology; Aquinas uses it in several different ways as knowledge is expanded through elicitation and analysis (Fig. 8.2).

Aquinas initially assumes that traits will be bipolar with ordinal ratings between 1 and 5. The expert is asked to rate each solution with regard to each trait, but the expert may specify different rating scales (unordered, interval, or ratio). Aquinas later assists in recognizing and changing types of rating scales.

```
IMPLICATIONS:
  GOOD GRAPHICS implies:
       LARGE PROBLEM CAPACITY ( 94)
  AERO-ELASTIC CAPABILITY implies:
       GOOD GRAPHICS (.88)
       LARGE DYNAMIC ANALYSIS (1.0)
       LARGE PROBLEM CAPACITY (1 0)
  LARGE DYNAMIC ANALYSIS implies
       GOOD GRAPHICS (.83)
       LARGE PROBLEM CAPACITY (1 0)
  NON-LINEAR ANALYSIS implies:
       LARGE PROBLEM CAPACITY ( 92)
  POOR GRAPHICS implies·
       LARGE PROBLEM CAPACITY (1 0)
       NON-LINEAR ANALYSIS (1 0)
  SMALL DYNAMIC ANALYSIS implies
       NO AERO-ELASTIC CAPABILITY (1 0)
       SMALL PROBLEM CAPACITY (1 0)
       LINEAR-ANALYSIS ONLY (1 0)
  SMALL PROBLEM CAPACITY implies:
       NO AERO-ELASTIC CAPABILITY (1.0)
       SMALL DYNAMIC ANALYSIS (1.0)
       LINEAR-ANALYSIS ONLY (1.0)
  LINEAR-ANALYSIS ONLY implies
       NO AERO-ELASTIC CAPABILITY (1.0)
```

Fig. 8.3. Implication generalizations between traits are generated based on grid ratings.

Aquinas also elicits information about the importance of each trait. This knowledge is used later in the decision making process.

3.2. Analyze and Expand the Initial, Single Grid

Once a grid is complete, an analysis is performed to show *implications* between various values of traits (Fig. 8.3). A method similar to ENTAIL (Gaines and Shaw 1985; Shaw and Gaines 1987a) derives implications: rating grid entries are used as a sample set and fuzzy set logic is applied to discover inductive implications between the values. This method uncovers higher-order relationships among traits and later helps build trait hierarchies. The expert can also use an interactive process (*implication review*) to analyze and debug this information; the expert may agree or disagree with each implication. If the expert disagrees, the knowledge that led to the implication is reviewed, and the expert can change the knowledge or add exceptions that disprove the implication (Boose 1986). Certain types of *implication patterns* are also uncovered.

Discovery of *ambiguous* patterns, for example, may mean that traits are being used inconsistently (Boose 1986; Hinkle 1965).

```
Go, Edit, Fill-in, List, etc...
AQU** ANALYZE-MATCHES

The two elements ATLAS and NASTRAN are matched at the 85% level. Can you think of some pair of traits which
would distinguish between them?
AQU** YES
What is that distinguishing trait?
AQU** CERTIFIED
What is that attribute's opposite?
AQU** NOT CERTIFIED

Please rate these things on a scale of 5 to 1, where 5 means more like CERTIFIED and 1 means more like NOT
CERTIFIED.
 CERTIFIED(5) NOT CERTIFIED(1)
   (E3SAP) ** 1
   (ANSYS) ** 5
   (STARDYNE) ** 1
   (GT-STRUDL) ** 5
   (NASTRAN) ** 5
   (ATLAS) ** 1
   (MARC) ** 1

The two elements GT-STRUDL and ANSYS are matched at the 88% level. Can you think of some pair of traits which
would distinguish between them?
AQU** YES
What is that distinguishing trait?
AQU** POWER UTILITIES
What is that attribute's opposite?
AQU** NOT POWER UTILITIES
                 :
```

Fig. 8.4. Aquinas analyzes solutions and traits for close matches. Here, the expert enters a new trait to help further distinguish between ATLAS and NASTRAN.

Aquinas also analyzes the discrimination power of a rating grid by checking for similarities between traits and solutions (Fig. 8.4).

3.3. Test the Knowledge in the Single Grid

The dialog manager recommends that the grid knowledge be tested by running a consultation (Fig. 8.5). The expert is asked to provide desirable values for the traits associated with an instance of the case under consideration. These values may be appended with a certainty factor and/or the tag ABSOLUTE to show an absolute constraint. Consultation questions are ordered according to a computed benefit/cost ratio that depends on both the generated system (e.g. entropy of a given trait (Quinlan 1983)) and the specified expert (e.g. cost of obtaining information) parameters. The questions may also be be ordered according to an arbitrary specification given by the expert. Performance is measured by comparison of experts' expectations with Aquinas consultation results.

```
Go, Edit, Fill-in, List, etc...
AQU** TEST

What is the LINEARITY attribute for this case (NON-LINEAR-ANALYSIS  LINEAR-ANALYSIS-ONLY) ?
AQU** NON-LINEAR-ANALYSIS
What is the PROBLEM-SIZE attribute for this case (LARGE-PROBLEM-CAPACITY  SMALL-PROBLEM-CAPACITY) ?
AQU** LARGE-PROBLEM-CAPACITY
What is the DYNAMIC-SIZE attribute for this case (LARGE-DYNAMIC-ANALYSIS  SMALL-DYNAMIC-ANALYSIS) ?
AQU** LARGE-DYNAMIC-ANALYSIS
What is the AERO-ELASTIC attribute for this case (AERO-ELASTIC-CAPABILITY  NO-AERO-ELASTIC-CAPABILITY) ?
AQU** AERO-ELASTIC-CAPABILITY
What is the GRAPHICS attribute for this case (GOOD-GRAPHICS  POOR-GRAPHICS) ?
AQU** GOOD-GRAPHICS .5

Test results for case number 101:

        1 : NASTRAN ( .85 )
        2 : ATLAS ( .83 )
        3 : MARC ( .53 )
        4 : ANSYS ( .24 )
        5 : GT-STRUDL ( .24 )
        6 : STARDYNE ( -.59 )
        7 : E3SAP ( -.89 )
```

Fig. 8.5. The rules are used to test grid knowledge by running a consultation in Aquinas.

Two methods are available in Aquinas for turning rating values in grids into solution recommendations. One approach involves mapping this information onto certainty factor scales (Fig. 8.6). Each rating in the grid is assigned a certainty factor weight based on its *relative strength* (a rating of 5 is stronger than a rating of 4), the *relative weight* the expert has assigned to the trait, and any *absolute constraints* that the expert has specified for the trait. In the test consultation, EMYCIN's certainty factor combination method (Adams 1985) is used to combine the certainty factors. The result is a rank-ordered list of solutions with certainty factor assignments. These certainty factors are also used when rules are generated for expert system shells.

Another approach available employs Saaty's Analytic Hierarchy Process to order a set of possible solutions (Saaty 1980). Grid information obtained through pairwise comparisons or through regular rating grid methods is mapped onto *judgment matrices*. The *principal eigenvector* is computed for each matrix; the eigenvectors are normalized and combined to yield a final ranking of the solutions. Each solution has a score between 0.0 and 1.0. In a knowledge base consisting of multiple grids, these values are propagated through the hierarchies.

3.4. Build Hierarchies (Solutions and Traits in Multiple Grids) from the Grid

Next, the dialog manager recommends that the expert expand the trait and solution hierarchies by performing a *cluster analysis*. Aquinas uses a method of single-link

```
--- GENERATING RULES ---

If: LINEARITY = LINEAR-ANALYSIS ONLY
Then: AERO-ELASTIC = NO AERO-ELASTIC CAPABILITY (1.0)

If: PROBLEM-SIZE = SMALL PROBLEM CAPACITY
Then: DYNAMIC-SIZE = SMALL DYNAMIC ANALYSIS (1.0)

If: PROBLEM-SIZE = SMALL PROBLEM CAPACITY
Then: LINEARITY = LINEAR-ANALYSIS ONLY (1.0)

If: LINEARITY = NON-LINEAR ANALYSIS
Then: PROBLEM-SIZE = LARGE PROBLEM CAPACITY (.8)
                    :

If: LINEARITY = LINEAR-ANALYSIS ONLY
Then: The analysis-package may be E3SAP (.4)

If: LINEARITY = LINEAR-ANALYSIS ONLY
Then: The analysis-package may be STARDYNE (.2)

If: LINEARITY = LINEAR-ANALYSIS ONLY
Then: The analysis-package may be ANSYS (-.2)

If: LINEARITY = LINEAR-ANALYSIS ONLY
Then: The analysis-package may be NASTRAN (-.4)

If: PROBLEM-SIZE = SMALL PROBLEM CAPACITY
Then: The analysis-package may be E3SAP (.5)

If: PROBLEM-SIZE = SMALL PROBLEM CAPACITY
Then: The analysis-package may be ANSYS (-.25)

If: DYNAMIC-SIZE = SMALL DYNAMIC ANALYSIS
Then: The analysis-package may be E3SAP (.3)

If: DYNAMIC-SIZE = SMALL DYNAMIC ANALYSIS
Then: The analysis-package may be GT-STRUDL (-.15)

If: GRAPHICS = POOR GRAPHICS
Then: The analysis-package may be MARC (.1)
                              :
68 rules were generated
```

Fig. 8.6. Rules are generated from rating grids and implications.

hierarchical cluster analysis based on FOCUS (Shaw and Gaines 1987a) to group sets of related solutions or traits. The junctions in the clusters can be seen as conjectures about possible new classes of solutions or traits. These more general trait or solution classes may be named and added to the hierarchies. *Laddering* is also used to find traits at varying levels of abstraction (Boose 1986; Hinkle 1965). "Why?" questions are used to find more general traits, and "How?" questions help find more specific traits.

3.5. Use Several Rating Value Types (Transform Ordinal Ratings to Nominal and Interval Ratings) to Represent Knowledge

Aquinas helps the expert convert several traits with ordinal values into traits with nominal and ratio scaled rating values. The expert rates the solutions again in terms of the new values and these values appear on the grid. Aquinas provides several forms of estimation help: START-&-MODIFY, EXTREME-VALUES, DECOMPOSITION, and RECOMPOSITION (Beyth-Marom and Dekel 1985).

Aquinas also helps the expert change trait scale types by checking values associated with particular kinds of traits. For instance, bipolar traits that receive only extreme ratings may be better represented with a nominal trait.

3.6. Test Knowledge in Hierarchies

A potential set of solutions is identified by the expert. A partial problem model is constructed, evaluated, and abstracted in a bottom-up fashion through the trait hierarchy of each solution level. Through this process the solution is refined as the children of the best solutions are chosen for continued evaluation. Bottom-up abstraction takes place again in the trait hierarchy at the new solution level, and the cycle continues until all remaining solution candidates have been evaluated. Then an ordered list of solution candidates is obtained. This information from a single case may then be combined, if desired, with information from other cases to derive a final ranking of solution candidates. Users may override this general model of inference propagation by specifying explicit inference paths and parameters.

3.7. Edit, Analyze, and Refine the Knowledge Base

Once the expert has entered information about one case, additional cases may be described. A list of relevant solutions and traits could be entered from scratch, but that would be inefficient if there were significant overlap in those required by a previously entered case and a new one (Mittal, Bobrow, and Kahn 1986). Aquinas allows an expert to copy pieces of hierarchies (and, optionally, their associated values) between cases. Information copied in this way can be modified to fit the new context. This facility may also be used to copy pieces of hierarchies between experts.

3.8. Further Expand and Refine the Knowledge Base

Hierarchies and rating grids continue to be used during the session to expand and refine the knowledge base (Fig. 8.7). Aquinas contains a variety of other tools to help accomplish this (e.g. analysis and comparison of different experts, incremental interviewing, trait value examination, trait range boundary examination, completeness checking, and combining similar traits).

3.9. Generate Rules for Expert System Shells

The expert is the judge of when the point of diminishing returns has been reached within Aquinas. When such a point is reached, a knowledge base is generated for an expert system shell, and development continues directly in that shell. Similarity and implication analyses allow experts to determine whether traits or solutions can be adequately and appropriately discriminated from one another. The system provides correlational methods for comparing the order of Aquinas recommendations to an expert's rankings.

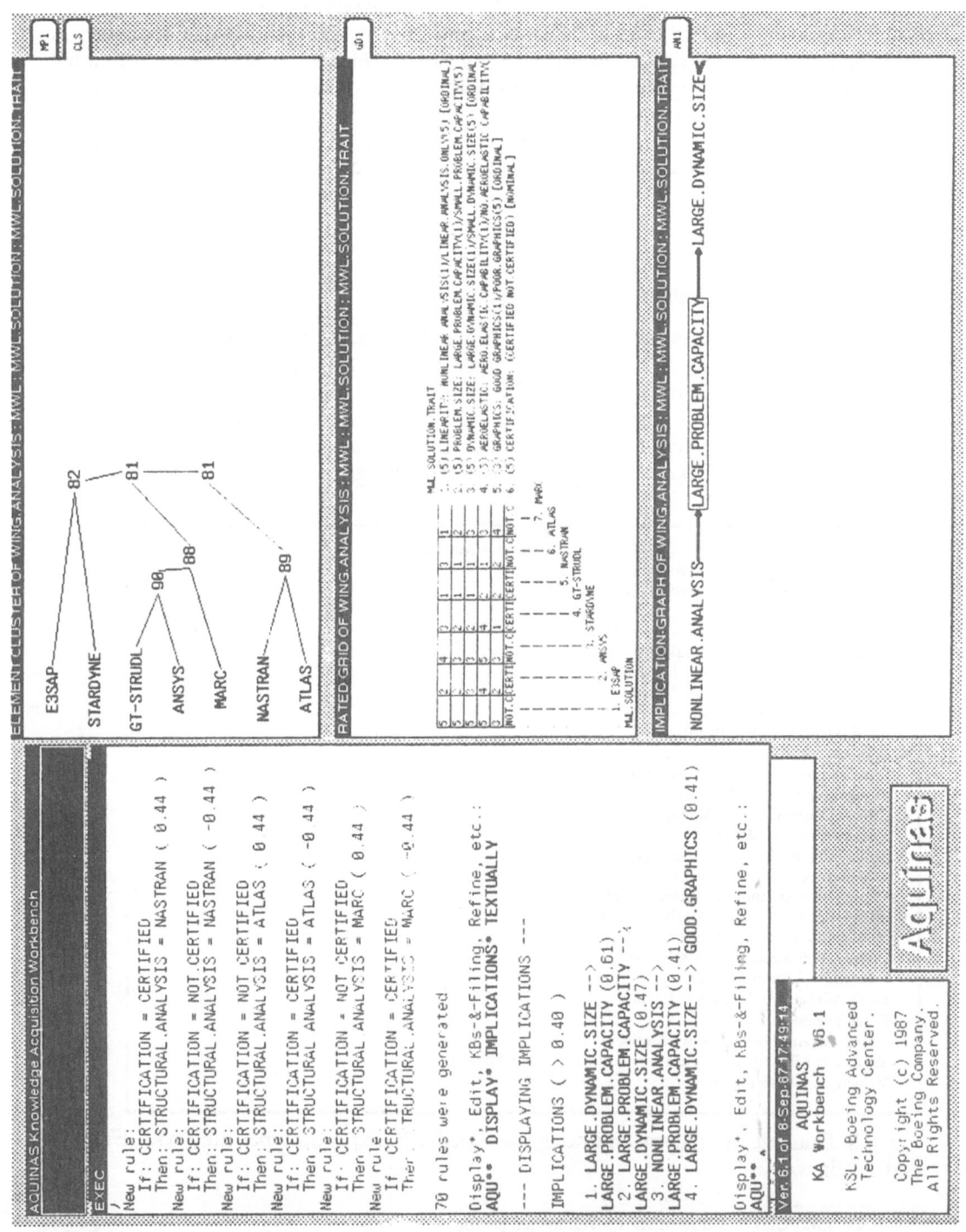

Fig. 8.7. Aquinas screen showing developing hierarchies, a rating grid, and an implication analysis graph for the Structural Analysis Advisor.

Aquinas can generate knowledge bases for several expert system shells (KEE, KS-300/EMYCIN[9], LOOPS[10], M.1, OPS5, S.1, and others). The knowledge contained in grids and hierarchies is converted within Aquinas into rules, and the rules are formatted for a particular expert system shell. Appropriate control knowledge is also generated when necessary. Rules are generated with screening clauses that partition the rules into subsets. An *expert clause* is used when expertise from multiple experts is weighted and combined together. A *case clause* controls the focus of the system during reasoning.

4. DISCUSSION

4.1. Advantages and Limitations of Aquinas

4.1.1. Improved process efficiency and faster knowledge base generation: Aquinas inherits the advantages of ETS: rapid prototyping and feasibility analysis, vocabulary, solution and trait elicitation, interactive testing and refinement during knowledge acquisition, implication discovery, conflict point identification, expert system shell production, and generation of expert enthusiasm (Boose 1986). It is much easier for users to learn knowledge-based system concepts through using Aquinas than through

---

Aerodynamic Analysis Software (PANAIR) - Geometry Advisor
Aerodynamic Analysis Software - Front End Advisor
Aerodynamic Analysis Software - Back End Analyzer
Aircraft Fault Isolator
Airplane Configuration Mass Property Estimation Risk Analyzer
Airplane Design Flutter Analyzer
Automatic Flight Controls Diagnostic Aid
ATLAS Structural Analysis Advisor
B1 Diagnostic Consultant
Bond Durability Consultant
Composite Materials Advisors
Documentation Update Consultant
Eigenvalue Solution Advisor
Energy Control System Model Evaluator
Energy Aircraft Advisor
Finish Advisor for Design Engineers
Finish and Corrosion Control Consultant
Flight Controls Human Factors Assistant
GT-STRUDL Structural Analysis Advisor
Molded Rubber Seal Advisor
Office Automation System Advisor
Propulsion System Advisor
Resin Advisor for Composite Parts
Rivet Selector
Structural Analysis Software Advisor
Transport Airplane Configuration Initial Selection Advisor
Velocity Analysis Advisor

Fig. 8.8. Example of analysis and design knowledge-based prototypes built with ETS and Aquinas.

---

[9] KS-300/EMYCIN is a trademark of Teknowledge, Inc.
[10] LOOPS is a trademark of Xerox Corporation.

reading books or attending classes (i.e. rules are automatically generated and used dynamically in consultations; new vocabulary is incrementally introduced).

ETS and Aquinas have been used to build dozens of prototype knowledge-based systems in the design and analysis area (Fig. 8.8).

4.1.2. Improved knowledge base quality: Aquinas offers a rich knowledge representation and reasoning environment. We believe that Aquinas can be used to acquire knowledge for significant portions of most structured selection design and analysis problems. Hierarchies help the expert break down problems into component parts and allow reasoning at different levels of abstraction. Varying levels of precision are specified, with multiple types of rating scales when needed. Multiple methods for handling combining uncertain information are available based on needed precision and convenience.

Knowledge from multiple experts may be combined using Aquinas. Users may receive dissenting as well as consensus opinions from groups of experts, thus getting a full *range* of possible solutions. Disagreement between the consensus and the dissenting opinion can be measured to derive a *degree of conflict* for a particular consultation. The system can be used for cost-effective group data gathering (Boose 1987).

Analytic tools help uncover inconsistencies and circularities in the growing knowledge base.

4.1.3. Better knowledge base maintenance and comprehensibility: Elicitation, structuring, analysis, and testing of knowledge is based on specific cases. When knowledge in Aquinas is updated, it is done so with respect to a specific case. Addition of new knowledge in this way can be strictly controlled by the expert; the tendency for local changes to degrade other cases is thus curbed.

The expert builds and refines knowledge in rating grids and hierarchies — not directly in production rules. As a result, knowledge at this higher level of abstraction is more compact, comprehensible, and easier to maintain.

The growing collection of old rating grids and case knowledge represents an important resource for building a variety of new knowledge-based systems. Knowledge is stored explicitly with associated problem cases, making knowledge bases easier to update and maintain.

Currently, a user may copy and change any portion of the Aquinas knowledge base during a consultation. In the future, each expert will be able to protect areas of knowledge. The expert may believe protection is necessary because some knowledge should not be changed or because the knowledge has commercial value.

4.1.4. Extensions to Personal Construct Theory methods: Aquinas significantly extends existing Personal Construct Theory methods. Rating grid knowledge can be tested and used interactively to make decisions. Rating grid information may be arranged and coupled in hierarchies. Multiple rating scale types are available (not just bipolar ordinal scales). Many grid analysis tools are available in a single workbench.

4.1.5. Process complexity: Aquinas is not as easy to use as was ETS using single grids. There are many elicitation and analysis tools. The decision-making process and inference engine can be set up to work in several different ways. We expect that continuing improvements in the dialog manager will help make the system more comprehensible and decrease the learning time for new users.

4.2. Theoretical Issues — Knowledge Elicitation

Personal Construct Psychology methods provide no guarantee that a *sufficient* set of knowledge will be found to solve a given problem. Aquinas attempts to expand the initial subset of solutions and traits based on problem-solving knowledge for specific cases. The goal is to solve enough cases so that the knowledge is sufficient to solve *new* cases. This is the methodology of knowledge engineering in general; Aquinas helps make the process explicit and manageable.

Hierarchical decomposition can be used to build intuitive, comprehensible models that seem to behave in reasonable ways. One disadvantage is that some problems do not easily fit the hierarchical model. It also may be true that a particular problem would best be represented by a *collection of conflicting hierarchies* (hierarchies for mechanical problems tend to model structure *or* function, not both, and both may be necessary).

The use of multiple rating value types provides more flexibility, convenience, and precision in representing knowledge. However, deciding which particular type of variable to use can be a complex task. The dialog manager offers some assistance, but the expert usually must learn appropriate usage of rating types through experience.

Experts develop Aquinas knowledge bases serially. In the future, we would like to build a participant system in which many experts could dynamically share rating grids and hierarchies (Chang 1987).

4.3. Analysis and Inference

Multiple analysis tools and elicitation methods in Aquinas help the expert think about the problem in new ways and tend to point out conflicts and inconsistencies over time. Lenat argues that knowledge representations should shift as different needs arise (Lenat 1983a; Lenat 1983b). This should lead to better problem and solution descriptions, and in turn, to better problem-solving.

Inference in Aquinas is efficient because the problem space is partitioned. Information in the trait hierarchies is attached to particular levels of solutions. Although no formal studies have been conducted consultation results using the methods described above seem reasonable.

Rule generation for expert system shells is straightforward. Development of the knowledge base can continue in an expert system shell that may offer advantages of speed, specialized development and debugging facilities, and inexpensive hardware.

5. FUTURE DIRECTIONS

We intend to build a knowledge acquisition environment that includes specific domain knowledge for specialized application areas and can acquire knowledge for synthetic problems.

Presently Aquinas works best on problems whose solutions can be comfortably enumerated (*analytic* or *structured selection* problems such as classification or diagnosis) as opposed to problems whose solutions are built up from components (*synthetic* or *constructive* problems such as configuration or planning). Simple classification can be thought of as a single decision problem (single grid). Complex structured selection problems may require a set of linked data abstraction/solution refinement decisions (multiple grids). The next step may be to generalize this process to acquire and represent synthesis knowledge for planning, configuration, and design problems where the order of linked decisions in solution hierarchies may represent precedence of events or goals rather than just solution refinement. In these problems hierarchies may be assembled at consultation time rather than constructed totally in advance as they are currently. Grid cells might sometimes contain an arbitrary computation rather than a rating. These would include results of functions (such as found in spreadsheets) or database retrievals. Deeper models of the structure and function of physical systems could be modeled.

An important step in expanding the knowledge acquisition workbench concept is the linking together of other specialized tools. At the Boeing Knowledge Systems Laboratory we are investigating ways of integrating diverse knowledge representations from different Laboratory projects so that this may be more easily accomplished. In the domain of knowledge acquisition, we feel that the approach used in SALT (Marcus, McDermott, and Wang 1985; Marcus and McDermott 1986; Marcus 1987) is particularly promising. SALT is a system that interviews experts to build knowledge bases for certain types of constructive problems. (The first use of SALT was to configure elevators.)

Development of the Aquinas workbench will continue in an incremental fashion. Techniques will be continuously integrated and refined to build a more effective knowledge acquisition environment.

ACKNOWLEDGMENTS

Thanks to Roger Beeman, Miroslav Benda, Kathleen Bradshaw, William Clancey, Brian Gaines, Cathy Kitto, Ted Kitzmiller, Art Nagai, Doug Schuler, Mildred Shaw, David Shema, Lisle Tinglof-Boose, and Bruce Wilson for their contributions and support. Aquinas was developed at the Knowledge Systems Laboratory, Advanced Technology Center, Boeing Computer Services in Seattle, Washington, USA.

REFERENCES

Adams, J. (1985): "Probabilistic reasoning and certainty factors," in *Rule-Based Expert Systems: The MYCIN Experiments of the Stanford Heuristic Programming Project*, Buchanan, B. and Shortliffe, E. (eds.), Addison-Wesley, Reading, MA, USA, pp. 263-271.

Beyth-Marom, R. and Dekel, S. (1985): *An Elementary Approach to Thinking Under Uncertainty*, Lawrence Erlbaum Associates, London.

Boose, J. H. (1985): "A knowledge acquisition program for expert systems based on personal construct psychology," *International Journal of Man-Machine Studies*, **23** , pp. 495-525.

Boose, J. H. (1986): *Expertise Transfer for Expert System Design*, Elsevier, New York.

Boose, J. H. and Bradshaw, J. M. (in press, 1987): "Expertise transfer and complex problems: Using AQUINAS as a knowledge acquisition workbench for knowledge-based systems," *Special Issue on the AAAI Knowledge Acquisition for Knowledge-Based Systems Workshop, International Journal of Man-Machine Studies*, **25** .

Boose, J. H. (1987): "Rapid aquisition and combination of knowledge from multiple experts in the same domain," *Future Computing Systems*, **1**(2), pp. 191-216, Elsevier.

Chang, E. (in press, 1987): "Participant systems," *Future Computing Systems*.

Clancey, W. (1986): "Heuristic classification," in *Knowledge-Based Problem-Solving*, Kowalik, J. (ed.), Prentice-Hall, New York, pp. 1-67.

Gaines, B. R. and Shaw, M. L. G. (1981): "New directions in the analysis and interactive elicitation of personal construct systems," in *Recent Advances in Personal Construct Technology*, Shaw, M. L. G. (ed.), Academic Press, New York, pp. 147-182.

Gaines, B. R. and Shaw, M. L. G. (August 1985): "Induction of inference rules for expert systems," *Fuzzy Sets and Systems*.

Hinkle, D. N. (1965): "The Change of Personal Constructs from the Viewpoint of a Theory of Implications," Ph.D. dissertation, Psychology Department, Ohio State University, OH, USA.

Kelly, G. A. (1955): *The Psychology of Personal Constructs*, Norton, New York.

Kitto, C. and Boose, J. H. (in press, 1987): "Heuristics for expertise transfer: The automatic management of complex knowledge acquisition dialogs," *Special Issue on the AAAI Knowledge Acquisition for Knowledge-Based Systems Workshop, International Journal of Man-Machine Studies*, **25** .

Lenat, D. (1983): "The nature of heuristics (Part 2)," *Artificial Intelligence*, **21** .

Lenat, D. (1983): "The nature of heuristics (Part 1)," *Artificial Intelligence*, **19** .

Marcus, S., McDermott, J., and Wang, T. (August 1985): "Knowledge acquisition for constructive systems," *Proceedings of the Ninth Joint Conference on Artificial Intelligence*, Los Angeles, CA, USA, pp. 637-639.

Marcus, S. and McDermott, J. (1986): "SALT: A Knowledge Acquisition Tool for Propose-and-Revise Systems," Technical Report, Pittsburgh, PA, USA.

Marcus, S. (in press, 1987): "Taking backtracking with a grain of SALT," *Special Issue on the AAAI Knowledge Acquisition for Knowledge-Based Systems Workshop, International Journal of Man-Machine Studies*, **25** .

Mittal, S., Bobrow, D. G., and Kahn, K. (November 1986): "Virtual copies: At the boundary between classes and instances," *SIGGPLAN Notices, Proceedings of the Object-Oriented Programming Systems, Languages, and Applications Workshop, Portland, Oregon*, **21**(11).

Quinlan, J. R. (1983): "Learning efficient classification procedures and their application to chess endgames," in *Machine Learning — An Artificial Intelligence Approach* **1**, Michalski, R. S., Carbonell, J. G., and Mitchell, T. M. (eds.), Tioga, Palo Alto, CA, USA, pp. 463-482.

Saaty, T. L. (1980): *The Analytic Hierarchy Process*, McGraw-Hill, New York.

Shafer, G. and Tversky, A. (1985): "Languages and designs for probability judgment," *Cognitive Science*, **9** , pp. 309-339.

Shaw, M. L. G. and Gaines, B. R. (in press, 1987): "Techniques for knowledge acquisition and transfer," *Special Issue on the AAAI Knowledge Acquisition for Knowledge-Based Systems Workshop, International Journal of Man-Machine Studies*, **25** .

Shaw, M. L. G. and Gaines, B. R. (in press, 1987): "PLANET: A computer-based system for personal learning, analysis, negotiation and elicitation techniques," in *Cognition and Personal Structure: Computer Access and Analysis,*, Mancuso, J. C. and Shaw, M. L. G. (eds.), Praeger Press.

9. MIND: A Design Machine
— Conceptual Framework —

M. Nadin and M. Novak

The Ohio State University, 1501 Neil Avenue, Cranston Center
Columbus, OH 43201, USA

Abstract: *We define ICAD (Intelligent Computer Aided Design) as the instantiation of the interactive relation between designer and a variable, category-based, extendible machine, based on a parallel configuration of Intelligent Processors. The paper provides an outline of a computational design theory, followed by a set of observations and recommendations as specifications for an ICAD system. Issues of methodology and configuration conclude the conceptual analysis. In place of 'expert systems for design,' implemented as partial solutions to the design problem within limited domains, we propose 'expert systems of design,' meta-expert systems containing knowledge of the design process. Expert systems of design can be built using current technology, and can be used to organize and guide the development of conventional, domain-specific, expert systems modules. The design team is the metaphor around which we developed the conceptual model of a multifunctional intelligent design (MIND) machine.*

Keywords: concurrency, intelligent processor, artificial intelligent processor, natural intelligent processor, visual processing.

1. INTRODUCTION

In contemporary society design plays an increasingly important role. The visual component has proven to be not just a way of embodying results but primarily the means for a new way of thinking. In spite of the very promising capabilities of computers for supporting the visual, design has benefited little from the advent of computers. The computer industry recognized early on the design potential of this technology but never took the time to address the specific needs of the very complex human activity that design is. General purpose machines were declared design workstations once software packages with graphic capabilities were made available.

CAD systems continue to be developed under the "general computing" paradigm, and while they offer an ever greater list of capabilities, they continue to be problematic because:

1. While design specifications are complex, people in the computer industry felt entitled to act on behalf of designers and made their specifications a standard for the trade.

2. Hardware has rarely been conceived in view of the specific purpose for which a particular computer is built.

3. The variety of design activities makes the determination of a common denominator quite difficult.

4. In view of the short term success strategy that the computer industry pursued, long term development (such as the type needed for developing a design station) was and still is being avoided.

Computer aided design systems currently available on the market are moving towards increased integration of functions operating on a single database, or on closely related databases. The GDS integrated system offered by McDonnell-Douglas, for example, combines modules for site-planning (as understood by the civil engineer), heating, ventilating and cooling (HVAC), heat load calculations, structural analysis, space planning, facility management — all built around a central architectural modeler. Information can flow back and forth between the modules. The modules contain a considerable portion of the knowledge that an architect's consultants bring to a large scale project or that the architect has available "in-house" for small to medium scope projects.

The following questions are raised by this:

1. What is it that the designer actually does? While it is clear what each of the engineering disciplines does, it is much less clear what the designer's 'territorial' task actually is. Insofar as design is not simply the sum of the parts, the increased availability of 'expert modules' makes pressing the question of emphasis and difference. Each specialized module that can be added to an integrated system clarifies what design is by precisely pointing out what design is not. Aesthetic and semantic issues become increasingly important as more and more of the technical and functional aspects of design approach resolution and automation. Accordingly, we should be able to evaluate these qualities and support them in a future design machine.

2. To a certain extent, the designer is an organizer of all the different disciplines involved in the realization of a project. Accordingly, two questions must be analyzed:

    a. Is this part of the 'necessary and sufficient' contribution of design?

    b. Should a design machine provide means which facilitate this aspect of design?

In this paper we will first describe the requirements for a computational theory of design and the characteristics of design that such a theory must address. We will then present a model for intelligent CAD systems, followed by a methodology for extracting information about design intelligence. Finally, we will discuss the configuration of intelligent CAD systems using present and future technologies.

1.1. Premises

A basic characteristic of the conceptual model we propose is that design is a major activity of the mind. Intelligence, which we define as generalized pattern recognition and the associated interpretation and transformation of patterns as forms of shared experience, participates in design in various stages: identification of needs, formulation of goals, critical evaluation of previous models and self-evaluation, as well as the ability to orchestrate various components participating in the design process. The interdisciplinary nature of design is made possible by intelligence insofar as it allows for elements often quite difficult to relate to each other (such as aesthetic and economic considerations, technological requirements and sociological prescriptions, high performance and ease of use, to name only a few) to belong to and actually constitute a design identifiable through its general qualities and through its originality. During our research, we established several premises:

a. Design is a dominantly visual activity, involving not only intelligence in general, but a specific form of intelligence related to the visual.

b. Design is an open-ended activity.

c. Although some design aspects can be represented in forms of linear relations, design by its nature is nonlinear.

d. Design is redesigning, i.e., it never starts from zero.

These premises established, we notice that quite often issues of design are treated in computer aided design by accepting restrictions which are not design-specific but computationally motivated. We make an epistemological decision which might affect actual implementation but which in the long run assures that we deal with design and not with some operations which are accidentally performed by designers.

This decision can be expressed as follows: Instead of dealing with a subset of design, assumed closed, complete, consistent, we deal with design as open-ended, as process, and in which inconsistencies are part of the final product. The kind of design we decided to cover, i.e., the problems of the mind it involves, imposes the need to conceive of a hybrid system. We have in mind an evolving structure, continuously reconfigured in the process of interaction between the machine intelligence as provided in programs and the natural intelligence of the designer. We call this a "tunable" design machine with an important autopoietic component.

2. TOWARD A DEFINITION AND THEORY OF DESIGN

First, let us examine a structural theory of design and then extend the requirements implicit in the theory into the area of design computing.

Design covers various fields of activity, such as architecture (landscape, interior, urban, monumental), visual communication, engineering, and industrial design. It is one of the most pervasive components of human activity. The following simplified representation of almost any kind of design evidences the relation between design, designer, and beneficiary (Fig. 9.1).

The diagram can be slightly improved if instead of defining the object of a designer's work as the product we deal with a higher level concept: the problem. In

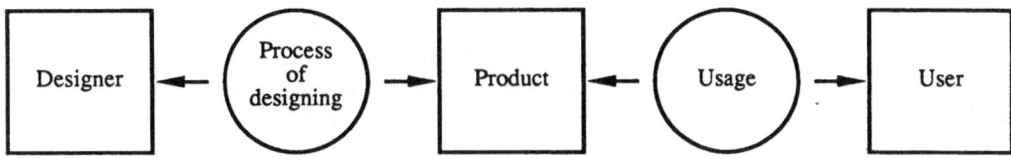

Fig. 9.1. Designer and user

this case, design is identified as problem-solving, one of today's dominant paradigms. Problem-solving, however, is not sufficient to cover the full extent of design.

The process of designing is quite difficult to describe due to the interdisciplinary nature of design. This interdisciplinarity can be represented, although the representation we have chosen (viz. Venn diagrams in set theory) is not necessarily exhaustive (Fig. 9.2).

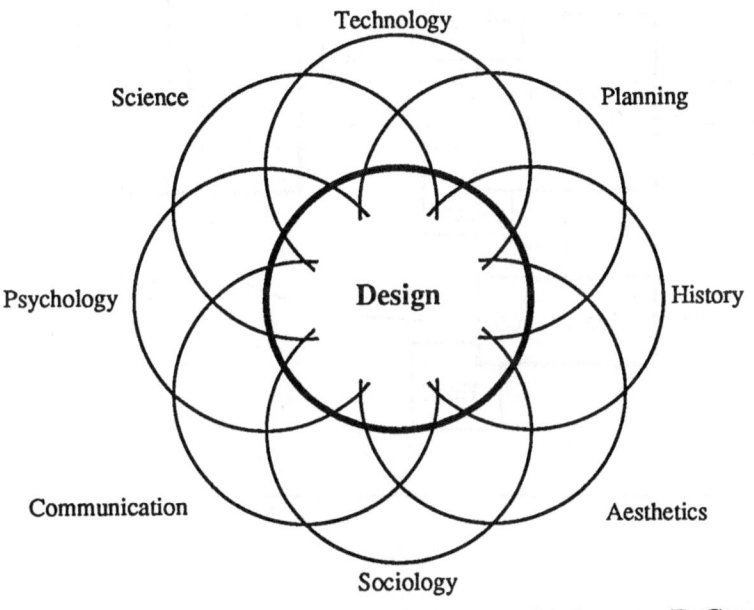

Fig. 9.2. Interdisciplinarity of design

The "specialized" components (such as the planning component, aesthetic quality, the social and psychological aspects of design and the product designed, communication, science, technology, etc.) require an integrated procedure (how to bring them together while maintaining the integrity of design or as a means to achieve it) as well as a self-critical moment. The components mentioned have elements in common. On a higher level representation dealing with these common elements we should be able to identify structural equivalences which are quite critical in the design of the intelligent component of the system which will. participate in the integration. The self-critical moment of design is represented by the fact that designers as well as users of design compare new designs to previous work and situate design in the broader context of culture and civilization. It is useful at this juncture to point, through an additional diagram to the environment (in a broad sense, i.e. social, natural, cultural) in which and in relation to which designs are generated, i.e. to the pragmatic framework (Fig. 9.3).

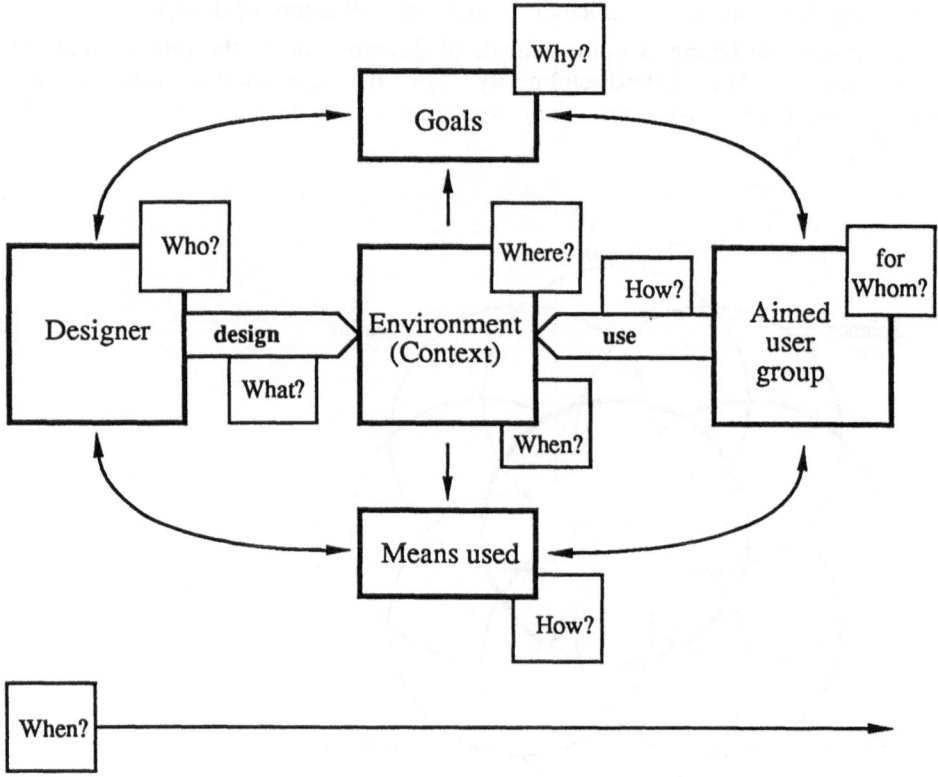

Fig. 9.3. Context and signification of design

In their work, designers use:

- specialized knowledge acquired through design education,
- general knowledge (belonging to culture),
- tools (simple to complex, such as pencils, rulers, production tools, and more recently, computers).

They also apply what is called "aesthetic sensitivity," i.e. knowledge concerning form and its expressive power according to ideals historically acknowledged. In the interdisciplinary design work, a hierarchical structure within which aesthetics occupies a high position is evident. By no accident, almost all design theories refer to this relation between the aesthetic component and the scientific and industrial aspects of design. Historically, design can be seen as the interaction of the aesthetic component, which encompasses art, craft, and the components of science and industry.

The design process, in its close relation to design products and their use, implies *design intelligence*, *cultural sensitivity*, and a *critical attitude*.

The designer works towards a goal (product) to be achieved with the help of the representation of this goal (Fig. 9.4). Representation can be in words, written statements, drawings, models, diagrams, etc. The dominant form of representation is visual. Some authors go so far as to speak of *visual thinking*, a metaphor that describes the participation of visual representations in thought processes.

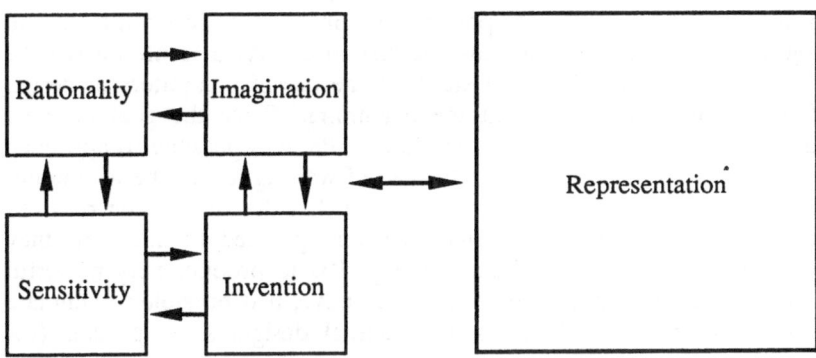

Fig. 9.4. Aspects of representation

3. DESIGN AND COMPUTING

To formulate a computational theory of design is not the goal in itself but the means necessary for accomplishing the global aim of providing a foundation for computer aided design. A computational theory of design must address the components of design in their most general expression and state the relationships between them in a way similar to the way communication theory deals with communication, for example. As with any theory, once one grasps the general elements and the fixed and variable relations of those elements, one should be able to enter values specific to the system so as to accommodate, explain, or predict the behavior of particular design efforts. A computational theory of design must be a theory of design foremost; it must be computational in that its expression allows the computer to "understand" design and to "behave" accordingly.

On its own, such a theory will be a contribution to all design fields. Most designers have noticed the need for a design theory as a prerequisite for design education, design criticism, and design evaluation. In the process of building such a theory, the contribution from other fields (communication, aesthetics, cognitive science, sociology, and others) is important. The theory should not be hardware-dependent, but should suggest ways of improving understanding of hardware requirements specific to different applications. Computational design, together with other computational disciplines, has become necessary precisely due to the challenges to design in today's society.

4. CATEGORIES FOR A COMPUTATIONAL DESIGN THEORY

Technology advances toward faster computers, increased memory, parallel processing, expert systems containing knowledge within specific problem domains, artificial intelligence techniques employing heuristics for problem-solving, natural language recognition and computer-generated speech, pattern recognition, vision systems, and others. It also makes available increasingly powerful graphics processors with 2D and 3D and even holographic capabilities and real-time responses. As a result we see the possibility for an alternative approach to the study of the use of computers in design along the line of the premises introduced at the beginning. Since the goal of these approaches is clearly to produce "intelligent processors," *IP*s, it is possible to approach this problem top down rather than bottom up. Instead of waiting for all the techniques implied in current research to be resolved, we can assume that they have been resolved (to any desired degree of resolution, as we shall see) and proceed to find how they should be employed. This is possible because, if the goal is to produce very powerful "intelligent processors" with the capabilities mentioned above, it it possible to simulate those artificial "intelligent processors," *AIP*s, with actual designers using tools (for efficient work), natural intelligent processors, *NIP*s.

The problem can be simplified to one familiar with a design studio: How does the designer in charge communicate with the design team 1) assuming that we already have a group of "intelligent processors" (simulated by a group of designers); 2) assuming that we have identified a (tentative, hypothetical) common language which can be arbitrarily complex but not infinite. What kinds of communication are necessary for these processors to design an object? What is the role of the components (speech, written language, drawings, models, animation) in defining a design language? Obviously we have in mind behavioral aspects characteristic of the design team work and we try to find ways to capture these in forms that would make intelligent computation possible.

We do not expect to identify the "absolute" primitives, contexts, codes, interactions, etc. Rather, we intend to arrive at an understanding of the conceptual categories, the basic elements pertinent to the pragmatics of design. The reason for this can best be understood through analogy. If we try to design a chair, we can either try to design one for the "average" (though non-existent) person and produce a chair with a tolerable degree of discomfort for most people; or we can design an adjustable chair which allows any individual to adjust the chair to one's own dimensions. In the first case, we attempt to design for the "ideal" shape, i.e. within the logical model of categories; in the second case, we attempt to identify the nature, locus, and magnitude of adjustments that will accommodate the full variability of human dimensions, the experiential component to which Lakoff (1987) makes reference in his recent work.

The second approach is in keeping with the goal of creating a "design machine." This machine should allow the designer to adjust it according to that designer's approach to design, which hopefully will not remain constant but will change as the designer "matures." The design machine must be useful to designers with different viewpoints and must allow, even encourage, the formation of new theories and approaches to design (some of which can be extended to several forms of design).

The creation of a hardware/software configuration for use in any task involves the embodiment of a theory, even if the configuration contains its own "inference engine." Thus any system relies on our ability to state explicitly what it is that we do when we do activity X. That is, the creation of any system relies on an act of communication which we evaluate through the behavior it leads to. In the case of design, this implies that the explanation and interpretation of the behavior as studied by previous protocol analysis approaches is irrelevant. What is important is our ability to communicate things characteristic of our interaction with computers: how do we program the computer for a design task or how do we use the program? or , in a more general way, what would an intelligent program have to accomplish in order to fulfil a design task? Research in this direction must try to uncover the *categories* of *communication*, the *categories* of *contexts* within which *communication* is meaningful, the *categories* of *conceptual primitives* that form the tokens of exchange in this communication, and so on. Ultimately, the creation of a "design machine" requires that we establish codes of communication between the designer and the computer, which means establishing the pragmatic framework of design.

Traditional protocol analysis applied to design has relied on the possibility of deriving an *explanation* of design processes from the *interpretation* of the observed behavior of the designer as well as from *concurrent explanations* given by the designer during design and from subsequent *recollections* of the thinking that prompted certain design decisions. These protocol analyses rely on the explanation or interpretation of the design behavior relevant to the problems of using computers in design. If we extrapolate from current research directions, it is possible to conduct research that does not rely on the explanation or interpretation of design behavior, *but that studies the actual relevant behavior itself*, as long as it can be studied at all. Furthermore, we conclude, this behavior is the only one that can be studied without studying the actual architecture of the human brain.

5. OBSERVATIONS AND RECOMMENDATIONS

We have defined the general character of design, and outlined the requirements of a theory of design that can be used to guide the development of intelligent CAD systems. We also suggested an approach for gathering information related to the way designers communicate in the process of designing. It is now necessary to look at some specific and critical characteristics of design as they result from the premises established at the outset of this paper. Based on the assumptions made we present observations and practical recommendations for an actual machine.

5.1. Concurrency: The Primacy of the Visual

OBSERVATION: The predominance of visual representations in design is well-established, and the effectiveness of visual interfaces, even in systems that are not directly related to design, is clear. The primacy of the visual representation in design applications stems from two elements fundamental to design: *concurrency of input and output* and *concern with form*. Visual representations, unlike verbal and other representations, are primarily concurrent rather than sequential. Beyond simply allowing the designer to see what something will "look like," this allows the state of the whole design, or at least several parts of it to be inspected at once, and progress and problems to be monitored continuously. Furthermore, the visual representation allows designers to concentrate on the form of the design, where form is used not to describe the surface characteristics of the evolving design but the abstract structure (as in the form of a novel, the form of an argument, etc.). In this way the visual can be seen as a multi-channel parallel means for input and output that has the capacity to handle both concrete and abstract information.

RECOMMENDATION: If the visual is understood in this way, then the primary interface between designer and machine will be visual, and specifications of high speed interactive graphics are natural, as are requests for full color for the rapid presentation and manipulation of complex information. The capacity to support multiple visualization techniques through fast mapping is critical to any ICAD system.

5.2. Analysis vs. Synthesis

OBSERVATION: While analysis is important at several points in the process of designing, the act of designing is first and foremost an act of synthesis. As the word synthesis implies, it is an act of bringing together, comparing, combining, fitting, assembling a multiplicity of parts into a whole. Characteristically, the parts *are not homogeneous*.

RECOMMENDATION: Not only must a system allow the designer to manipulate many different elements, but it must also put few restrictions on what those elements may be and how they are to be manipulated. Special tools must be provided for the retrieval, superimposition, combination, and transformation of different elements.

5.3. Normative and Procedural Aspects of Design

OBSERVATION: Procedural theories of design fall short of explaining design to any depth. Normative considerations ("ought to be" theories) guide the design of architecture, products, communications, and other fields where a semantic dimension is important. Even in engineering design, as in science (Kuhn 1962) there are dominant paradigms which guide what should be done. There is evidence that such normative concerns offer clear benefits by providing the designer with a conceptual platform from which to launch the first attack against the problem (Rowe 1987).

RECOMMENDATION: Although normative concerns are *de facto* parts of much of the generation and evaluation of design and form a great part of the context within which designs are understood they have been ignored almost completely by computer aided design systems developers. An ICAD system must allow the designer to formulate normative as well as procedural theories and must provide tools for their application to the emerging design. Such tools must allow the designer to construct design schemata easily, check for consistency of application of design principles, allow the designer to make reference to previous solutions and ideas, etc.

5.4. Parallel and Episodic Design Process

OBSERVATION: As soon as they are given a problem, before collecting information, analyzing it, and arriving at a reformulation of the problem, designers begin to generate and evaluate solutions. Later, even as the project nears completion they speak of finally understanding what the "real problem" within the problem was. The design process that designers follow is usually presented as a rigid-state flowchart with arrows that denote backtracking, and a sequence of phases such as: problem statement, programming, data collection, data analysis, problem restatement, alternative generation, alternative evaluation, design development, evaluation, and communication.

This conception of design, even with backtracking, is essentially sequential and assumes rigid boundaries between stages.

RECOMMENDATION: We propose an alternative view: a parallel design process with soft boundaries between tasks and traversed in an unpredictable, episodic manner. The designer keeps all "stage files" open all the time, simply shifting attention from one to the other or assigning information to one stage or another. The design is not complete until all the stages have run their course. This view of the design process accounts for the "aha!" phenomenon of sudden realization, as a piece of information that is literally out of phase falls into place in another stage in the process.

5.5. Personal, Evolving Approaches

OBSERVATION: Not only is the design process not sequential, not only does it not have hard boundaries, but it is not even constant. The process a designer uses "matures" with the designer, just as the designer's normative and procedural theories of design evolve. The process may vary both over the short range and over the long range. In addition, the process varies from designer to designer, and even from mood to mood.

RECOMMENDATION: An ICAD machine needs to have a variable architecture which recognizes and even encourages the evolving approach. This is particularly important in view of the various levels of the visual at which designers work.

5.6. Streams of Association: The Flow State

OBSERVATION: The episodic traversal of the parallel process stages, and the variability of the design process itself follows the designer's "train of thought," the stream of associations made during the effort to understand and solve the problem. Psychologists describe a state where the mind is free to move fluidly from concept to concept, and during which a person's sense of the passage of time is diminished, as a "flow state." Given flexible tools, such as tracing paper and pencil, a designer can enter this state with ease in the process of concentrating on the problem. Present CAD systems scarcely recognize (let alone respond to) this issue.

RECOMMENDATION: An intelligent CAD system must provide means for encouraging the designer to enter this state quickly and easily, by providing means of interaction that are sufficiently fast so that they can follow the stream of associations the designer is working on. Moreover, the ICAD system must be sensitive to the designer's *pace*, actively provoking the designer by inquiries, suggestions, recommendations, in addition to passively responding to the designer's requests.

5.7. Problem Reformulation and Alternative Generation: Multiple Representations

OBSERVATION: Designers often attempt to solve problems by shifting their viewpoint with respect to the original design specification and by playing games of hypothesis of the form "what-if-y-were-z?" When generating alternative solutions they often apply the "what-if?" form to certain parts of the design but not to others, or not to the same degree, setting up, in effect, informal generative systems. Finally, if the situation in which they must design is in itself problematic — for example, if there is not enough time or there are not enough funds to complete a project — they apply the "what-if?" form to the design process itself, thus moving to a meta-design level. We see that in the reformulation of the problem, the generation of alternatives, and the design of an approach to design, multiple representations are maintained and mappings are constantly being revised and updated.

RECOMMENDATION: An ICAD system must allow multiple representations of problems, solutions and processes, and provide a panoply of tools for moving from one to the other.

5.8. Management of Partial or Previously Rejected Solutions

OBSERVATION: Designers frequently return to solutions that were sketchy, partial, or rejected. Traditional media leave a physical record that allows for this kind of "hill-climbing." This can occur either diachronically or synchronically. The following is a good example of diachronic use of partial solutions:' Alvar Aalto, a leading figure of modern architecture is said to have designed on a continuous roll of paper which was set up so that, using cranks, the history of a design's development could literally roll before his eyes. The synchronic use of partial design solutions is evident in the multiple "tracings" designers make, where selective copying allows information from previous attempts to be recovered easily. Both approaches have implications for ICAD that are relatively easy to implement.

RECOMMENDATION: The diachronic record of partial solutions is simply a history of commands; the synchronic requires provisions for recognizing and extractings partial solutions of elements that, as we mentioned previously, may not be homogeneous. Further intelligence could be added if the system could analyze partial solutions and bring them up as particular conditions were met.

5.9. Problem/Solution Relationship

OBSERVATION: Many design problems involve the adaptation of previous solutions to similar problems. It is often possible, therefore, to gather previous solutions and classify them into different types, creating a typology of solutions. Solving a problem involves searching for an appropriate type of solution and modifying it to suit the requirements of the problem at hand.

Many design problems do not have direct precedents, and one cannot simply modify a pre-existing type. However, since the solution to a problem stands to the problems it addresses as figure stands to ground, it can be seen that a parallel, invisible knowledge of problems is implicit in our knowledge of the world. When the designer is faced with new problems it is this knowledge that provides access to the solution space. This knowledge is usually implicit and unarticulated, yet it is indexed by partial choice strategies. These strategies mediate between problems and solutions.

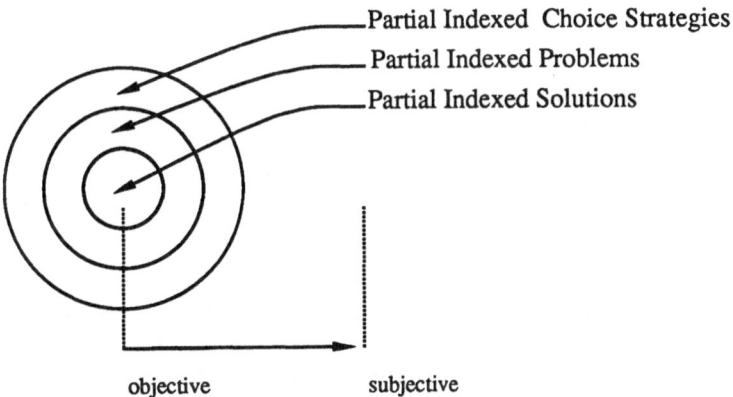

Fig. 9.5. Hierarchical relation of solutions, problems, and choice strategies

RECOMMENDATION: A typology of problems can be constructed that allows the designer to manipulate the problem itself, as well as to have access at "indexed partial solutions" (Brown and Chandrasekaran 1985), indexed partial problems and indexed choice strategies (Fig. 9.5). Indexed solutions represent objective answers while choice strategies embody "intelligence" and support subjective components.

5.10. Roles and Performances

OBSERVATION: Designers very often "cast" either the developing design or themselves in "roles," relying on accumulated experiences of typical situations to provide backgrounds for visualization or simulation of situations, and as means for achieving empathy with the users of designs. Designs are often judged by their "performance."

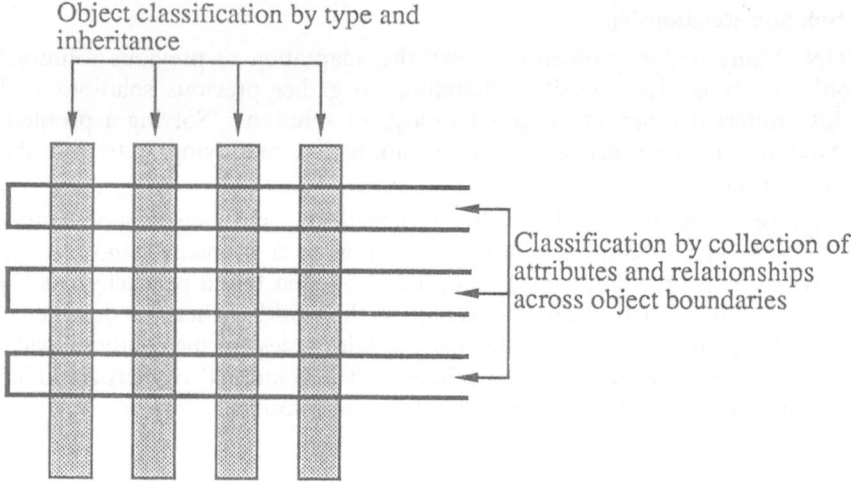

Object classification by type and inheritance

Classification by collection of attributes and relationships across object boundaries

Fig. 9.6. Classification by attributes and relationships

RECOMMENDATION: The frame and script representation of knowledge familiar to artificial intelligence researchers should be extended to allow the representation of knowledge of "roles" and "performances"; a key operation using this representation must be "casting" and "performing." Unlike contexts, which passively illuminate meaning, *frame/script/role/performance* representations can allow the designer to place objects in situations dynamically.

5.11. Classification

OBSERVATION: The abilities to classify, and act upon, selected features of iconic and symbolic aspects of design problems are characteristic of design intelligence. Suppression and enhancement of features, feature extraction, and recognition of objects, attributes, relations and actions are classification acts central to design cognition.

RECOMMENDATION: It is necessary to provide efficient visual classification tools via semantic networks. At a more general level, classification within problems has to be supported through frames, scripts and "roles" and "performances," as described above. Action upon features of the iconic and symbolic realm has to be supported by adequate computer architecture. In addition to vertical, hierarchical classification through inheritance or other means, which can be restricting, a flexible, horizontal, scheme of classification through reuse of attributes across object descriptions is desirable. Objects must allow for an open-ended, extensional (Veth 1987) description. New 'attribute set' objects should be created automatically by the system (Fig. 9.6).

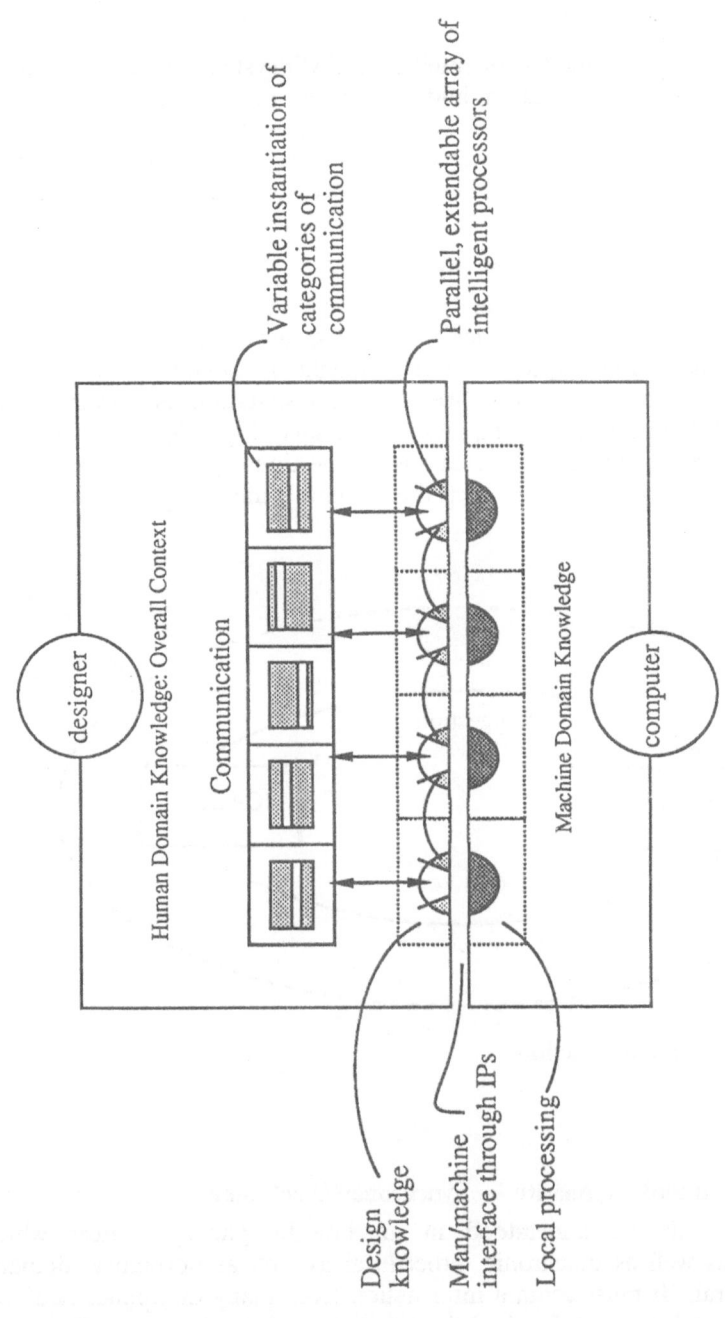

Fig. 9.7. System overview

6. A MODEL FOR ICAD

We can now suggest a model for intelligent CAD systems based on a structure uniting IPs, categories of communication, and the concept of a variable and extendible machine (Fig. 9.7). More precisely, the model proposed consists of several structural components: flow information metaphor, context interpreter, process and problem management, variable interface, intelligent processors.

6.1. Information Model: The Flow State

The primary goal an ICAD system must strive to achieve is the maintenance of a state in which concentrated mental activity occurs, the sense of the passage of time is diminished and uninterrupted, "fluid" thought is possible. To achieve this goal, maximum transparency of interface and representation is required and real time response with high resolution graphics is necessary (Fig. 9.8). Furthermore, the system must be active in making observations, suggestions, provocations, associations, and otherwise providing stimuli that keep the designer's mind engaged with the problem.

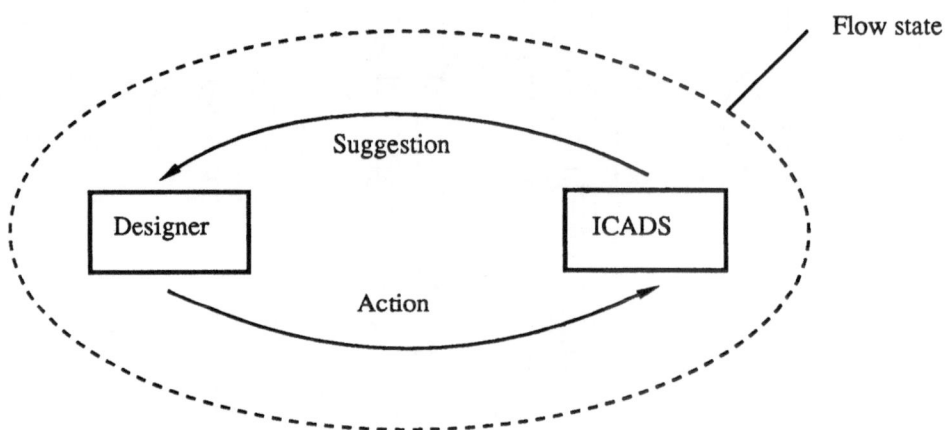

Fig. 9.8. Flow state interaction

6.2. Context: Interdisciplinarity — Synchronic/Diachronic

Design can only be understood in an interdisciplinary context which must be synchronic as well as diachronic, procedural as well as normative, domain-specific as well as general. It must contain information from many disciplines and must be easily expandable, both through local editing of its knowledge base and through the addition of modules representing other disciplines. Some of the modules need only be virtual modules — the actual information provided by electronic network information or other databases. It can be visualized as consisting of a modular but interrelated knowledge

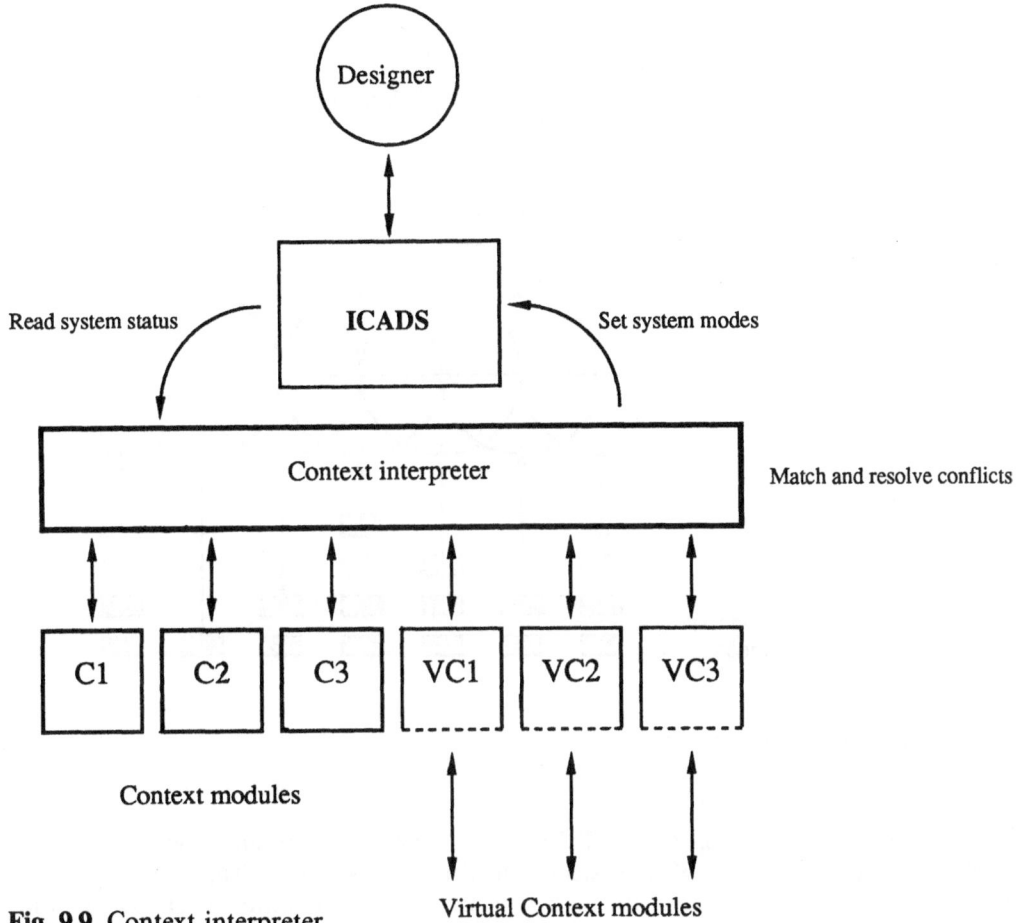

Fig. 9.9. Context interpreter

base and a context interpreter. The context interpreter's task is to attempt to find as many matches as possible between the user/system interaction and the context knowledge modules and to use this information to monitor and control the system's variable configurations (Fig. 9.9).

6.3. Process: Process Management — Parallel, Episodic, Individual, Evolving Process

The design process in this model is a parallel process. All stages of the process commence at the onset of the design effort and remain active throughout the duration of the project. Even though the designer may focus attention on one aspect of the design at a time, observation indicates that, at any point, the designer may divert attention to another part of the process, apparently following a stream of associations. In any case it seems clear that when one works on one task one does not necessarily stop thinking of another task. Moving from task to task therefore needs to happen as quickly as possible and with the least restriction in terms of timing, task completion, and so on.

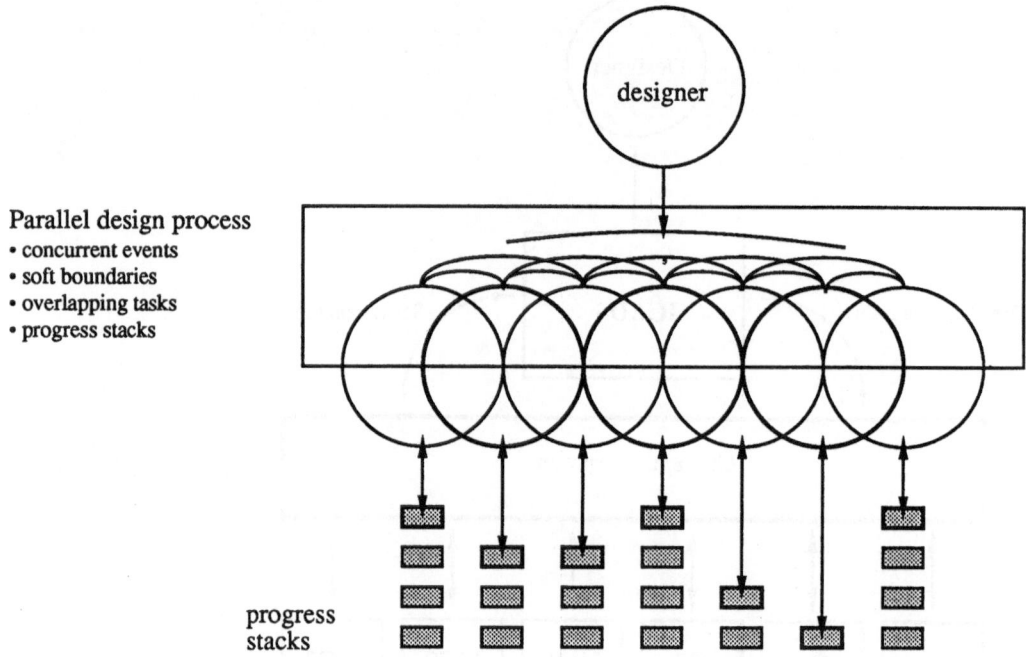

Parallel design process
• concurrent events
• soft boundaries
• overlapping tasks
• progress stacks

Fig. 9.10(a). Parallel design process

As shown in Fig. 9.10, the parallel process involves overlapping, concurrent steps. Each step may be thought of as a stack that needs to be filled. The idea of *progress stacks* is used to indicate that the design is not completed until all stacks are full, though each stack may have items removed or inserted at any point along the design period. Thus, for example, the designer may make some progress in problem definition, think of some alternative solutions, consider communication options — adding tokens to the progress stacks — then realize that the problem definition is inadequate — removing tokens from that stack — return to ideas about communication, and perhaps not complete the problem definition stack until the design is almost complete. Visual feedback could allow the designer to monitor the overall progress of the project.

Unlike previous models, this model assumes that the stages of the design process are not rigid, but that they have soft boundaries. In addition, the design process is seen as being episodic (Rowe 1987), individual, and evolving. An ICAD system cannot impose a fixed process on the user, but must instead provide tools for inspecting, evaluating and modifying it as needed. In that sense, a meta-process level is implied which will allow the designer to re-design the design process.

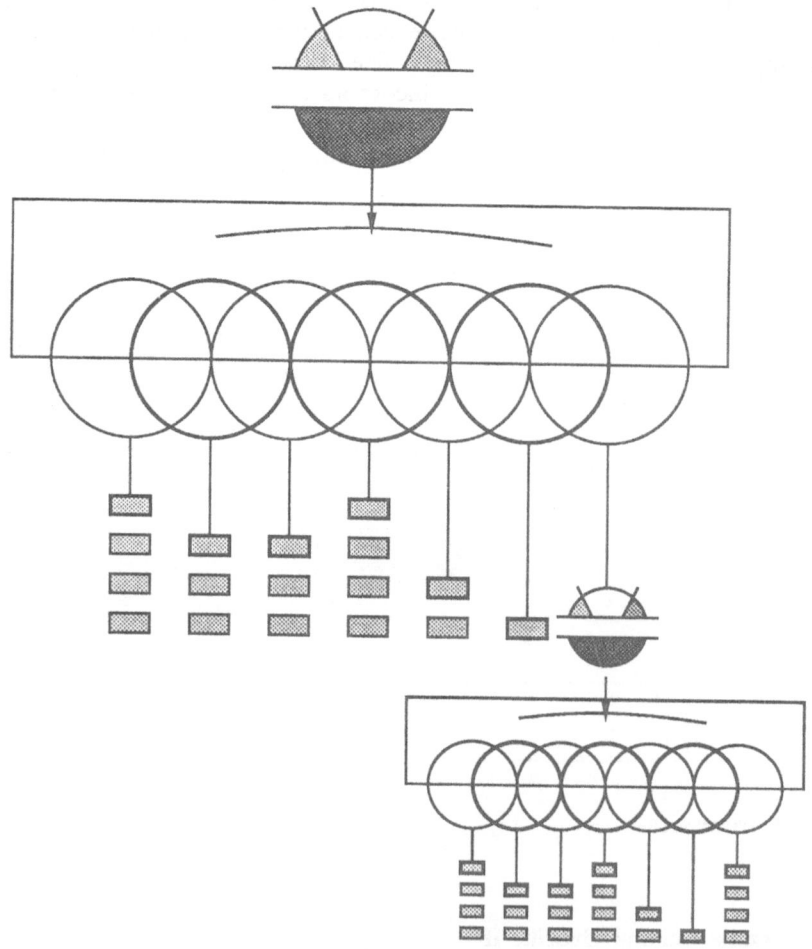

Fig. 9.10(b). IPs in parallel design process

6.4. Interface: Interface Management — Variable Machine

Based on what is already known about designers and how they work, it is expected that the Design Machine will be a variable machine. As such, it must not only be adjustable by the designer, but must also be able to "recognize" the designer's "state of mind," the place of the designer's actions in the design process, the context appropriate to understanding these actions. It must "adjust" its interface as necessary to support the designer, different designers, at different times, facing different design problems, and similar changes in the context of design. We shall especially consider the role of information related to the transitions between states (of thinking, of the design process, of scope, scale, category, association, etc.).

A human intelligent "processor" learning design (in college, in a studio, etc.) can follow a train of thought suggested by the experienced designer (procedural learning) and often will attempt to anticipate the next step. The knowledge that allows this to happen will also be researched, formalized, and incorporated in the Design Machine. This knowledge is expected to be relative, and to vary from designer to designer, from task to task, from time to time, from domain to domain, and from context to context.

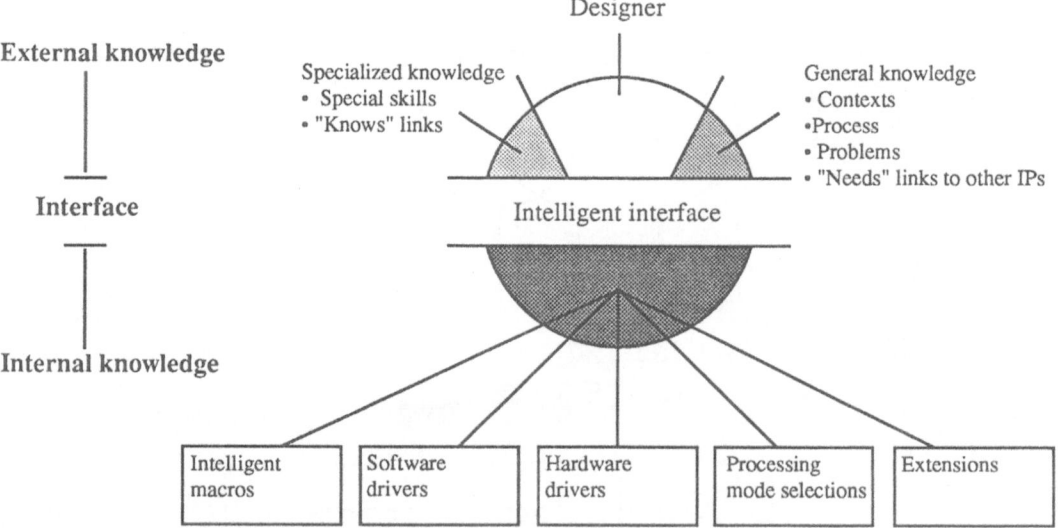

Fig. 9.11. Intelligent processors

6.5. Intelligent Processors: General/Specific IP

The following model (Fig. 9.11) illustrates the architecture of the intelligent processors themselves: each processor has external, internal, and interface knowledge. The external knowledge is divided into general, specialized, and designer characteristic knowledge, while the internal knowledge contains information about the computer, peripherals, and software. Knowledge connecting the external knowledge and the internal knowledge is contained in the interface section of the IP's and provides an intelligent connection between the designer and the lower level software and hardware. The general knowledge of the IPs is made up of information allowing access to the context knowledge, access to other IPs, and knowledge of the design process and its progress stacks. The specialized knowledge component allows each IP to act as a specialist when needed, while still acting as a normal member of the IP design team under most circumstances. Knowledge of the designer allows it to recognize idiosyncrasies, anticipate questions, and adjust to different users. Knowledge of the computer, peripherals, and software constitute the other major portion of an IP. Between the external and the internal knowledge is a section containing intelligent interface knowledge, knowledge which provides the designer with a mapping of actions which is not one-to-one, but one-to-many, so that one action by the designer may elicit many actions by the system.

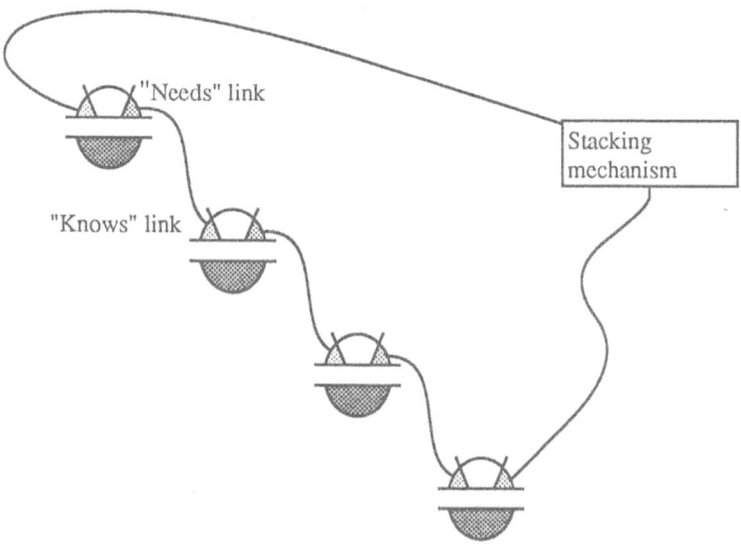

Fig. 9.12(a). IP link/loop configuration

The relationship between the IPs is reconfigurable, so that parallel, sequential, and hierarchical connections can be established as needed (Fig. 9.12). A stacking mechanism allows recursion to occur by keeping track of partial results, so that ring and loop connections can also be established. Such connections are evident in design activity, especially during problem definition, ideation, and alternative generation, when IPs interact rapidly with the designer and with each other.

6.6. Problem: Problem Management

A problem management utility is provided. This utility contains knowledge about problem solving approaches, i.e. top-down, middle-out, bottom-up, as well as heuristic tools, rules of thumb, rules for divergence and convergence. It also contains other problem solving strategies particular to the designer and tools for choice of strategies. It is used to monitor the designer's interaction and recognize the approach taken. At times it can interrupt the designer and propose another strategy, or suggest another viewpoint. It is responsible for maintaining the "flow state" mentioned earlier.

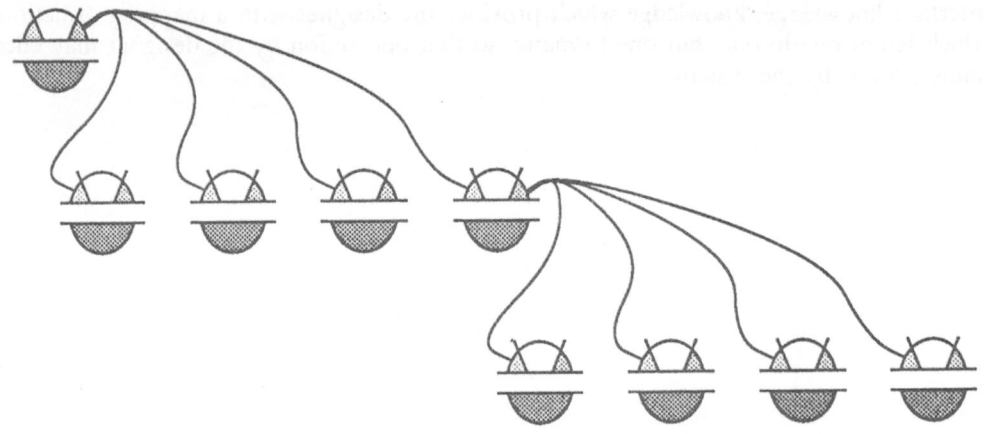

Fig. 9.12(b). IP hierarchical/networked configuration

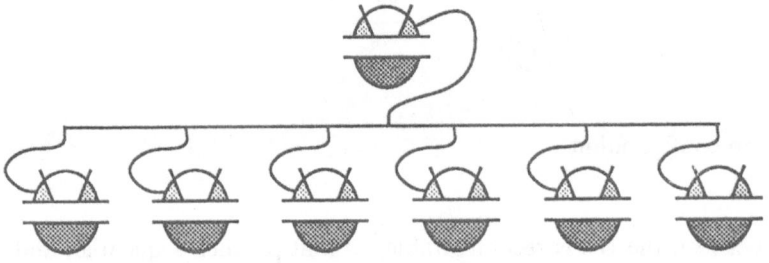

Fig. 9.12(c). Concurrent IP configuration

6.7. Representation: Representation Management

The representation management utility works in parallel with the problem management utility (Fig. 9.13). It provides the designer and the IPs with a broad selection of representations that allow the designer to look at the information in a variety of different ways. In order to maintain the flow state, and based on the progress being made and the designer's place in the design process, the representation management utility can suggest alternative representations to the designer. An integral part of the representation management subsystem is to maintain partial solutions for review and reuse, and to provide tools for their manipulation and combination.

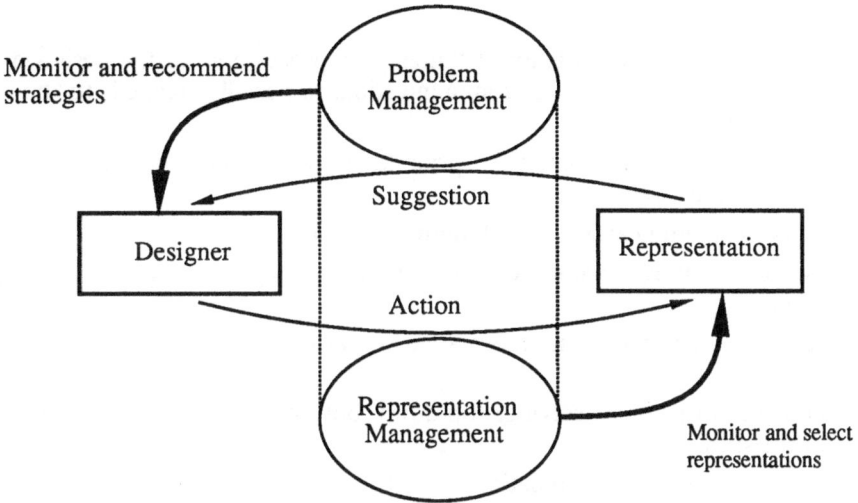

Fig. 9.13. Problem and representation management

7. METHODOLOGICAL APPROACH

By focusing on NIPs it is possible to collect the information needed for the implementation of the AIPs and to arrive at the necessary formalization for the creation of a system. The following methodology forms the basis for research aimed at identifying the categories of communication used in design.

7.1. Analysis

The results of the analytical stage may be of only limited usefulness as synthesis is likely to be very different from analysis. Nevertheless, the following approach can be suggested:

- Identify possible communication categories by involving designers in explaining why designs acknowledged as "good" and "valid" are "good" and "valid"; identify conceptual primitives in each category;
- Identify contexts relevant to design within which categories and primitives are meaningful;
- Identify prior knowledge implicit in contexts;
- Identify classes or representations used: iconic, symbolic, indexic, and their relation to semiotic categories of interpretation;
- Identify how all these are communicated, the degree of precision (i.e. vagueness), the influence of precision, etc.

7.2. Synthesis

- Instruct the designer to "lead" a team of "intelligent processors" in the design of simple objects; observe aspects of communication, especially related to how designers "externalize";
- Identify representations, their relation to prior knowledge, conceptual categories, primitives, and contexts used;
- Identify means of communication: verbal, mimetic, visual;
- Identify ways of thinking employed: association, analogy, metaphor, simile, synecdooche, etc.;
- Identify modes of communication: affirmative, negative, declarative, imperative, interrogative;
- Identify approaches to problem-solving: top-down, bottom-up, breadth first, depth first, special heuristics;
- Identify manner of problem reformulation;
- Identify aesthetic considerations and their formulation in the design process;
- Identify choice strategies for knowledge acquisition and distribution;
- Combine results with those of analytical phase.

7.3. Hypothesis and Testing

- Create a hypothetical "language" model and use it for specifying design programs;
- Instruct a group of designers and the design leader on this model; clarify which elements of the hypothetical language can be used; observe discrepancies, assumptions, shared knowledge, etc.; test using programs;
- Improve language model;
- Repeat as necessary.

7.4. Extract Patterns: Operational Knowledge

In order to facilitate implementation on computers, we need to operationalize the knowledge acquired. The steps to follow are:

1. Identify categories of conceptual adjustments;
2. Formalize these categories in the hypothetical design language;
3. Test and improve the language using intelligent processors and prototypical software/hardware configurations;
4. Search for patterns of frequency, distribution, interconnection, proportion, and especially transition in the ranges of variables identified; search also for constants, minima and maxima, averages, and other such quantities that may be related to quantifiable measures of performance;

In general, analyze the information gathered from the use of prototypical systems in order to find what settings are important for different phases of a design task, for different tasks, and for different task domains.

A rapid hypothesis/test cycle should be set up to allow fragments of hypothetical languages to be rapidly formalized and tested.

1. The designer communicates with IPs using an initial hypothetical language derived from the analysis of existing designs, other interviews with designers, and conventional design concepts and terminology. NIPs carry out commands manually. Hypothetical language is augmented by communication directed towards synthesis.

2. The designer communicates with IPs. IPs act as front ends for fast interactive workstations.

3. Designer/IP communication is formalized into a sketch "language" on artificial intelligence/expert system workstations.

4. The designer uses sketch language to design. Discrepancies and inadequacies are noted. Hypothetical language is updated. Process is repeated with new hypothetical language.

8. CONFIGURATION

In addition to the requirements discussed in the recommendations mentioned above, we have to add requirements pertinent to the visual processing specific to design. There are two levels at which visual processing takes place; the iconic level (realistic images) treated mainly in bitmapped two- and three-dimensional representations and a symbolic level of abstracted images. The two levels require different kinds of processing in order to support optimum interaction. One should be highly vectorizable, supported by single instruction multiple data (SIMD) architectures, dealing with image processing, the other supported by multiple instruction multiple data (MIMD) architectures, in the symbolic domain. Since the designer goes from one level to the other frequently, the IPs should support the change in mode. We suggest here a configuration appropriate for this purpose (Fig. 9.14).

The iconic domain of our real time design machine requires intense computation. Conventional computer architectures, based on the von Neumann machine, are limited in their speed. However, they are flexible. SIMD uses direct memory access and thus avoids time consuming MOVE instructions.

The two-dimensional arrays used in image processing are well suited for the vector processing of SIMD. The only condition is to keep the pipeline operation uninterrupted. The MIMD architecture is a multiprocessor system useful where the SIMD architecture with pipeline processors or systolic arrays is not recommended because of the predominantly scalar nature of the problem. When we need multiple small window processing, as in design applications of a symbolic nature, the MIMD architecture is better suited (Giloi 1986).

Having this architecture in mind, we have to define different levels of software, starting with the operating system and all its utilities, drivers, communications software, file management, etc. Since it is supposed to be a real time machine, we have to provide real time capabilities.

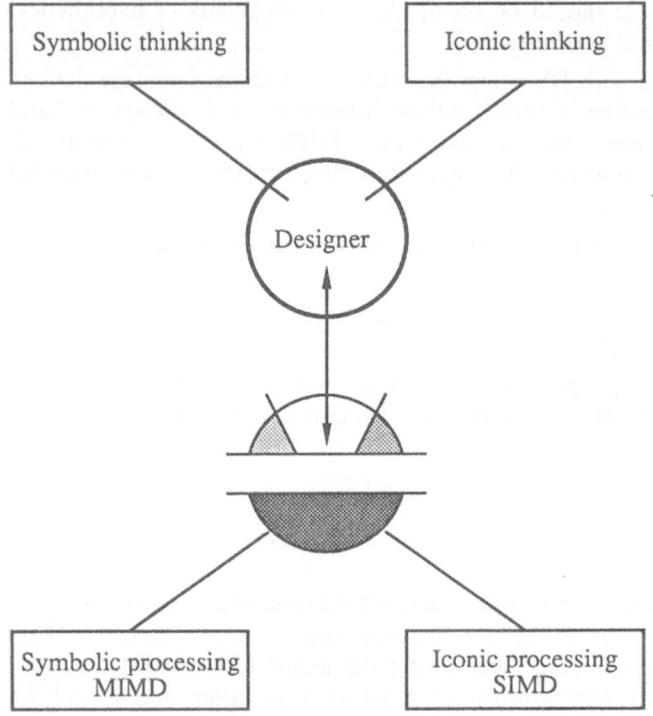

Fig. 9.14. Processing configuration

The applications software level is designed having in mind programming languages in which design applications are programmed. Together with appropriate languages, we have to provide a library of generic and specific routines for image analysis and image synthesis. In view of the machine's variability ("tunable"), the user interface will be built up by the user. It seems quite probable that the programming style will be the result of using languages with new data types characteristic of iconic image processing and symbolic image processing pertinent to the designer's activity. These types are provided in addition to conventional data types such as integers, strings, and arrays.

9. CONCLUSION

The complexity of design makes the attempt to build ICAD machines unlikely to result from a piecemeal approach. We discovered that what is necessary is an encompassing, but not enclosing, concept. This is represented by the suggested variable, category-based, extendible machine based on IPs, for which this paper proposed a model. We do not try to emulate ways in which designers solve partial problems but rather suggest an environment in which design intelligence is supported in the interaction between designers and the design machine.

REFERENCES

Brown, D. C. and Chandrasekaran, B. (1985): "Expert systems for a class of mechanical design activity," in *Knowledge Engineering in Computer-Aided Design, Proceedings of the IFIP WG5.2 Working Conference 1984 (Budapest)*, Gero, J. S. (ed.), North-Holland, Amsterdam, pp. 259-290.

Giloi, W. K. (1986): Functionality and Architecture of Machine Vision Systems, Lecture Course (notes for the USLA computer vision course).

Kuhn, T. (1962): *The Structure of Scientific Revolutions*, Chicago University Press, Chicago, IL, USA.

Lakoff, G. (1987): *Women, Fire, and Dangerous Things: What Categories Reveal about the Mind*, Chicago University Press, Chicago, IL, USA.

Rowe, P. (1987): *Design Thinking*, MIT Press, Cambridge, MA, USA.

Veth, B. (1987): "An integrated data description language for coding design knowledge," in *this volume*, Springer-Verlag.

Report on Session 4

Chair: P. Bernus †
Cochair: Zs. Ruttkay ‡
Edited by: V. Akman †

The two papers in this session have addressed different scopes of intelligent CAD.

1. *What is design?*

   Boose's paper suggested that multiple criteria selection and methods related to it are important parts to be included as design tools. One may interpret the message of the Design Machine paper by Nadin and Novak in this respect — a strong stress on the multidisciplinary nature of design. Even if it sometimes seems to deal with a single discipline only, the wholistic aspect of creativity must be predominant. If design is exercised by a group and not by an individual, then design has a specific social aspect, too.

2. *Role of CAD in future:*

   Boose's paper about a specific yet important class of knowledge acquisition functions makes it clear that today prospective CAD systems users can find themselves in a difficult situation: Even though there exist a large number and type of tools developed for different subareas of a design subject, it is hard to judge what tool to use (or whether to use one at all) for given design situations. We have to face the knowledge representation problem. Future CAD systems, by the Design Machine concept, need to support the visual aspect — the designer's way of thinking. This is true for the different iconic representations of the target of design and the "flow" of design as well.

3. *Role of intelligence in CAD:*

   In Nadin and Novak's presentation, it is mainly the parallel and multilevel nature of design that makes intelligence a necessary feature of CAD systems. The proposed architecture, with self-similar levels of detail, is considered as a natural architecture for intelligence. Parallelism on the same level of detail is supposed to play a role similar to the cooperation of designers in a design studio.

---

† Centre for Mathematics and Computer Science, Kruislaan 413, ı098 SJ Amsterdam, The Netherlands.

‡ Computer and Automation Institute, Hungarian Academy of Sciences, Kende ut. 13-17, 1111 Budapest, Hungary.

During the Q&A period, Schramel made the comment that an expert must know all the tools before using them. His question is "How long does it take to get them known in Boose's system?" Boose replied that the user is not expected to know anything about the content of the system. It operates as an expert system which operates with certainty factors, but it is hidden from the user. If one thinks of the empty system as an expert system shell, one can use it for different application domains. One can learn how to use it within a couple of hours, Boose added.

Akman wanted to know Boose's view about certainty factors. Boose accepts that there is a debate in the AI community about how to handle uncertainty, e.g. the way certainty factors were incorporated into MYCIN. It is too easy to handle uncertainty by applying certainty factors. In Boose's case, it is the system which actually computes the certainty factors and they are not used later on by the reasoning engine.

Koegel asked Nadin and Novak if they think that CAD systems should take into considerations aesthetics in evaluating a design. They see it is an important aspect which cannot be automated in a precise sense. All the same, it has to be taken into account. Tomiyama had doubts about how effective their approach can be, even in a time frame of five years. Nadin and Novak's reply was that they do not want to make any prophecy. However, they reminded the audience of the exciting developments in computer graphics and added: "Now you have on your screen what you had not dreamt about before."

Koegel questioned if they incorporate a model of how designers move from one task to another. Their reply was that they have to analyse the protocols of human design. Yet they think that association has the main role in the control of design and in all kinds of problem solving.

Session 5

10. Modular Design As Algebraic Composition

M.P. Fourman and R.M. Zimmer

Abstract Hardware Workshop, Department of Electrical Engineering
Brunel University, Cleveland Road, Uxbridge, Middx UB8 3PH, UK

Abstract: *The aim of this research is to study how high-level mathematical abstractions could be used for reasoning about hardware design. These abstractions will be used as the foundation for a CAD tool for VLSI. Of course, different algebraic abstractions are appropriate on different levels of the design hierarchy. Uncovering, formalising and integrating these will be difficult but is necessary for the development of future CAD tools. We have developed a general algebra, applicable at all levels of the hierarchy, for composing small modules into large designs. Having a common composition algebra throughout allows transformations between levels (both refining transformations downwards and verification and abstraction transformations upwards) to be algebraic morphisms. In this paper, we present a sketch of the composition algebra, and closely follow a design at the behavioural level. It is hoped to give an indication both of the use of composition and of the kind of verification that can be done at the highest levels.*

Keywords: CAD, hardware verification, VLSI, category theory.

1. INTRODUCTION

The design of Very Large Scale Integrated (VLSI) Circuits is a complex task. One useful tactic for handling this complexity is to reason about a design as a conglomeration of several more or less independent units, or *modules*. Another tactic is to design circuits *hierarchically* — the designer must be able to translate his design through a sequence of design levels going from a high-level behavioural description down to physical layout. An intelligent CAD system for VLSI design should incorporate the abilities, first, to compose several small designs into a large design and, secondly, to verify that a design at one level of the hierarchy meets its specification at a higher level.

We believe that the control of these processes is best achieved by the use of formal mathematical abstractions throughout the design process. Of course, many different abstractions are employed in system design. If all the abstractions used are instances of the same general algebra, then it is clear what shape the top-down (refinements) and bottom-up (abstractions and verifications) transformations should take — they should be morphisms in this algebra. This paper concerns some of these algebraic applications to CAD systems. A general algebraic system for modular composition is given. This composition algebra is quite general and has been used to study modular composition on various levels of the design hierarchy from stretchable cells for layout up to Petri net models of high level behaviour.

As well as giving an exposition of the composition algebra, this paper closely follows an example of a design at the high-level behavioural stage. Verification of high-level behavioural descriptions should be supported even in an automated logic synthesis system. If the high level description is in error then "correctness by construction" merely leads to a correct implementation of a faulty specification. Presently, errors made at this level may not be uncovered until the design has progressed far enough to enable traditional functional simulators to exercise the design; indeed, some errors can never be detected by simulation. An intelligent CAD system for VLSI should provide a tool enabling the designer to verify some aspects of the behavioural description.

The aim of this paper is to give an indication both of the use of composition and of the kind of verification that can be done at the highest levels.

2. THE GENERAL ALGEBRAIC FRAMEWORK

Abstract composition is a structured synchronisation. To compose objects it is necessary to say in what ways they synchronise. This synchronisation information is given by a partial map (sometimes just a relation) from the objects themselves to a set of labels.

Fibred Categories: It is only the labels that are used in the synchronisation. Therefore, it is only the labelled parts of the objects that can be synchronised. In the simple case of circuit level composition, some of the wires are labelled, and synchronising two circuits involves saying which labels of the one are related to which labels of the other — i.e. which wires get connected to which wires. The wires without labels are internal wires that cannot be connected to other modules, they are said to be *hidden*.

The algebraic framework for reasoning about such compositions is that of a *fibred category*. The general mathematics will be fully developed in a separate paper (Fourman and Zimmer 1987). For the purposes of the paper, we will give a brief flavour of the general algebra and then concentrate on a high-level behavioural example.

Let P be an object. An *E-labelling* of P is a partial function from P to E.

Given $f: E \to F$ and an E-labelled algebraic object, P, we obtain an F-labelled object, f_*P, by composition. We can also obtain an E-labelled object, f^*Q, f^*Q, from an F-labelled one, Q, by "pulling back." This object, f^*Q, comes equipped with a universal morphism to Q, "over f" (Fig. 10.1). When we say we have a fibration, we are merely asserting the existence of this structure.

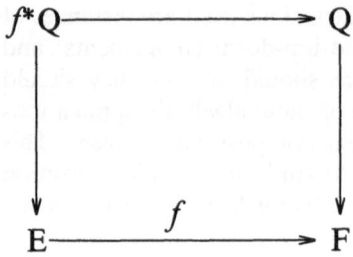

Fig. 10.1

If f is a partial map giving a subset $F \subset E$, (note that the morphism is in the opposite direction from the inclusion), f^* gives a trivial extension of the label set, while f_* gives the notion of hiding. With other morphisms, it gives the notion of relabelling. When we consider the inclusion as a morphism $g: F \to E$, extension arises as g_* and g^* is a form of restriction.

Synchronising two labelled objects is a matter of saying how the labels of one relate to the labels of the other. This information is given in the form of an event set, E, and partial functions from E to the two labelling sets. The composition is then the limit of a certain diagram in the fibred category.

Abstraction and Refinement Morphisms: One of the major advantages of having this uniform treatment of composition is that the composition forms a skeleton on which to flesh out the notion of abstraction and refinement morphisms. For example, because we know how to compose circuit descriptions and Petri nets within this framework to define an abstraction morphism which takes a circuit description in terms of wires and gates to a Petri net description we simply need to define the abstraction function for a few basic circuits. Then the Petri net abstraction of a circuit composed of several basic one is the composition of the Petri net abstractions of the basic circuits.

That is if we have Petri net abstractions of an AND gate and an inverter, PN(and) and PN(inv), respectively, then we automatically have the Petri net abstraction for a NAND gate by forming the appropriate composition.

3. PETRI NETS AND THEIR COMPOSITION

For the remainder of this paper, we will concentrate on the Petri net instance of the composition algebra. Petri nets have the advantages of being, first, useful on various levels of the design hierarchy (here they will be used to model high-level behaviour) and, secondly, of being familiar to practicing engineers. We do not mean to suggest that Petri nets are the only, or indeed the main, example of the algebra — we only hope that the Petri net example will serve to elucidate several ideas from the general theory.

A Petri net is simply a set, P, of *places*, a set, T, of *transitions* and two relations from T to P giving, for each transition, the preconditions and postconditions of that transition.

A *marking* of a Petri net is a subset (possibly a subset with multiplicities) of the set of places. The elements of that subset are said to be *marked*. A marking indicates what commodities are available at a given time.

A transition is said to be *firable* in a given marking if all of its preconditions are marked. The effect of firing a firable transition is to unmark its preconditions and to mark all its postconditions.

A *labelling* of a Petri net, PN, is an assignment of names to some of the transitions of PN.

Winskel (1984) has shown how Petri nets should be regarded as algebras in the category of (multi)relations. This viewpoint provides a natural category of Petri nets. We replace his synchronisation algebras by categorical constructions which provide further unity to this approach. In this paper we restrict ourselves to safe Petri nets, which may be viewed as algebras in the category of relations between finite sets, with labellings taken from the category of finite sets and partial maps. A more general (and more uniform, but more technically abstruse) treatment, using multirelations throughout will be explored elsewhere. For our present purposes, these simplifications will not be a problem.

Fig. 10.2

We consider Petri nets whose transitions are labelled by events from a set L shown in Fig. 10.2. A Petri net is a pair of arrows in the category of relations. These relate each transition to its pre and post places. The labelling is a partial map from the transitions T to the set of events.

Given two Petri nets labelled using the same set of events, there is an obvious notion of morphism between them (Winskel 1984). Given by the commuting diagram of Fig. 10.3 (where the morphism on places is a relation, and that on transitions is a partial map), for each labelling set, L, we have a category of L-labelled Petri nets. Given two Petri nets labelled from different sets of events, we can only compare them if we have a comparison morphism between the sets of events as shown in Fig. 10.4.

We say this morphism of labelled Petri nets is "over" the morphism on labels. Technically, this gives a fibration of labelled Petri nets over the category of finite sets and partial maps. In this context, various constructions have a natural categorical representation. We give only the examples we need here, leaving the general theory to another paper.

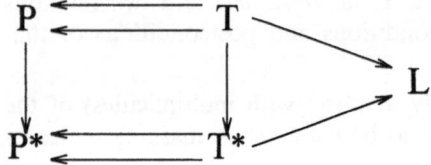

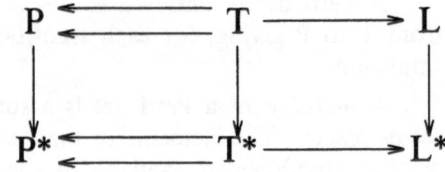

Fig. 10.3   **Fig. 10.4**

Composing Petri nets: The synchronisation information necessary for the composition, at the Petri net level, is given by functions to the various labelling sets. A transition from one Petri net synchronises with a transition from another in the composition if the labels of the two transitions get mapped to by the same element of the synchronisation or event domain. For example, a car cannot enter an intersection unless the light corresponding to that car's position is green — a car leaving must synchronise with a light staying green. There is a basic asymmetry here — a car cannot leave without the light being green but the light can be green independently of car positions. This is modelled in the Petri net synchronisation by having one element of the event set that is mapped to both "light stays green" and "car leaves" and another which just gets mapped to "light stays green."

The composition may be characterised as a universal solution to the problem of giving a Q-labelled Petri net with a morphism to each component, respecting the morphisms on labellings. (One reason for our use of partial maps for labels and transitions is that, if we generalised these to relations, synchronous composition, defined in this way, would not always exist. A better solution to this problem is actually to use multirelations everywhere (Fourman and Zimmer 1987).)

4. VERIFICATION USING TEMPORAL LOGIC

The logic used for verification on the Petri net level is a species of temporal logic called CTL, or *Computation Tree Logic* (Clarke, Emerson, and Sistla 1983). CTL was developed to reason about state diagrams. The basic terms are those of propositional logic plus some others referring to things being true at some future time. This type of language is good for expressing conditions like:

1. "if there is a car waiting at the eastern corner, then the eastern light will eventually turn green" — a liveness condition.

2. "whenever the northern light is green, the eastern light is red for sufficiently long to avoid collisions" — a safety condition.

Computation Tree Logic is a path logic of state graphs. The basic semantic relation of satisfaction is written

$$s \mid = j.$$

(Read, "*j* is valid in state *s*".) The propositional connectives have their standard, classical interpretation. In addition, we have modal operators including:

$$s \mid = \mathbf{A} j$$

iff, for every state, t, immediately succeeding s,

$$t \mid = j,$$

and

$$\cdot s \mid = j \, \mathbf{EU} y$$

iff, there exists a path, *p*, from *s*, to a state *t* such that

$$t \mid = y$$

and for each predecessor, *v*, of *t* in *p*, we have

$$v \mid = j.$$

The operator **EF** is defined by:

$$\mathbf{EF} j \in T \, \mathbf{EU} j.$$

The operator **G** is defined by:

$$\mathbf{G} j \in \mathbf{EF}(j).$$

Computation Tree Logic also allows for fairness constraints which are needed to express the design requirement in more complex examples.

The general outline of verification on this level is to generate from a Petri net, PN, a state machine whose states are the possible markings of PN. A transition in the state machine corresponds to the firing of a transition of the Petri net. Then using CTL we express statements like the liveness and safety conditions written above and the system tells us if the conditions hold in every state reachable from the start up conditions.

5. AN EXAMPLE: A TRAFFIC-LIGHT CONTROLLER

The problem is familiar (Fig. 10.5). We must design a mechanism whereby traffic can continue to flow (liveness) but collisions are avoided (safety). Firstly, the problem must be described formally. There are two parts to this description. The first gives a formal model of the possible behaviours of the crossing and cars. This model must tell us which events in the system are subject to our control. The problem is to restrict these behaviours in an orderly fashion. The second part of the description specifies what properties we require of the controlled system. (In our example, safety and liveness conditions.) The unconstrained system is described as a Petri net which is built from smaller Petri nets.

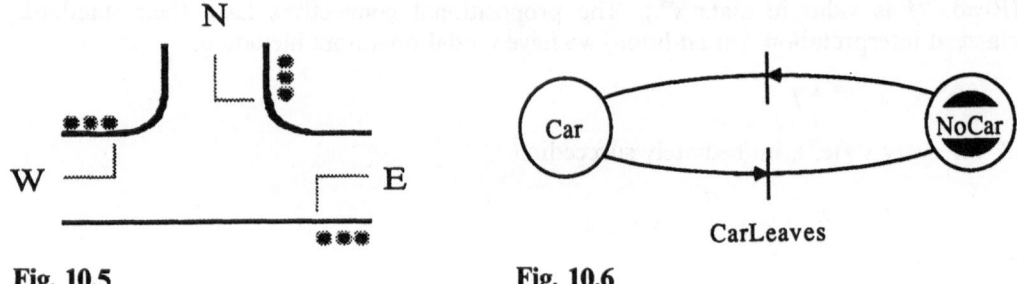

Fig. 10.5   **Fig. 10.6**

The arrival and departure of cars at each light is described in Fig. 10.6. The behaviour of a traffic light is given by the Petri net shown in Fig. 10.7. The traffic regulations (which we assume are obeyed) tell us how these two nets must synchronise. A car can only leave if the light is green. Otherwise, transitions can happen independently. We specify this by introducing a higher-level event, CarLegallyLeaves, which may be viewed as the synchronisation of the lower-level events CarLeaves and IsGreen. We also allow any of the lower-level events except CarLeaves to occur asynchronously. This gives the Petri net shown in Fig. 10.8 where a "traffic light" has events

$$TE = \{IsGreen, GoGreen, GoAmber, GoRed\}$$

and a "stream of cars" has events

$$CE = \{CarLeaves\}.$$

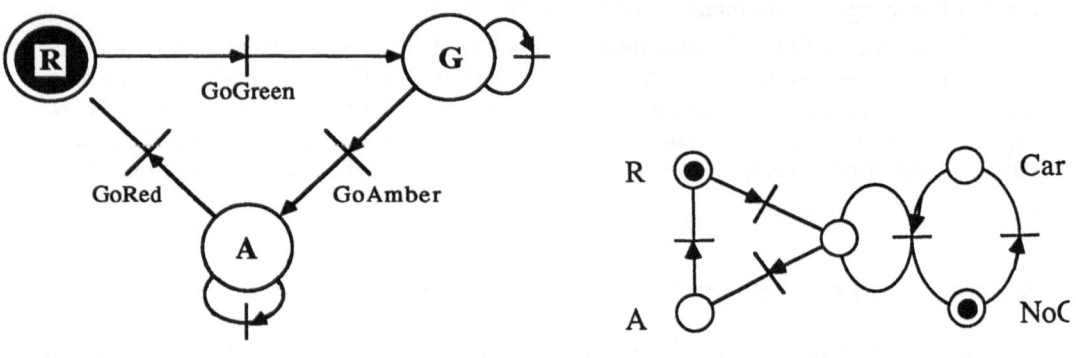

Fig. 10.7   **Fig. 10.8**

We synchronise one of each with the new event set

$$Q = \{CarLegallyLeaves, IsGreen, GoGreen, GoAmber, GoRed\},$$

with the partial maps,

1. Q → CE taking CarLegallyLeaves to CarLeaves, and

2. Q → TE taking CarLegallyLeaves to IsGreen, and each other event in Q gets mapped to to itself.

We now hide some of these events by mapping Q to {GoGreen, GoAmber, GoRed}. (Remember, we are using partial maps.) We form three copies of the resultant Petri net parametrised by {N, E, W}. Finally, we build a controller with the event set

R = {EA, ER_NG, NA, NR_EG_WG, EG_WG, WA, WR}.

We synchronise this with the three Petri nets to form a new Petri net with event set R, using four partial maps, the identity on R, and the obvious maps from R to the event sets of the three parametrised Petri nets.

The physical system which must be controlled is modelled as three such Petri nets, which are initially independent. Our system allows us to create parametrised instances of a given Petri net automatically. We use the prefixes N, E, W, to distinguish the three copies. In fact, to be complete, we must also specify which parts of the system can be controlled directly — we have no direct control over the arrival and departure of cars (bringing in a policeman is not an admissible solution), we can only control the changing of the lights.

Thus, the designer must produce a control Petri net and say how it synchronises with the events,

N_GoGreen E_GoGreen W_GoGreen
N⁻GoAmber E⁻GoAmber W⁻GoAmber
N⁻GoRed E_GoRed W_GoRed,

which change the various lights.

Here is a putative controller. The transitions are labelled to indicate which lights they change, as shown in Fig. 10.9. This controller is to be a very naïve one which does not — and in this statement of the problem may not — look to see if cars are waiting. If we wished to allow this possibility we should make events SeeCar and SeeNoCar and make them visible to the controller.

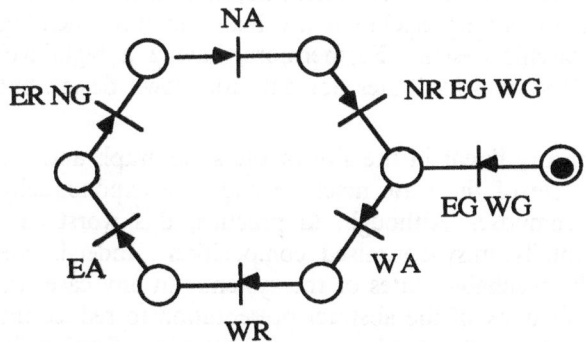

Fig. 10.9

We synchronise this controller with the three Petri nets representing the intersection, and verify automatically that the specification is satisfied. We can then go on to implement the controller using various standard methods. (This is intended as a pedagogic rather than a practical example. In real life, we would have a more demanding requirement and a more complex implementation.)

Verification of the controller: A first attempt at expressing part of the safety requirement might be

$$G \text{ (green\_W} \supset \text{ red\_N)}.$$

CTL formulae are, in general, true in some states and false in others. The "G" says "globally." $G(j)$ is true in some state, s, iff j is true in every state reachable from s. The rest of this formula should be familiar. (Classical propositional logic is a part of CTL). We really want to say more, however. We want to ensure that if in some state the west light is green then the north light is red and it will still be red in the next state (this will give any car crossing time to get safely across).

$$G \text{ (green\_W} \supset \text{ red\_N} \angle \text{ A (red\_N))}.$$

"A" stands for "all" and means "for all immediately following states." There are two other safety conditions for the traffic light controller. The one above with E replacing W and a corresponding condition for the times when N is green. Of course, so far, there is no guarantee that any light will ever be green. The simplest way of guaranteeing safety is to leave all lights always red! The liveness conditions will preclude this. We want to ensure that no light ever gets stuck on red. For example, the term, G(EF green_W), expresses that the west light eventually turns green. This introduces the operator EF, "exists finally." $EF(j)$ is true at s iff there exists a path from s which reaches a state where j is true. The problem has now been formally described.

6. CONCLUSIONS

A coherent algebraic theory could provide a useful foundation for a CAD system able to handle the complexity of chip design. The Petri net system outlined in this paper is a case in point — knowledge of the underlying algebra is invaluable in implementing strategies such as modular and hierarchic design. Furthermore, having a high-level mathematical underpinning to a CAD system facilitates not only top down design but also bottom up verification.

The Petri net checker has complexity linear in the size of the state graph and the formula in question. However, the size of the state machine may rise exponentially with the number of Petri nets we compose. Although, in practice, this worst case should not occur as it represents totally unsynchronised composition. Indeed, one function of a controller is to limit the reachable states of the system. In any case, we hope we will be able to exploit the hierarchy of the abstract presentation to reduce the complexity of checking large designs. This will introduce an element of verification (in the sense of formal proof) to our system.

We must not only provide tools to enable high-level design entry (synthesis tools), but also provide tools for design validation at the same level. Before we can build such tools, we must formalise the abstractions the designer uses (or provide abstractions he can use). To be useful, such abstractions must appear natural to the engineer. A tool which requires lengthy retraining will not be used. More research into appropriate abstractions of hardware and into the manipulations of these abstractions, like the composition algebra described in this paper, is needed.

REFERENCES

Clarke, E. M., Emerson, E. A., and Sistla, A. P. (1983): "Automatic Verification of Finite State Concurrent Systems Using Temporal Logic: A Practical Approach," Research Report No. CMU-CS-83-152, Carnegie Mellon University, Computer Science Department.

Fourman, M. P. and Zimmer, R. M. (in preparation, 1987): An Algebraic Theory of Composition.

Winskel, G. (1984): "A New Definition of Morphism on Petri Nets," *Lecture Notes on Computer Science*, **166** , Springer-Verlag.

11. Intelligent Computer Aided Synthesis of Digital Signal Processing Systems

D. Genin, A. Dardenne, and J. de Moortel

Tektronix AI Research Center
Kapeldreef 75, B-3030 Leuven, BELGIUM

Abstract: *The goal of our research is to open new perspectives in CAD by identifying different techniques and theories of Artificial Intelligence (AI) that can be used to create the next generation CAD tools. Therefore the design task has to be better understood and models of it must be built. Our work is based on studies we made in a specific design domain, but our model of the designer will be at a high level of abstraction and will cover a wide class of problems. Good designers think about problems from a global viewpoint. Starting from assumptions they build different worlds containing different possible solutions. They evaluate and compare that limited set of possible solutions by using common sense and domain knowledge. Truth maintenance systems and time reasoners have been used to construct such a model of the designer. This paper relates our way towards a solution to this problem.*

Keywords: CAD, knowledge representation, common sense reasoning.

1. INTRODUCTION

Our research focuses on the designer and the design task. Design strategies have to be identified, and the way the designers put them into practice must be better understood. In order to get a deep understanding of the designing process, we chose a specific domain, observed designers, and analysed existing CAD tools. This led us to define the specification of the new intelligent CAD tools for that specific domain. The information and knowledge we have collected from that particular application led us to think that the design methodology we have modeled could be used in a much broader domain and could cover a larger class of applications. Problems dealing with the choice of resources to perform some work within a certain period of time, such as job shop scheduling, could be tackled by our system methodology.

The problem domain of this project is the automatic synthesis of synchronous multiprocessor system chips starting from a high level behavioral description of a digital signal processing system (DSP). The designers from the Interuniversity Micro-Electronic Center (IMEC) in Leuven, Belgium, the methodology they use (de Man 1986), and the CAD tools developed by the Center (de Man et al. 1986; Rabaey and de Man 1987) were taken under study.

The high level language used to describe DSP is an applicative language called Silage (Hilfinger 1985). The target architecture is a predefined set of execution units identified and designed at IMEC. Such an architecture is especially suited for the so-called third generation DSP algorithms involving large blocks of sample data subject to complex decision making algorithms. Typical applications include: speech synthesis and analysis, modems, digital audio (compact disk signal processing), matrix based processing, and ISDN (Integrated Services Digital Network).

2. THE DESIGNING TASK

It is important to be aware of the complexity of the design of DSP chips in order to understand our approach to solve the problem. An AI approach to any problem starts with a study of the human ability to solve the problem. Designing is an art. It is very difficult to capture the essence of it. Moreover there are no real experts (in the sense used in expert system theories) for this particular mapping problem because the design methodology and strategy are still under development. Therefore we had to observe people who have some experience and trust our common sense.

The methodology developed at IMEC is called *"Meet-in-the-middle design strategy"* (de Man et al. 1986). It refers to the fact that one opts for a strict separation between system and silicon design levels. The system designer will call parameterized modules designed by silicon specialists. The choice of a minimal set of *primitive silicon modules* is the result of series of studies undertaken by the IMEC DSP specialists. The primitive silicon modules were designed by *silicon specialists* in a bottom-up way while the *system engineers* work top-down from the high level specification down to the "middle line" where they call the parametrisable silicon modules.

The problem faced by the designer is to map a set of operations onto an architecture. The operations are defined by an algorithm written in Silage. The target architecture is constructed by composing a set of modules selected among *predefined execution units* (EXU). The designer is allowed to use different instances of the same EXU. Each EXU is able to perform different operations, some more efficiently than others. This means that a specific operation can be performed in many different ways. The designer has to choose for each operation the best suited EXU. Therefore he has to evaluate two main cost functions: the cycle count and the area. Choosing specific execution units for each particular operation is called the *mapping problem*. Ordering the execution of the operations on the chosen EXU's in the time domain is called the *scheduling problem*.

The system specification limits the freedom of the designer. The ideal resulting chip must have the smallest size possible and be fast enough. Being fast enough means that the circuit must execute the DSP algorithm in a number of cycles less than a certain maximum. Therefore we do not try to find the fastest solution.

The cost functions are always conflicting as is the case with all interesting problems. Minimizing the size of the circuit by using a small number of EXU's results in a slow system. Using parallelism to speed up the circuit gives rise to large chips. One may wish to reduce the number of buses between EXU's, but this may cause collisions during data transfer that can only be solved by reducing the speed of the system. Many examples of this kind could be listed. But the important thing here is to understand that the designer must find compromises between a large number of conflicting constraints. All the implications of the choices the designer can make are not always directly obvious and it is difficult to cope with all of them at the same time.

A good designer will optimally use the resources in order to generate an interesting architecture. How can he make the best choices? One can argue that the design is performed in many steps, each iteration optimizing the solution. In fact, the reason why the designer is able to improve his previous sketch is because he collected more information about the problem, leading him to some improvements. A good designer will be able to come up with a solution in less iterations. This is due to the heuristics he uses or from a better understanding of the problem which allows him to collect more information at each iteration.

The DSP designer has to solve two problems with conflicting cost functions: the mapping problem and the scheduling problem. The order in which those problems are addressed can have an important impact on the quality of the solution. Our intuition and experience tell us that conflicting constraints should be handled in parallel to avoid the problem of finding a local minimum.

One can use those observations to implement different models of the designer. The fundamental questions we have to answer are: What kind of information is important to the designer? Which design level does that information belong? What is the domain of the knowledge he uses? When and how does he collect the relevant information? When and how does he use it?

3. POSSIBLE MODELS OF THE DESIGNERS

We first will go through strategies that are or could be used to model the reasoning process of a good designer. Each of the following strategies answers differently to the questions mentioned above. The mapping of a Silage statement onto some EXU corresponds to a step in the design. We thus have to model that succession of choices and mappings. The design process can be represented by a tree of decisions. For this particular problem the search space is very large, even for the mapping of very simple algorithms. Each decision or choice has a large number of implications. Two different classes of algorithms can be differentiated through the way they evaluate design decisions and thus by the way they explore the search space: depth-first and breadth-first searches.

In this section three types of depth-first search will be briefly presented. The differences between those systems lie in the degree and type of "intelligence" they involve. At first glance the breadth-first solution seems to be less interesting because of the size of the search space. However, a closer look at the breadth-first technique reveals interesting properties of it.

3.1. Chronological Backtracking

In a depth-first search one solution to the problem is examined at a time. If some cost evaluation rejects the current solution the system will backtrack to the point where the previous choice was made and another decision will be made. Prolog or an Emycin-like system could easily be used to implement that model. But it is clear that we are far from an intelligent way of doing it. The new decision the system will make after the backtracking will often have nothing to do with the problem encountered in the solution including the previous choice. Indeed the origin of the problem may lie at another level in the decision tree and a lot of backtracking and recomputing will often be necessary to discover the right level at which the wrong decision is undone.

This previous strategy is called *chronological backtracking* because the system backtracks one level at a time to undo the last decision. A solution to this problem implies the need for some intelligence in the backtracking procedure so that one can address the right cause of the problem. It is clear that a designer uses more intelligence during his walk through the search space.

3.2. Dependency-Directed Backtracking

Instead of using a chronological backtracking system we could model our designer as a *dependency-directed backtracker* (non-chronological backtracking). This implies that the system must be able to keep track of the place where decisions were made, so that when a problem occurs it is possible to jump back to the origin of the problem. All the drawbacks of the chronological backtracking approach are not solved here. Even if we are able to identify the place where the wrong choice was made, we still need a way to identify a better choice.

3.3. Cathedral II system

Cathedral II (de Man et al. 1986; Rabaey and de Man 1987) is a system built at IMEC to aid the designer to design a DSP chip. This system is intended to be used as a computer aided design tool and not as a tool to completely automate the design process.

Cathedral II is based on the principle that design is an iterative task. After each iteration the designer should be closer to the solution or should know that a particular path is a dead end. Cathedral II helps the designer view the implications of his choices.

The solution chosen at IMEC consults the designer during the backtracking. The depth-first search helps the designer to see the consequences of choices (designer's own choices or system choices) and the interaction gives the designer control of the system. Entry points are defined through which the designer can access the system to modify its choices. Between these points the system is left on its own. If the system arrives at an user entry point without a solution, the user has to discover the reason for this failure by studying the choices made and the inferences drawn. This means that Cathedral II relies on the intelligence and the expertise of the user to make the right choices.

This system has been implemented in Prolog for the mapping problem (Vanhoof, Rabaey, and de Man 1987) (choice of an architecture). The scheduling of that particular architecture was first done by using an Integer Linear Programming technique. This approach has been replaced by a heuristic scheduling technique implemented in Pascal (Goossens et al. 1987)

The evaluation of the cost functions has been split into two parts. First the area of the chip is considered during the mapping problem, then the cycle count is evaluated during the scheduling of the architecture. If problems are encountered during one of these two phases the system will ask for a new architecture. The designer is able to help the system by introducing pragma's to the system. A *pragma* is a hint given to the system to help it in making decisions. In fact a pragma forces the system to map certain operations onto certain EXU's. For example if the time constraints are not met, one can tell the system to map all the multiplications on multipliers. If the area constraints are violated one can force the system to use an ALU to execute multiplication, this will slow down the system but uses a general purpose unit more efficiently.

In summary, Cathedral II employs an interactive approach to solve the mapping problem. The user works at a high level and can see the implications of his choices. This is certainly valuable compared to more conventional CAD tools where the designer has to work at each level in the design.

Nevertheless, we believe that we can further expedite the design process by utilizing AI techniques in the feedback loop, thus relieving the designer from all but the specification task. To accomplish this we must develop a comprehensive model of the designer.

4. A MORE COMPLETE MODEL

We began by studying the DSP designer. Designers use tools in a design environment. It is our experience that they are conditioned by their tools and often cannot see beyond them. It is sometimes difficult to understand what their design methodology would be in the absence of tools or work environment. The architecture of our designer model has been conditioned by the study of *the good choice*. What does a good choice mean in the specific problem of mapping Silage equations onto a customizable architecture, or what constitutes a good choice? A good choice must be based on reliable information. This means that the designer must know what the algorithm consists of and understand the DSP he wishes to build a chip for. In other words he must have a global view of the DSP specification, instead of a set of local views. He must therefore extract as much information as he can from the specification. Part of that information is explicit, but most of it is implicit and must be inferred by appropriate tools. Good designers or experts have the capacity to deeply understand the specification. They know what kind of information they need. Based on that knowledge, they can consider design issues together with their implications. Good designers are able to "visualise" solutions before they start working on them. This fundamental capacity permits them to make the right decisions at the first attempt. This saves them from having to iterate many times through the design before an acceptable solution is found. A good understanding of the problem allows them to select and use appropriate heuristics to better prune the search space by making good choices among the alternatives.

We also believe that designers build *different hypothetical worlds* each of which corresponds to a set of *assumptions* they make after the specification has been understood. When working on real problems the best possible understanding cannot lead to the discovery of all the parameters that will uniquely specify the best solution. When dealing with problems where the designer must relax conflicting constraints, it is very hard to cope with all the implications of every assumption. After having fixed the value of certain parameters, the designers specify ranges of values for the remaining parameters. In our system, a specific value within such a range is called an assumption. Each assumption can be considered as the origin of a hypothetical world. That world will become true if its origin value is chosen among the other possible values. The maximum number of hypothetical worlds is thus equal to the product of the size of the range of the "undefinable" parameters. This product is a maximum because all the possible combinations of assumptions will not lead to a feasible solution; some combinations will be contradictory. The number of feasible solutions can still be quite high and it is obvious that designers will only consider a small number of those worlds. Therefore designers must choose the most plausible alternatives. That capacity of selecting the most promising alternatives is decisive and has an important effect on the quality of the solution and on the number of iterations necessary to find it. These worlds are not entirely disconnected and have parts in common. When a designer is working in one world, he may infer things that are valid in others so that these deductions are only made once. While working towards a solution, the designer collects more information that will allow him to make responsible choices among the assumptions, thereby limiting the number of worlds.

This explains why we have rejected the depth-first approach. In a depth-first search each Silage statement is considered individually and a choice is made. We need a global view of the algorithm, therefore we think that a local analysis of each Silage statement will never be able to generate a good mapping of the algorithm. A good mapping of a statement must take other mappings into account in order to make the best use of the resources.

5. THE GLOBAL VIEW

The need for a global view was stated in the previous section without really defining on what basis that global view should be specified. Although we will reason in the context of the mapping of DSP algorithms onto a multiprocessor architecture, we think our view applies to a broad class of problems.

The designer must find a minimal set of execution units (resources) and schedule the execution of the Silage algorithm within certain time constraints. The choices the designer makes thus belong to two different domains: an architectural domain and a temporal domain. An Artificial Intelligence approach starts from the premise that the power lies in the knowledge. It is therefore mandatory for the designer to collect information about both domains. This will be useful for all further reasoning the designer will do on the specification.

5.1. Global Architectural View

A good mapping of a Silage statement must take other mappings into account in order to make the best use of the resources. One should know before starting the mapping what kind of execution units and how many of them will be available. But those questions can only be answered after the mapping is completed. A good designer, in fact, has a "feeling" about what the architecture will be and can predict the need for some execution units before starting the mapping.

Consider for example the mapping of a multiplication onto an architecture where the time constraint is critical. From a local point of view the best mapping would appear to be on a multiplier. But it is possible that when that specific multiplication takes place, a slower unit (ALU) will be free. If this can be used without loss of time, the solution will be more compact (and therefore, better).

How can we know at the beginning of the algorithm that we could "reuse" a unit that will only be generated to fulfil a need at the end of the mapping? To know that we can reuse a unit, we must be aware of the fact that a need for it will appear, and that the unit will be free when we need it. Therefore we only can make the assumption that such a unit will be available and act as if it were already present.

In a global architectural view we must consider all the possible interesting mappings for each operation. This way unexpected opportunities can emerge, while local analysis of each statement would have missed them.

5.2. Global Temporal View

Everyone unconsciously manipulates time concepts everyday. Time is part of our world and we use common sense to reason about it. We easily classify and arrange events in the time domain.

Naturally, this holds for designers, too. They are aware of the temporal implications of the choices they make in the architectural domain. This is mostly a subconscious process; they do not really schedule the selected units *a priori*. This means they do not compute the start and finish times of operations while constructing their global temporal view. Later on, a precise schedule is developed which completely specifies the arrangement of operations in time.

The type of temporal computation used during the first phase of the design process could be characterized as high level symbolic computation — symbolic because numerical values are not considered and high level because the temporal objects in consideration are not points in time.

We believe it is possible to model high level temporal reasoning by making temporal intervals primitives (Allen 1983). Relationships between temporal intervals (i.e. interval A overlaps interval B) can be described in a hierarchical manner and analysed using constraint propagation techniques. The notion of duration can also be introduced to limit the number of possible relations between intervals.

As for the architectural domain, the designer considers different temporal worlds. Each temporal world corresponds to a selection of relations between intervals. These worlds are again contradictory because the relations between two intervals can be different in two different worlds. But it is also true that worlds have parts in common.

This must be taken into account in order to obtain a fast implementation of the temporal reasoning mechanism.

5.3. Strategy

Having established his global views, the designer can now define sub-domains of the architectural and temporal domains. Those sub-domains restrict the search effort in each domain to a limited area. Moreover, the existence of constraints relations between the architectural and the temporal domain further reduce the number of possible solutions to be investigated (Fig. 11.1).

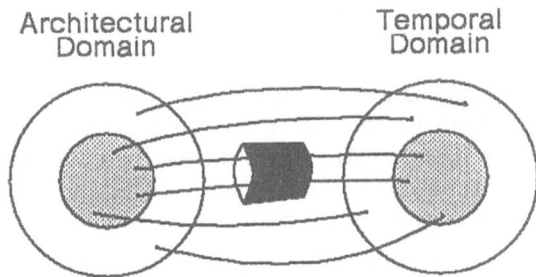

Fig. 11.1. Solution space

Two important ideas derive from these discussions. First in order not to work locally a breadth-first search would be more suited. A good strategy is certainly to avoid taking decisions too early. We must have enough information to generate the good choices at architectural and temporal level. Second a good decision can only be taken in the presence of sufficient criteria. This means that the area and cycle count cost functions should be evaluated simultaneously.

A first version of this model has been implemented with a hard-wired scheduler, i.e. without the symbolic temporal reasoner (the number of cycles required by a partial solution was computed incrementally). The architectural reasoner was able to prune the search space reasonably well.

Figure 11.2 shows the shape of that search space. If all possible solutions are considered its size grows exponentially from the start (S) to the finish (F) of the problem solving. This is represented by the region in gray. The central region (striped) shows the shape of the search space in the first version of our system (hard-wired scheduler). The point of discontinuity (D) corresponds to the point where the timing rules become active (the number of cycles required by a partial solution is higher than the maximum permitted number allowed in the specification).

Evaluating all the cost functions on all the partial solutions is certainly not feasible. An algorithm consisting of only five statements can be mapped on about 25,000 different architectures (one statement can be mapped onto seven or eight different units). Moreover if we consider that the arguments of any operation can be interchanged this leads to a number of solutions about 850,000!

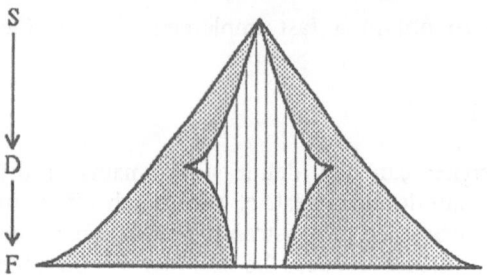

Fig. 11.2. Search space

Clearly the search tree has to be heavily pruned if we want a breadth-first solution. To assist us in this purpose, we have implemented a version of de Kleer's *Assumption-based Truth Maintenance System* (ATMS) (de Kleer 1984). Based on the manipulation of assumption sets, this system allows us to search the space of solutions efficiently, by maximizing the information transferred from one point of the search space to another.

The overall reasoning system consists of two components: a problem solver which draws inferences in the architectural and time domain and an ATMS which records these inferences. The assumptions are the primitive data from which all other data are derived. Until a final solution is found, sets of assumptions define partial solutions and represent different *worlds*. Assumptions can be contradictory, but there is no necessity that the overall database be consistent. Therefore the ATMS can handle multiple worlds. It is easy to refer to worlds and to move to a different point in the search space (different worlds) requires little work. Comparing solutions is therefore easy.

6. TRUTH MAINTENANCE SYSTEMS

In our context solving a problem means selecting among plausible alternatives. At first all the alternatives appear equally likely. One method for exploring the search space is simply to enumerate every point in it and test whether it contributes to a solution. In our case this strategy would be extremely inefficient because of the size of the search space. Moreover, we have already mentioned that the different worlds that form the search space have parts in common. Thus, it is advantageous to have the ability to have access to results obtained in one region of the search space from other regions.

A world is defined by a set of assumptions and defines a possible solution. Our goal is to find the best solution. Although we may not find the best one, we should be able to find solutions that meet some criteria of acceptability. This implies we must be able to compare (partial) solutions. An assumption has been defined as a specific value within a range a parameter can take. Different assumptions on the same range are thus contradictory. This means that the database that will record the different worlds will be inconsistent. Therefore we need a truth maintenance system to work safely with an inconsistent database.

6.1. Conventional Truth Maintenance Systems

A TMS serves different goals. It records the deductions made by the problem solver and then avoids repeating the same inferences. Once a contradiction has been discovered the world it belongs to is no longer considered. The TMS allows the problem solver to make non-monotonic inferences. By using dependency-directed backtracking techniques a TMS can remove contradictions from the database and rebuild a consistent world.

In a conventional TMS (Doyle 1979) only one world can exist at a time. Comparisons between parts of worlds are possible but are computationally expensive. For the same reason dependency-directed backtracking should be avoided.

A world containing only a subset of assumptions contains other worlds. If we compare a solution of the problem to a point in a hyperspace a world can be viewed as a point, a line, a plane, a hypervolume, etc. Inference should therefore be made in the most general world possible and be inherited by the subworlds.

6.2. Assumption-Based TMSs

By keeping track of the relations between data and the worlds to which they belong, the ATMS removes the necessity for a contradiction-free database. Inconsistencies between worlds, and therefore between data belonging to different worlds, are recorded in a *nogood set*. This implies that data that belong to different worlds can coexist in the database. Thus it is possible to compare and combine them at no cost. Worlds are not represented as disjunct entities. Only the incompatibilities between them are explicitly recorded in the *nogood*.

Instead of explicitly representing the worlds, the ATMS constructs a data structure which makes consistency checking very fast. The ATMS builds this data structure by associating description of world with data, instead of the association of data with world as in conventional TMS. Every datum in the ATMS receives a compact description (*Label*) of every world in which the datum holds. This lets the problem solver to work in all worlds at once.

Normally, it is extremely rare for a problem solver to ask for the contents of a world. During the problem solving, the number of hypothetical worlds can be very large, but if we do not construct them, it does not matter. At the end of the problem solving, when enough contradictions between assumptions have been found, the number of worlds (the number of solutions) is relatively small and it is then realistic to construct them by combining the mutually consistent assumptions.

As worlds are not explicitly distinguished, a result obtained in a world is automatically available in the other worlds. When a derivation is made by the problem solver, the ATMS records it in the most general way, so that it covers as large a region of the search space as possible. When a contradiction is recorded the ATMS finds its most general form in order to rule out as much of the search space as possible.

6.3. Basic Data Structure

An *assumption* designates a decision to assume without any commitment as to what is assumed. An ATMS *environment* is a set of assumptions. Logically an environment is a conjunction of assumptions. The *nogood* is the set containing the inconsistent environments in their most general form. For example if the environment {ABE} is in the *nogood*, while the inconsistent environment {AB} is discovered (or *vice versa*), only {AB} will be kept in the *nogood* because it is more general than {ABE}.

An ATMS *world* (or context) is the set formed by the assumptions of a consistent environment combined with all nodes derivable from these assumptions.

The *node* has the structure:

$$node = \;<datum, label, justification>.$$

The ATMS node contains the datum inferred by the problem solver, a label computed by the ATMS and a justification the problem solver has created for it. The node

$$a = \;<E_1 \, \text{Before} \, E_2, \{\{AB\}, \{C\}, \{D\}\}, \{(b, (c, d))\}>$$

represents the fact "event E_1 took place before event E_2" and is derived from either the node '*b*' or the nodes '*c*' and '*d*'. The label states that '*a*' holds in the environments {AB}, {C} and {D}. Its computation is based on the justifications provided by the problem solver.

A node with an empty justification and a label containing the empty environment holds universally and is called a *premise*. An assumption is represented by a special kind of node.

6.3.1. Remarks: The ATMS is incremental, updating only the changed worlds. Before the problem solver makes a derivation, the label of the possible new node is computed by the ATMS and checked against the *nogood*. This prevents the problem solver from working in inconsistent worlds. There are two inference procedures in the overall reasoning system (Problem solver, ATMS) both operating on the same expressions but treating them entirely differently. For more details the reader should refer to de Kleer (de Kleer 1986, 1986a, 1986b; de Kleer and Williams 1986).

7. THE TEMPORAL REASONER

The problem of representing temporal knowledge and temporal reasoning arises in a wide variety of domains. In our specific problem of coordinating different resources we utilize a high level symbolic representation of time. Our temporal representation uses the notion of a temporal interval as primitive. Relationships between intervals are computed by using constraint propagation techniques. Our temporal reasoner is based on Allen's work (Allen 1983, 1984). Extensions have been made to cope with the notion of interval duration and a truth maintenance system is used to check the consistency between temporal relations and between temporal relations and duration.

7.1. Interval as Primitive

The temporal reasoner provides an additional mechanism for reducing the search space. After understanding the specification, the designer limits the number of possible solutions by making different choices. A number of alternatives remain open. Choices about these alternatives will be made when more information will be available. To make the first choices, and during the problem solving the designer uses common sense. He makes inferences about time, which appear to be made automatically and effortlessly. The events the designer considers have certain durations. During the first simple inference he makes, he considers possible relations between position of events in time, but the start and finish times of these events are not precisely defined. We think he considers an event as an entity instead of being composed of imprecise time points. It is easier to think about 't' and 's' by saying 't overlaps s' instead of

$$(t- < s-) \angle (t+ > s-) \angle (t+ < s+)$$

where '$-$' means start point and '$+$' end point.

Later on when the designer has a better understanding of the problem to be solved and of the possible implications of his assumptions, he can reason with exact time. A good model for the temporal entities used by the designer is the concept of time interval.

Table 11.1. Temporal relations

Relation	Symbol	Symbol Inverse	Pictorial Example
X Before Y	<	>	—— ·········
X Equals Y	=	=	··············
X Meets Y	m	mi	——-·········
X Overlaps Y	o	oi	—— ··········
X During Y	d	di	········—·····
X Starts Y	s	si	·············
X Finishes Y	f	fi	·········——

7.2. Temporal Relations

Allen (1983, 1984) defines 13 relations between intervals (Table 11.1). Transitivity relations are also defined, e.g.

```
IF A Ri B
  AND B Rj C
THEN A Rij C
```

where Rij can be a set of relations; for example

IF A Meets B
 AND B During C
THEN A Overlaps C
 OR A During C
 OR A Starts C

By using this temporal algebra it is possible to compute all the relations between intervals given a set of relations between some intervals. Rules linking relations between intervals and duration are also defined. For example:

IF A During B
 OR A Starts B
 OR A Finishes B
THEN Duration (A) < Duration (B)

IF A During B
 OR A Starts B
 OR A Finishes B
 OR B equals B
 AND Duration (A) = Duration (B)
THEN A Equals B

These rules are very simple but extremely useful, because they assert further constraints which reduce the solution space.

Reference intervals are introduced to limit and to control the propagation of constraints. They are used to group together clusters of intervals. These clusters are related to the rest of the intervals only indirectly via the reference interval. For example a loop in the algorithm is described to the rest of the algorithm through a reference interval.

7.3. Consistency Maintenance

Instead of giving one relation between intervals to the system, we generate a set of possible relations for some intervals, each corresponding to an assumption. A number of possible values for the duration are also given to the time reasoner. This number is usually two or three.

In order to maintain the consistency between these constraints the ATMS records the consistent and contradictory environments. At the same time the ATMS handles the assumptions made by the architectural reasoner. It acts as a link between the two problem solvers. The goal is to find the best combination of assumptions that produces a set of intervals with as little overlapping as possible. The graphical representation (Fig. 11.3) shows that the chains of intervals should end in the region in gray. This means we search for the slowest solutions that still meet the temporal specification. This is based on the heuristic that the slowest solutions require, in general, the smallest area.

Figure 11.4 shows the shape of that search space (region in black). With the temporal reasoner we are able to eliminate from the beginning the fastest and the slowest solutions.

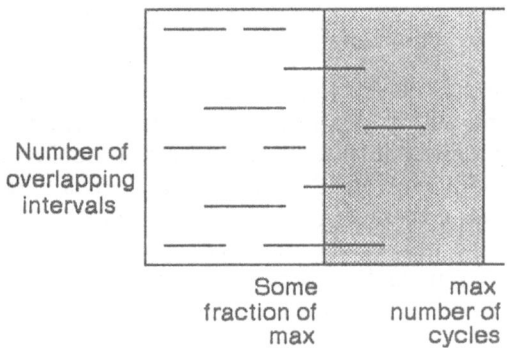

Fig. 11.3. Interval configuration

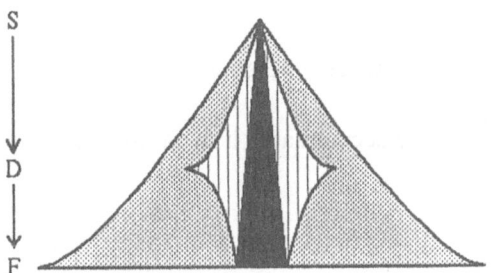

Fig. 11.4. Search space

8. THE ARCHITECTURAL REASONER

This part of the system is domain-specific. The temporal reasoner can be used in any problem dealing with time and the assumption-based truth maintenance system is very helpful in all the problems confronted with the necessity to select among plausible alternatives.

The knowledge used in the architectural reasoning part of the system was collected during interviews with IMEC DSP specialists. The architectural reasoner is frame-based. This means that the knowledge is articulated around objects. For example, in our system the operator '+' is an object that inherits the general properties of operators, but it also knows that it can be mapped onto different units such as an ALU (Arithmetic and Logic Unit), an ACU (Address Calculation Unit) or a Multiplier. Each of these different units offers advantages and disadvantages. On the other hand, an ALU knows what kind of operations it can implement, and it also knows that the number of instances of itself should be limited to a maximum of three.

199

The architectural reasoner uses the ATMS to store the assumptions it makes and to communicate with the temporal reasoner.

Operations can be viewed from different perspectives. The algorithmic view defines the type of the operation, its argument, and its dependences to other operations. The architectural view records the possible units onto which the operation could be mapped. The temporal view considers the different possible schedulings. It is clear that the assumptions from the two views constrain each other. Therefore the architectural reasoner bases its choices on the first results obtained by the temporal reasoner. Once the first assumptions have been made, both reasoners work in parallel.

8.1. Rules

The task of the architectural reasoner is to generate the most promising assumptions concerning the mapping of operations onto units and also to test the quality of the combination of assumptions. Selecting between assumptions should be made as soon as enough information is available to make a good choice. Examples of rules used by the architectural reasoner are listed below:

- The maximal number of ACU's is a function of the depth of nested loops and of the number of address calculations in the algorithm.

- If there exist different units of different types and an operation can be mapped onto these units then select the most specific one.

- An addition should be mapped onto a multiplier-accumulator if both operands are the results of multiplications.

9. CONCLUSIONS

We have proposed an architecture for a dedicated silicon compiler. This system merges different concepts and theories from the domain of artificial intelligence: knowledge-based systems, constraint propagation, common sense reasoning, and knowledge representation.

The system is implemented is Smalltalk to take advantage of its object-oriented properties. This system is still under development. Its architecture seems to fulfil our needs. Nevertheless we still need to augment the knowledge present in the system before testing it on real examples.

REFERENCES

Allen, J. (1983): "Maintaining knowledge about temporal intervals," *Communication of the ACM*, **26**(11), pp. 832-843.

Allen, J. (1984): "Towards a general theory of action and time," *Artificial Intelligence*, **23** , pp. 123-154.

de Kleer, J. (August 1984): "Choices without backtracking," *Proceedings of the fourth National Conference on AI*, Austin, TX, USA, pp. 79-85.

de Kleer, J. (1986): "An assumption-based TMS," *Artificial Intelligence*, **28**(2), pp. 127-162.

de Kleer, J. and Williams, B. (August 1986): "Back to backtracking: Controlling the ATMS," *Proceedings of the Fifth National Conference on AI*, Philadelphia, PA, USA, pp. 910-917.

de Kleer, J. (1986): "Problem solving with the ATMS," *Artificial Intelligence*, **28**(2), pp. 197-224.

de Kleer, J. (1986): "Extending the ATMS," *Artificial Intelligence*, **28**(2), pp. 163-196.

de Man, H., Rabaey, J., Six, P., and Claesen, L. (December 1986): "Cathedral-II: A silicon compiler for digital signal processing," *IEEE Design and Test*, **3**(6), pp. 13-25.

de Man, H. (September 1986): "Evolution of CAD-tools towards third generation custom VLSI-design," *Digest European Conference on Solid-State Circuits, ESSCIRC*, Toulouse, France, pp. 256-256c.

Doyle, J. (1979): "A truth maintenance system," *Artificial Intelligence*, **12** , pp. 231-272.

Goossens, G., Rabaey, J., Vandewalle, J., and de Man, H. (November 1987): "An efficient microcode compiler for custom multiprocessor DSP-systems," *submitted to the IEEE ICCAD-87 (International Conference on Computer Aided Design)*, Santa Clara, CA, USA.

Hilfinger, P. (May 1985): "A high-level language and silicon compiler for digital signal processing," *IEEE CICC Conference*, Portland, OR, USA, pp. 213-216.

Rabaey, J. and de Man, H. (May 1987): "Cathedral II: Computer aided synthesis of digital signal processing systems," *IEEE CICC-87 (Custom Integrated Circuits Conference)*, Portland, OR, USA.

Vanhoof, J., Rabaey, J., and de Man, H. (August 1987): "A knowledge-based CAD system for synthesis of multi-processor digital signal processing chips," *VLSI-87*, Vancouver, Canada.

Report on Session 5

Chair: F. Arbab †
Cochair: W. Eshuis ‡
Edited by: V. Akman ‡

Papers in this session were concerned with the design of VLSI components and systems.

Zimmer presented a paper whose premise is that the language of mathematics (specifically, category theory) is appropriate for expressing the intention and specification of a VLSI component. His view on design is that the initial specifications, expressed formally, are partial descriptions of the desired behavior. The result of the design process is a component whose behavior implies the initial specifications. Thus, design should be a sequence of refinements that takes the initial specifications to more detailed lower levels. Zimmer discussed a general algebra for composing modules into larger units. He regards communicating sequential processes (of Tony Hoare) and Petri nets as two instances of his algebra. Because Petri nets can be composed cleanly, he uses them to represent a design at its various stages of refinement. He sees his algebra as an appropriate formalism for modeling the process of design. For example, both refinement (top-down) and verification (bottom-up) transformations can be modeled as algebraic morphisms.

Genin presented a work whose primary application domain is automatic synthesis of synchronous circuits from high-level behavioral description. His premise is that designers approach design problems from a global point of view. Starting with assumptions, they build different worlds which may contain alternative solutions. They then investigate a subset of this set of worlds to check if they meet the specifications and are consistent. Each assumption is the origin of new hypothetical world. The worlds may not be disjoint and deductions made in a particular world may hold in other worlds as well. There are two types of worlds: architectural and temporal. Using a truth maintenance system and temporal reasoning the created worlds are related to each other. Surely, designers may create a multitude of alternative worlds. They nevertheless use common sense and domain knowledge to delimit their concerns.

† Computer Science Department, University of Southern California, SAL 200, MC 0782, Los Angeles, CA 90089, USA.
‡ Centre for Mathematics and Computer Science, Kruislaan 413, 1098 SJ Amsterdam, The Netherlands.

During the Q&A session, Akman asked Zimmer about the role of algebra: Is it just a notational tool? Zimmer emphasized that it is more than just a notation and helps in fact with the reasoning. Furthermore, the user does not know that there is a category-theoretic framework underlying his system. Arbab remarked that Zimmer's paper makes two proposals. First, Zimmer has a formalism to model design as successive refinement and second, he contends that his formalism is better than a silicon compiler approach. Arbab then wandered if Zimmer's formalism can also be used to build better silicon compilers. It turned out that Zimmer doubts the value of silicon compilers and thinks that they will face serious problems in the near future. His approach can deal with lower level optimizations that silicon compilers ignore or simply cannot handle.

Takala asked Genin about the inconsistency problems that arise within Genin's framework. Genin thinks that this is why there is a need for truth maintenance. Since creativity is a mixture of order and chaos, he first creates in his system some kind of confusion via assumptions and later confirms or denies them. When more information becomes available, verification of the assumptions induces a sense of order. Markov reminded Genin that mechanisms reminiscent of his system exist in Prolog and asked if he ever seriously considered Prolog. Genin replied by citing a major weakness of Prolog in regard to his intentions: Only one environment (world) exists at a time. Nadin found the user interface of Smalltalk (Genin's implementation environment) confusing and asked: Is this the kind of environment we are promising to our users? Both Genin and Bernus objected to this remark and vigorously supported the elegance and usefulness of the multiple windows, menus, pointing devices, etc., citing the well-known reasons for their importance.

Session 6

12. A Theoretical Model for Intelligent CAD

J. F. Koegel

Dept. of Mathematics and Computer Science, University of Denver
Denver, CO 80208, USA

and

Institutes for Information Processing Graz (IIG), Technical University of
Graz and Austrian Computer Society, A-8010 Graz, Austria

Abstract: *We present a model for future CAD systems which aid the designer by combining conventional CAD techniques with expert problem solving and design planning tools. Dialog with these tools is directed at the application level and all tools provide explanation. We describe how two models of the design process, the goal tree model of Mostow and problem solving by debugging almost right plans of Sussman and Steele, Jr., can be integrated to support both high and low level planning of engineering design. We suggest a set of paradigms for integrating the model.*

Keywords: CAD, knowledge-based systems, object-oriented systems, logic programming.

1. INTRODUCTION

The goal of current research in intelligent CAD systems is to further design automation by incorporating more of the designer's role into the system. This has two aspects:

1. A higher level of interaction between the designer and the design tool.
2. More powerful design problem solvers that are integrated into the design tool.

In order to achieve these goals, future CAD systems may include the following features:

1. More application-specific knowledge.
2. Design aids implemented symbolically as well as numerically.
3. Database interfaces that support both geometric data and design knowledge.
4. Extensive, graphically oriented simulation facilities.

One approach to increasing the intelligence level of the user interface is to increase the meaning of the symbols and commands provided by the interface. Rather than having the designer deal with objects at the graphics or drafting level (lines, curves, planes), the system should provide an interface that is appropriate for the given application. The architect deals with walls, plaster, floor plans, brick and timber. The

VLSI engineer deals with wires, transistors, and other electrical components. Each of these units of design have a conceptual as well as graphical meaning. Consequently, providing a more intelligent interface to the designer does not mean only providing appropriate graphical representations of these objects, but also making the system interact with the designer at this same conceptual level.

A second approach to improving the level of interaction of the system with the designer is found in those expert systems which not only emulate complex decision making but also can support their conclusions with explanations showing causal reasoning. This not only gives the user more confidence in the system but also allows interaction between the two at a higher level.

Special programs performing engineering analysis of designs are available in current CAD systems. In areas where optimization of a design is not easily quantified, either because of complexity or because of the inherent non-numeric nature (e.g. aesthetics), symbolically oriented systems could become a powerful design aid. Tools of this type could be implemented as expert systems for design. The knowledge bases for such systems could be constructed from the rules and heuristics used by experienced designers.

Many aspects of the design process would be candidates for this type of implementation.

1. Fundamental rules of design for the given application area could be incorporated into the system and applied either automatically or on demand.

2. A designer could specify a rough design and have the system refine it automatically.

3. A knowledge base of classic design examples could be used to automatically provide examples to the designer during the design process.

4. A tool might be designed to evaluate designs interactively or after a design is complete. It might give a textual appraisal of the design, listing both strengths and weaknesses.

If knowledge about the design and the design process is built into the system then it is possible to envision simulation facilities which make use of these tools. These facilities must deal with models that include both numeric and symbolic criteria.

There are many theoretical and methodological issues to be addressed here, including:

1. Representation of graphical objects and knowledge about these objects in a uniform and easily programmed manner.

2. Acquisition and representation of knowledge about the design process.

3. Selection of programming methodologies which are both efficient and suitable for symbolic and numeric processing. If more than one methodology is used, then their integration is a concern.

4. Program control, particularly the interactive application of design knowledge. Many production systems do not easily support interactive application of rules whereas interaction is critical in the CAD environment.

5. Balancing the system's knowledge with the user's knowledge so that the user feels aided and not controlled.

A model is proposed for evaluation with respect to the previous issues. The model combines conventional CAD techniques with expert problem solving and design planning tools and a powerful user interface. We suggest that the model can be integrated by applying the object-oriented paradigm to logic programming, data modeling, databases, and the user interface.

2. A MODEL OF A CONVENTIONAL CAD SYSTEM

Requicha (1980) presents a functional model of contemporary mechanical CAD systems shown in Fig. 12.1. In this model, the designer enters graphical application data using the geometric input system. This data is transformed into the geometric model for storage and manipulation by the system. The designer can also enter other application data using the non-geometric input system. This data is transformed into the non-geometric data model.

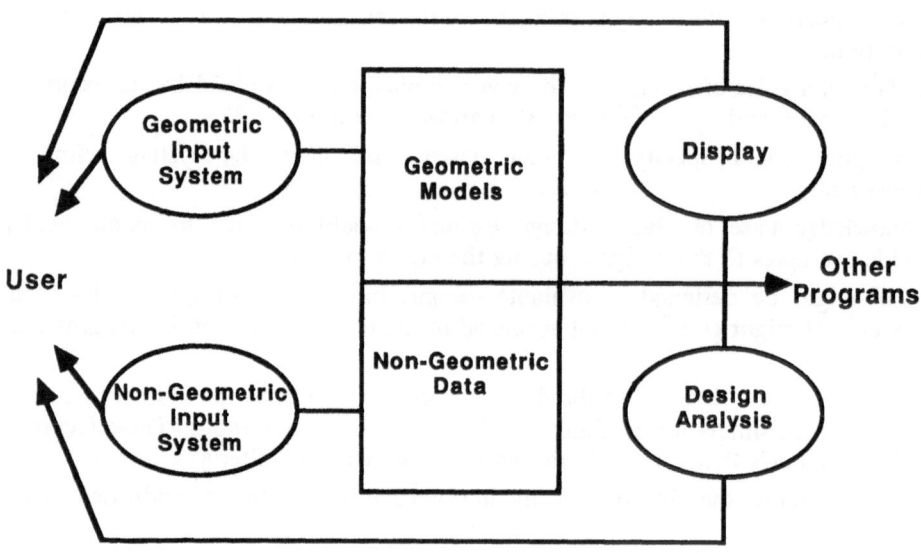

Fig. 12.1. A functional view of a mechanical CAD system (Requicha 1980)

The designer can evaluate the modeled data visually using the display interface. The designer can use specially written application software for design analysis. In conventional CAD systems, this analysis is almost entirely numerical. Those parts of the design process which cannot be evaluated quantitatively (either because quantitative measures are not available or because of complexity) must rely on the designer's expertise and use of heuristic knowledge and design experience.

The user interface in conventional CAD systems depends upon the application and the Geometric Modeling System (GMS) used. In general this interface is at the level of GMS rather than the application. For example, a GMS based on Constructive Solid Geometry (CSG) provides the designer with a set of primitive objects and operations for manipulating these objects. The objects are simple polyhedra such as cones, cylinders, and cubes. The operations are geometric in nature — for example, union and subtraction of solids. The designer constructs complex objects starting with the primitive objects and the operations. Some GMSs offer an interaction level that is relatively close to the application. For example, sweeping has a close analogy with many machining operations. However, in general the designer is required to interact at the level of GMS.

The designer may have access to a library of designs that have been built up over time by many designers. Reuse of these designs can significantly shorten the design cycle. Conventional design databases often use the relational data model for organizing part descriptions. The relational data model is not well suited for representing part hierarchies. Object-oriented data models provide a more natural representation for object relationships such as aggregation and generalization (Hardwick and Spooner 1987). These models can also support the automatic identification of reusable designs by providing a more suitable organization of engineering databases. nature of searching a large database.

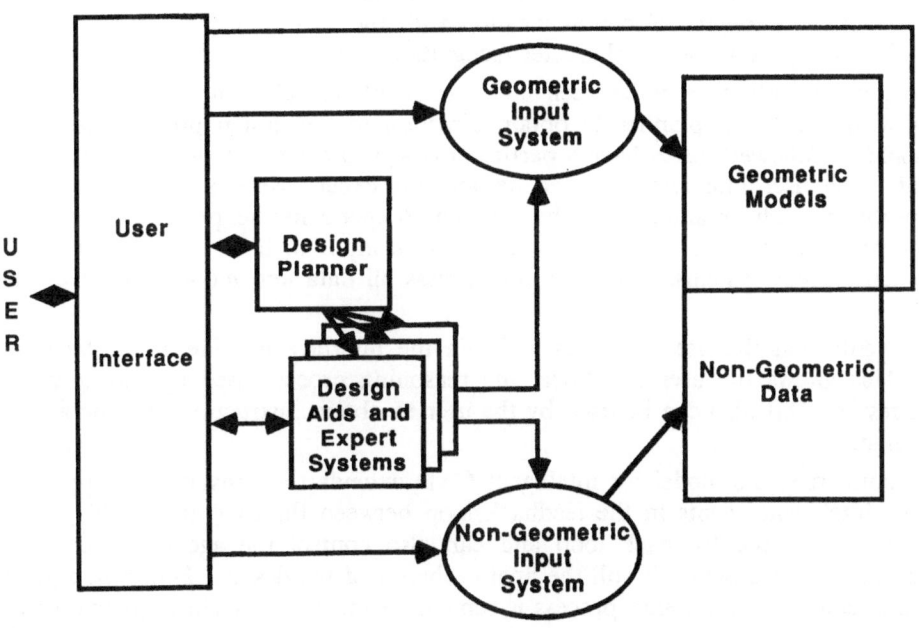

Fig. 12.2. A model for intelligent CAD

3. A MODEL FOR INTELLIGENT CAD

Figure 12.2 shows a model for intelligent CAD. The model contains some of the same elements found in Fig 12.1 but adds some components that make it significantly different. These are:

1. Design Planner: .The planner captures knowledge about the design process for use in guiding or evaluating the design activity. The planner is controlled interactively and makes plans from the current state of the design.

2. Design Aids/Expert Systems: These provide support to the designer for solving specific design problems. They can be used interactively or as batch procedures, but both types of interface must be supported in order that assistants/experts may be used both by the user and the planner. The may be application-specific (specifying tolerances on parts) or general purpose (a package for symbolic mathematics).

3. User Interface: The first difference between the two models is the greater number of interaction paths available to the user. The second is that the user interface uses application specific models to mediate all interactions with the other components of the system. This mediation is an important part of providing the right level of abstraction for the user. As part of controlling the level of interaction, the interface allows the user to specify how much feedback and guidance the system provides.

The user interface is reconfigurable both by the user and by the system. The interface should maximize the degree of communication with the user by exploiting techniques such as windowing, icons, online character recognition, etc.

The communication paths between the components are also important. The user interacts with the design planner to obtain direction in the design process (i.e. what steps should be followed) or to have a particular design procedure followed by the user evaluated. The user interacts with the design aids/expert systems to solve specific design problems. The planner uses the assistants/experts in the process of creating plans for design tasks. The aids/experts can provide input to both the geometric and non-geometric input systems. They can also access all data and models stored in the system.

All agents that the user interacts with via the interface provide an explanation facility. This allows the user to inspect the reasoning process used by the agent to obtain a result. This also can be used by the user to obtain instruction in some aspect of the design.

To summarize, the model for intelligent CAD extends the conventional approach by placing intelligent agents in the feedback loop between the user and GMS. Since the user is still in the feedback loop and can also control the agents through the interface, the human designer is still the final evaluator of the design. However, a great deal of the tedium of the design process is removed from the user since qualitative as well as quantitative decision making can be performed by the agents. The approach has the additional advantage that the user can use the system for training. It is also conceivable that the system could have a learning or knowledge acquisition mode where it would extend its knowledge base by observing the designer and evaluating the design library.

4. DESIGN PLANNER

4.1. Two Models of Design

The design planner embodies knowledge of the design process which provides the organizational framework for interaction between the designer and system. In this section we discuss two models of the design process and then describe how these might be mapped to a conventional planner.

Problem solving by debugging almost right plans (PSBDARP) (Sussman 1975; Sussman and Steele, Jr. 1980): In this model the designer starts the process by forming a high-level plan of the design. This plan partitions the design into subparts which can be dealt with somewhat independently. As the design proceeds, each subpart may be further partitioned until primitive elements are reached. The identification of the primitive elements and the specification of their parameters is dealt with iteratively using constraint propagation.

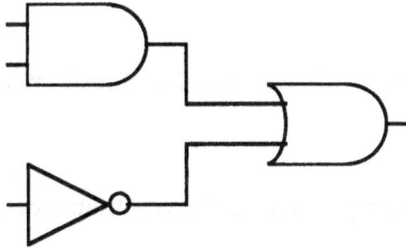

Fig. 12.3(a). Simple sum of products circuit

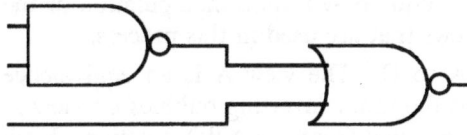

Fig. 12.3(b). Equivalent circuit with reduced chip and gate count

The constraints model the functional behavior of the set of elements currently under consideration. This locality is used to reduce the complexity of the design. Since a given set of constraints may not account for certain global effects of interest, different *views* can be attached to the design. The views allow alternate sets of equivalent

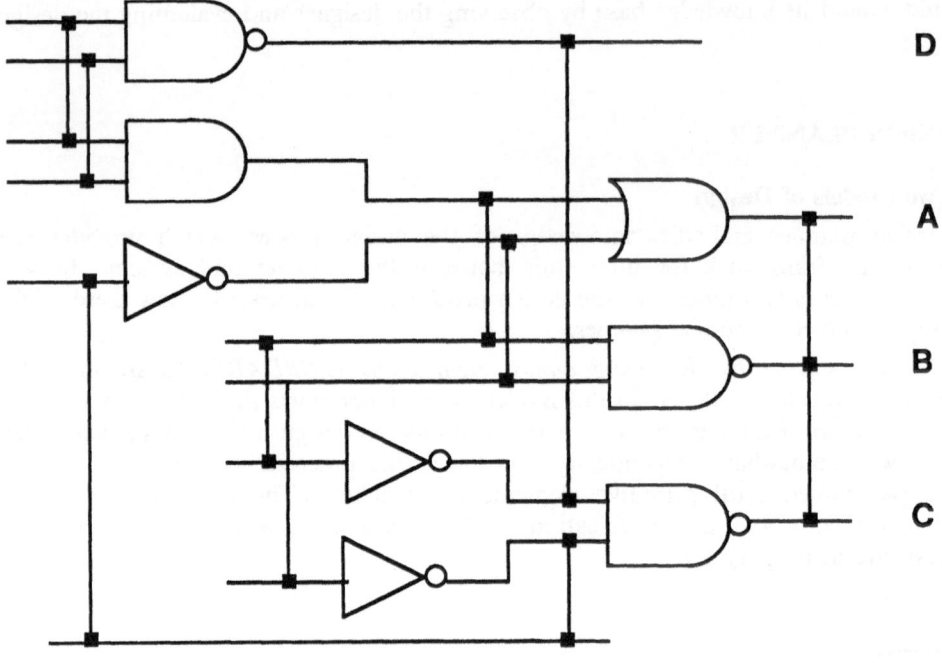

Fig. 12.3(c). Multiple views with equivalent constraints which are needed to transform circuit from AND-OR to NAND-NAND logic

constraints to be used as needed. In addition, the views as well as the constraints can be manipulated at different stages in the design.

Figure 12.3 shows an example from digital design of how multiple views of an object can be used to transform the object to a desired state. Figure 12.3(a) shows a simple sum of products circuit consisting of three different types of gates. It is desirable to transform this circuit to that shown in Fig. 12.3(b) which has a reduced chip and gate count. One way this might be done is by algebraic manipulation. However the designer can use a much easier technique based upon manipulation of the schematic. Figure 12.3(c) shows the different views that are used in this process.

Figure 12.3(c) shows four views, labeled A to D. The view A is an equivalence based on de Morgan's theorem. View B equates the input inverting bubbles on view A with explicit inverter gates so that they can be separated from the AND function. View C equates two serial inverters with a single wire (ignoring delay). (Notice that this view is able to reference a component within view B. This has implications on how views are modeled. This will be discussed later.) View D equates the AND gate and INVERTER gate with a NAND gate.

Of course having these views is just a part of the solution. One must know which views to use, in what order to apply them, and how to demonstrate their equivalences. These issues must be resolved at a higher level in the design process.

The premise of this model is that expert designers *know the form of the answer* when they begin a design. The design then involves manipulation of these forms until the problem constraints are satisfied. We believe that this is a suitable model for many areas of engineering design.

Goal tree model of design (Mostow 1985): Transformational models of design describe the design process as a series of transformations applied to the initial problem statement until the final design is achieved. The purpose of the goal tree model is to make more of the design process explicit, so that it can be reasoned about, verified, documented, justified, explained, etc. Those areas of design that it attempts to make explicit include:

1. State of the design.

2. Goal structure of design process.

3. Design decisions.

4. Rationales for design decisions.

5. Control of the design process.

Design is then modeled as a goal tree where each level of the tree consists of subgoals which provide more detail regarding the actions the designer wants to perform. The leaves of the tree are the primitive goals or transformations to be applied to the problem specification in order to produce the design. By making the goal structure explicit it is possible to abstract the design process to very high level planning. This would enable the system to aid the designer when the form of the design is not already known.

4.2. Evaluation

Perhaps the strongest case for the PSBDARP model is that it is a methodology used by expert designers in engineering disciplines. We can also make the following observations from the perspective of machine implementation:

- *Dynamic use of views*: Views, however domain-specific, can be manipulated dynamically. This flexibility is critical for problem solving, and requires that the underlying data model be able to support this dynamic definition and use of views.

- *Support for design reuse*: One area that needs to be explored is how views can be manipulated to support *reuse of designs* as a design criteria. For example, a certain view may correspond to a design in the database while other views may not. We could introduce reuse as a criteria in the application of views.

- *Design history*: Recording of the design steps could be useful for documentation, maintenance, redesign, and machine learning (Mostow 1985). The views and use of constraints could be part of this record.

- *Simulation*: Evaluating the performance of a design can be an important part of the design process. Constraints provide the functional definition needed for simulation.

In the goal tree model specifying goals explicitly allows more of the design process to be made visible to the system and therefore to the user. It also allows the system to interact with the user at a higher level of planning. On the other hand this model does not specify the problem solving approach(es) that the goals will direct.

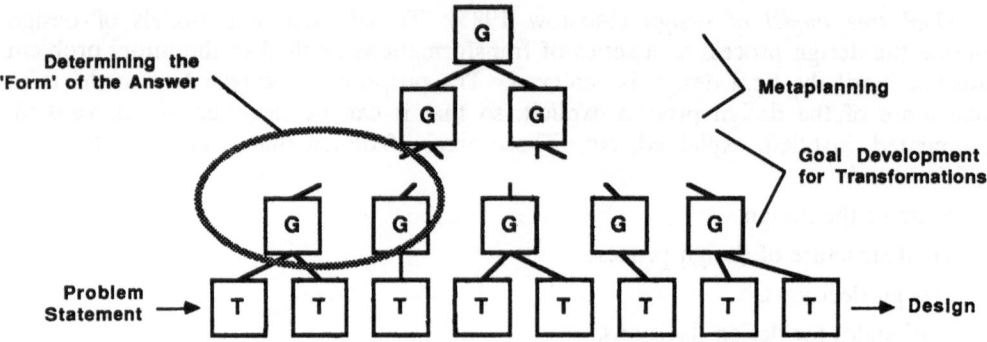

Fig. 12.4. Goal tree for goal driven problem solving by debugging almost right plans

We therefore see the two models as being complementary. The two approaches can be combined into one model which we call Goal Driven PSBDARP (GDP). We illustrate this model by means of the tree shown in Fig. 12.4. The top level of the tree contains the most abstract goals which represent the designer's general thoughts about design (here called metaplanning). The lower layers of the tree contain successively more concrete goals which also become more and more bound by the problem solving approach implicit in the transformations to be applied. The leaves of the tree are again the transformations to be performed. This model also makes explicit the process by which the designer determines the form of the answer and the transformations of the problem statement that produce this form.

4.3. GDP using a Hierarchical Planner

A hierarchical planner operates in a task space that is organized according to levels of abstraction. This permits the planner to deal with general and presumably important issues before getting enmeshed in details. Examples of hierarchical planners are ABSTRIPS (Sacerdoti 1974) and NOAH (Sacerdoti 1975). Since GDP is organized as an explicit goal tree with different layers of abstraction, the use of a hierarchical planner to implement GDP seems natural. We identify two important issues:

1. *Goal interaction*: Subproblems may interact. HACKER (Sussman 1975), a non-hierarchical planner, attempts to solve this problem by backtracking and patching the solution. NOAH uses critics and a strategy of least commitment.

2. *Use of domain knowledge*: How to make planning as domain independent as possible so that domain knowledge can be added modularly? One approach is to use plan generators which produce plans for specific tasks identified by the planner. The plan generators encapsulate the domain specific knowledge needed to evaluate a task. An additional layering could be used to have the plan

generators make use of the expert systems and design aids. This reduces overlap between the planner and the expert systems and may permit more interaction with the designer during the design process.

5. DESIGN AIDS/EXPERT SYSTEMS

Expert systems for engineering and CAD have been demonstrated for a variety of different kinds of tasks (Faught 1986). These systems have a number of common characteristics:

- Expert level problem solving skills in a narrow and well-defined area.
- Explanation of results and decisions.
- Use of hybrid representation techniques.

There are also *design aids* which may not be described as having an expert level of performance but which relieve the designer of some otherwise manual functions. These also must provide explanation. Both expert systems and design aids are self-contained and do not depend on other tools. The expert systems have separate knowledge bases.

These tools must be integrated with the system so that both the design planner and user interface can make use of them. This requires:

1. Separation of problem solving from input/output since the input/output needs of the user interface and the design planner differ.
2. Protocols to support interaction with both user interface and design planners.
3. An event driven design since interactions with the design planner and user interface may overlap.
4. Interfaces such that all tools have access to design and application data as needed.

We expect tools which support the design planner (e.g. for suggesting different views or directing use of constraints) to be available directly to the designer as well.

The system must also be extensible in that it should gracefully accept the removal and addition of tools. Well-defined protocols for interfacing tools with the system is one aspect of this. The system must know whether a tool exists which solves a particular function. When such a function is not available, one default is to direct the problem to the designer.

6. USER INTERFACE

The user interface mediates all interactions between designer and the system. It includes the following features:

1. Facilities for the designer to enter in a top level problem statement. The format of this statement could be natural language, key words, or a specification language.
2. User-controlled windows for each tool so that the designer can interact with tools in parallel.
3. Access to online reference materials (e.g. design handbooks) and the design databases with rapid indexing to topics and designs that are relevant.

A key characteristic of the user interface is that it is directed at the application level. Part of this will be obtained by the tools that are available (i.e. problem solvers for application specific problems). Additionally, the interface must be able to associate certain properties with the geometric objects the designer creates. These properties must be defined so that the designer can interact with the objects at the application level and obtain meaningful and realistic feedback. These properties also control how objects interact.

One way to specify object properties that would be useful for simulation and would also integrate with the design planner is to define the properties as constraints. A difficulty in associating properties with objects is that many CAD systems use modeling techniques which are convenient for geometric purposes but have no functional relationship with the object being defined. The system could infer these properties by identifying the object type from the problem statement. A library of primitive objects and their properties might be used for this inference.

7. PARADIGMS FOR AN INTEGRATED MODEL

7.1. Issues and Requirements for an Integrated Approach

In order to insure that the preceding model is integrable, we must consider the following:

1. Geometric models for GMS: There are a number of viable GMS alternatives each having certain advantages and disadvantages depending on the application. Not only must the GMS be appropriate for the application, but its geometric model should be integrated with the data models used in the rest of the system.

2. Data models for efficient storage and access of CAD data: CAD data are characterized by their large volume. In addition, conventional DBMS data models have been found insufficient for providing the needed support for CAD. Some of the issues include support for views, versions, and configurations.

3. Knowledge base models for organizing and controlling the knowledge used by the planner and design aids/experts: It is expected that the agents for intelligent CAD will require large knowledge bases. In addition, currently the integration of multiple expert systems is difficult.

Integration of the components of the intelligent CAD model depends upon the data models employed. Each of the three areas listed above has specific needs regarding representation of data or knowledge. In addition, the user interface will also need an appropriate data model. A modeling approach is required that will allow these different components to integrate, particularly when the model is scaled up to large applications.

A basic modeling approach that could unify these different needs has been available in object-oriented programming languages. More recently this approach has been applied to data modeling, databases and user interfaces. We propose the use of object-oriented logic programming, data modeling, and user interface management as a way to satisfy the integration problem as outlined in the following sections.

7.2. Object-Oriented Logic Programming

Logic programming as a powerful tool for symbolic processing has been described by Kowalski (1979). Currently the most practical implementations are limited to first-order Horn clause forms. Logic programming in general and Prolog in particular have been compared with other approaches to symbolic programming. Additionally, Prolog as a basis for graphics design has been discussed (Ogawa et al. 1984; Pereira 1983). Logic programming has a distinct implementation of pattern matching and uses deductive inference as its computational mechanism. It has been used successfully to implement planners and expert systems, key parts of our model.

Object-oriented programming has also received a great deal of attention due to its intuitive style of programming and utility in structuring a software system. A number of researchers have combined this paradigm with various of programming languages including Prolog (Chikayama 1984; Gallaire 1985; Zaniolo 1984). This has the additional advantage of adding modularity to the otherwise unstructured style of logic programs.

We have extended this concept by developing the language POOL (Parallel Object-Oriented Logic) where parallelism is obtained by message passing between concurrently executing objects and objects are synchronized by using monitor-like constructs (Koegel 1987b; Koegel 1987c). Object methods are first-order Horn clauses executed sequentially using backtracking in a depth-first manner. We believe that this language retains the advantages of object-oriented logic programming listed previously while providing the additional benefit of parallelism — an essential part of making future CAD systems practical to implement. In comparison with committed-choice non-deterministic languages (the predominant approach to parallel logic programming today) we believe that the POOL model offers a more natural structuring and synchronization style and an abstract machine more suited to conventional multiprocessor architectures.

7.3. Object-Oriented Data Models

Researchers studying the data modeling needs for CAD databases have identified the following characteristics as being essential (see, for example, (Katz, Anwarrudin, and Chang 1986)):

1. Aggregation: Representing composite objects as a collection of their components.

2. Version Control: Supporting multiple versions of functionally equivalent objects.

3. Multiple Views: Representing an object from a number of different design perspectives.

The molecular object model developed by Batory et al. (Batory and Buchmann 1984; Batory and Kim 1985; Buchman and Perez de Celis 1985; Chou and Kim 1986) supports the first two characteristics. In addition it defines a mapping to the relational data model which we believe is of practical significance. We have compared this model to a number of other object-oriented data models and found that other models either: 1) didn't provide for version control, 2) didn't explicitly map to the relational data model, or 3) were specialized to a specific application such as VLSI CAD (Afsarmanesh et al. 1985; Ketabchi, Berzins, and March 1986; Koegel 1987a).

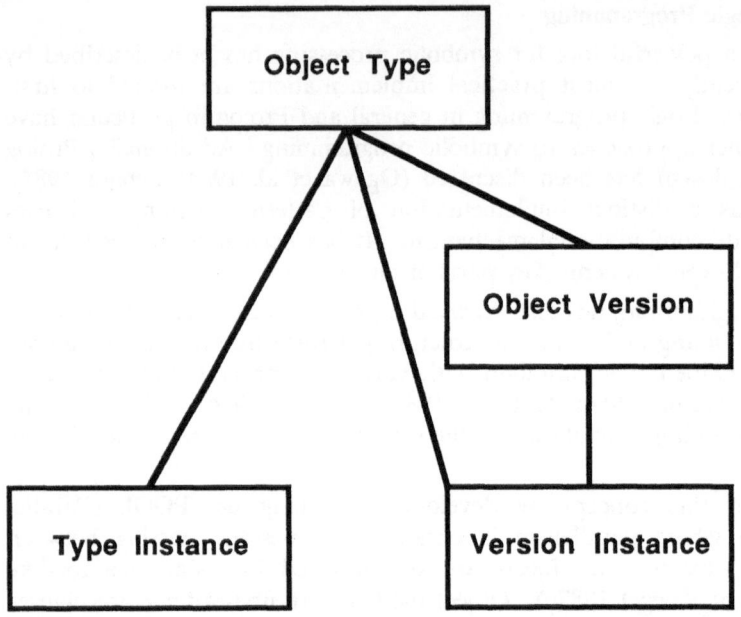

Fig. 12.5. Object type, version, and instance attribute inheritance in the molecular model (Batory and Buchmann 1984; Batory and Kim 1985)

Figure 12.5 shows the inheritance hierarchy for the molecular data model. The *type* defines those functional characteristics that are unique to all instances and versions of the type. There can be zero or more *versions* which vary by some non-functional attributes. Both types and versions can have *instances* which represent the actual design artifacts. The types, versions, and instances are connected in a strict inheritance (no default overriding) hierarchy in order to preserve semantic integrity in the database. This does not preclude use of a lattice hierarchy to link objects for other purposes.

Figure 12.6 shows a modified Entity-Relationship (E-R) diagram of a molecular object (Batory and Kim 1985). This diagram can be used to derive the mapping of an object to the relational data model (Batory and Kim 1985; Koegel 1987a). Figure 12.7 shows an example of type, version, and instance for molecular objects implemented in POOL (Koegel 1987a). This example defines objects geometrically. However, other object attributes can be described by adding the appropriate predicates to the object state.

The molecular object model does not support views, the third desired characteristic cited previously. In order to support views we should identify that two object descriptions are equivalent. This can be done as in (Sussman and Steele, Jr. 1980) by adding an equivalence predicate that operates on two objects: *eq(object1, object2)*. These two objects must be of the same type to be equivalent. Once defined as equivalent, they can be treated as different views of the same element.

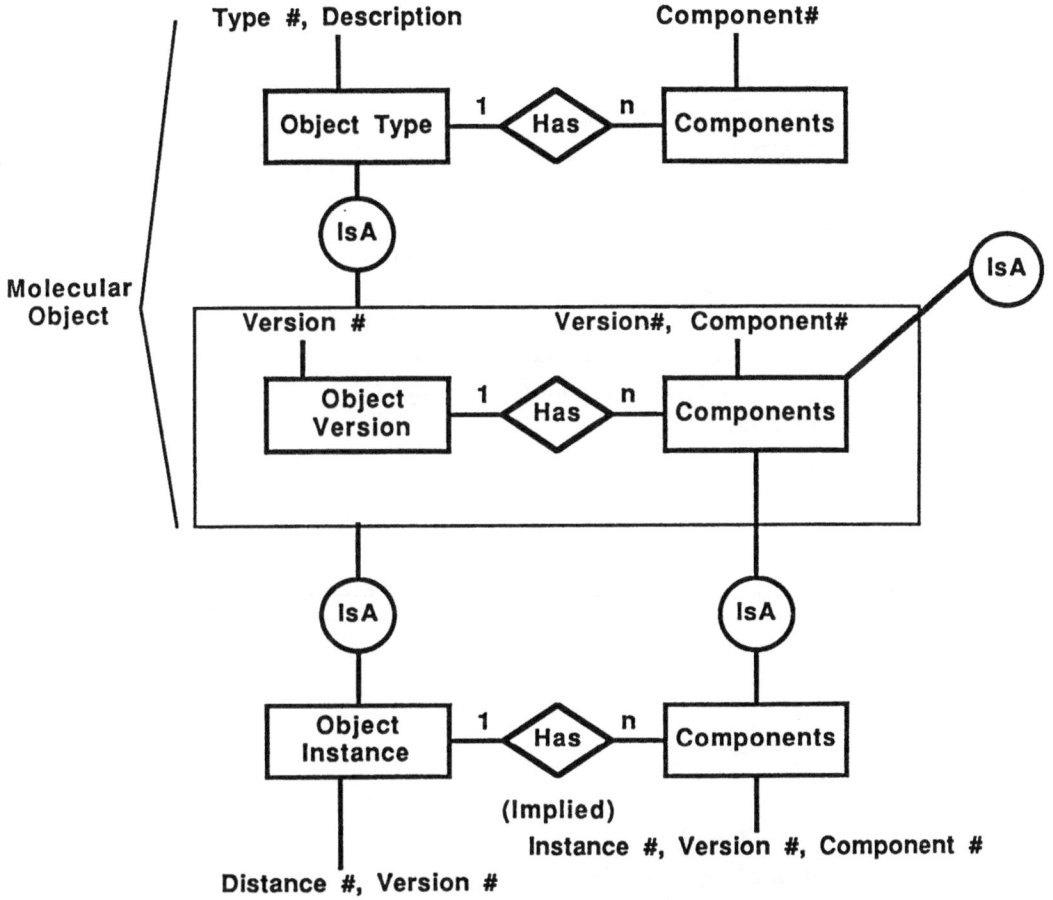

Fig. 12.6. Modified E-R diagram for molecular modeling (Koegel 1987a)

7.4. Object-Oriented User Interfaces

The object-oriented paradigm has been used to structure a user interface to provide modularity and extensibility (Sibert, Hurley, and Blesser 1986). However we believe that this paradigm can be the basis for adding the object properties mentioned in Section 6. In particular, the properties can be defined as object methods. If the methods are defined so that inputs and outputs are invertible, then the type of manipulation used for propagation of constraints can be performed.

7.5. Integration

The basis for integration of the intelligent CAD model is the use of the object-oriented paradigm. We have described how this paradigm can be used in symbolic

```
molecule[

        gate_4_and,
        metaclass(molecule),
        supers([]),
        state,
                type(1),
                has_part(terminal, c1, [input]),
                has_part(terminal, c2, [input]),
                has_part(terminal, c3, [input]),
                has_part(terminal, c4, [input]),
                has_part(terminal, c5, [output]),
        methods
           ...
].
```

Fig. 12.7(a). 4-input AND gate object type definition

```
molecule[

        gate_4_and,
        metaclass(molecule),
        supers([gate_4_and(type, 1)]),
        state,
                version(1),
                has_part(gate_2_and, c1, []),
                has_part(gate_2_and, c2, []),
                has_part(gate_2_and, c3, []),
                has_part(wire, c4, [gate_2_and(c1), terminal(c3),
                        gate_2_and(c3), terminal(c1)]),
                has_part(wire, c5, [gate_2_and(c2), terminal(c3),
                        gate_2_and(c3), terminal(c2)]),
        methods
           ...
].
```

Fig. 12.7(b). 4-input AND version based on 2-input AND type

programming, data modeling, and the user interface. The integration of these different areas with the object-oriented paradigm depends on the definition of protocols for message-passing. Object implementation is otherwise transparent, giving the implementation flexibility. Within different parts of the system different classes of objects will be needed to tailor objects and object hierarchies to different functions.

8. SUMMARY

We have presented a model for intelligent CAD based on combining conventional CAD techniques with tools for design planning and expert problem solving. We have described how two existing models of the design process, the goal tree model and problem solving by debugging almost right plans, could be integrated to model both low and high level planning of engineering design. The combined model can then be mapped to a hierarchical planner.

Expert systems and other design aids fit naturally into the model, provided certain structuring guidelines are followed and protocols are devised. Three accomplishments of the model in this area are: 1) integration of the planning function with the expert systems/design aids, 2) extensibility of the system by allowing addition and removal of expert systems/design aids, and 3) support for high interaction of the designer with the tools.

We have identified the need for the user interface to be directed at the application level. Object properties need to be defined and accessible at the user interface. An object-oriented approach to representing objects could be used here. If the object methods are implemented as constraints, the interface could assist in simulation.

We have described how an object-oriented approach could be the basis for integration of the components of the system. To this end we described how this paradigm applied to parallel logic programming, data modeling, and user interface management. Final integration depends upon structuring the objects and object hierarchies appropriately for the system and defining protocols for message passing between these objects.

ACKNOWLEDGEMENT

The author would like to acknowledge the assistance of Dr. Herman Maurer of Technical University of Graz.

REFERENCES

Afsarmanesh, H., McLeod, D., Knapp, D., and Parker, A. (August 1985): "An extensible object-oriented approach to databases for VLSI/CAD," *11th International Conference on Very Large Data Bases*, Stockholm, pp. 13-24.

Batory, D. S. and Buchmann, A. P. (August 1984): "Molecular objects, abstract data types, and data models: A framework," *Proceedings of the 10th International Conference on Very Large Data Bases*, Singapore, pp. 172-184.

Batory, D. S. and Kim, W. (September 1985): "Modeling concepts for VLSI CAD objects," *ACM Transactions on Database Systems*, **10**(3), pp. 322-346.

Buchman, A. P. and Perez de Celis, C. (1985): "An architecture and data model for CAD databases," *Proceedings of the 11th International Conference on Very Large Data Bases*, Stockholm, pp. 105-114.

Chikayama, T. (November 1984): "Unique features of ESP," *Proceedings of International Conference on Fifth Generation Computer Systems 1984*, Tokyo, pp. 292-298.

Chou, H. T. and Kim, W. (1986): "A unifying framework for version control in a CAD environment," *Proceedings of the 12th International Conference on Very Large Data Bases*, Kyoto, Japan, pp. 336-349.

Faught, W. S. (July 1986): "Applications of AI in engineering," *Computer (Special Issue on Expert Systems in Engineering)*, **19**(7), pp. 17-27.

Gallaire, H. (July 1985): "Logic programming: Further developments," in *1985 Symposium on Logic Programming*, pp. 88-96.

Hardwick, M. and Spooner, D. L. (March 1987): "Comparisons of some datamodels for engineering objects," *IEEE Computer Graphics and Applications*, **7**(3), pp. 56-66.

Katz, R. H., Anwarrudin, M., and Chang, E. (1986): "Organizing a design database across time," in *On Knowledge Base Management Systems*, Brodie, M. L. and Mylopoulos, J. (eds.), Springer-Verlag, Berlin, Heidelberg, New York, Tokyo, pp. 287-295.

Ketabchi, M. A., Berzins, V., and March, S. (1986): "ODM: An object-oriented data model for design databases," *1986 ACM Computer Science Conference*, Cincinnati, OH, USA, pp. 261-269.

Koegel, J. (March 1987): "Molecular Objects in POOL," Technical Report No. MS-R-8702, University of Denver, Department of Computer Science.

Koegel, J. (June 1987): "POOL: Parallel object-oriented logic," *Proceedings of the 1987 Rocky Mountain AI Conference*, Boulder, CO, USA.

Koegel, J. (January 1987): "POOL: Parallel object-oriented logic," Technical Report No. MS-R-8701, University of Denver, Department of Computer Science, Denver, CO, USA.

Kowalski, R. (1979): *Logic for Problem Solving (Artificial Intelligence Series 7)*, North-Holland, Amsterdam.

Mostow, J. (Spring 1985): "Towards Better Models of the Design Process," *AI Magazine*, **6**(1), pp. 44-56.

Ogawa, Y., Shima, K., Sugawara, T., and Takagi, S. (November 1984): "Knowledge Representation and INference Environment: KRINE, — An Approach to the Integration of Frames, Prolog, and Graphics," *Proceedings of International Conference on Fifth Generation Computer Systems 1984*, Tokyo, pp. 643-651.

Pereira, F. C. (June 1983): "Can Drawings Be Liberated from the von Neumann Style?," Technical Note No. 282, SRI International, Palo Alto, CA, USA.

Requicha, A. A. G. (1980): "Representation of Rigid Solid Objects," in *Computer Aided Design — Modelling, Systems Engineering, CAD-Systems (Lecture Notes in Computer Science)* **89**, Encarnacao, J. (ed.), Springer-Verlag, Berlin, Heidelberg, New York, Tokyo, pp. 2-78.

Sacerdoti, E. D. (1974): "Planning in a hierarchy of abstraction spaces," *Artificial Intelligence*, **5** , pp. 115-135.

Sacerdoti, E. D. (1975): "A Structure for Plans and Behavior," Technical Note No. 109, SRI International.

Sibert, J., Hurley, W., and Blesser, T. (1986): "An object-oriented user interface management system," *ACM Computer Graphics (SIGGRAPH '86)*, **20**(4), pp. 259-269.

Sussman, G. J. (1975): *A Computer Model of Skill Acquisition (Artificial Intelligence Series 1)*, North-Holland, Amsterdam.

Sussman, G. J. and Steele, Jr., G. L. (1980): "Constraints — A language for expressing almost-hierarchical descriptions," *Artificial Intelligence*, **14** , pp. 1-39.

Zaniolo, C. (February 1984): "Object-oriented programming in prolog," *1984 International Symposium on Logic Programming*, Atlantic City, NJ, USA, pp. 265-271.

13. Systematic Design in Intelligent CAD Systems

J. Duhovnik

Faculty for Mechanical Engineering, University Edvarda Kardelja
Murnikova 2, 61000 Ljubljana, YUGOSLAVIA

Abstract: *The paper describes design phases and types of design. Design phases define the function and functionality related to graphics presentation. In a design process four characteristic graphic presentations are produced; a scheme, a rough drawing, a drawing (project drawing), and a manufacturing drawing. Four design modes define the designing activities; new design, innovative design, variation design and adaptation design. Connecting the design phases with the design modes makes systematic design possible.*

Keywords: CAD, knowledge engineering, systematic design.

1. INTRODUCTION

CAD systems supporting the design process consist of separate activities and cannot solve problems in complicated applications. It is the knowledge of the design process and the processing systematics installed into CAD systems that will fundamentally increase their application. An interaction between the function and the shape is present during the whole process of developing a new product. By means of iteration, we try to achieve the optimum of both conditions in order to satisfy the economical purposes.

The paper represents the design systematics which has to be installed into intelligent CAD systems. The investigations carried out are a part of the research project within the frame of the RSS (Research Council of Slovenia) and have been applied and tested in industry.

2. FUNCTION AND FUNCTIONALITY

A designer unites elements and parts into a certain assembly in such a way that the fulfillment of a certain function is enabled. A function is represented by a natural phenomenon. A natural phenomenon can be physical, chemical, physiological, etc. A

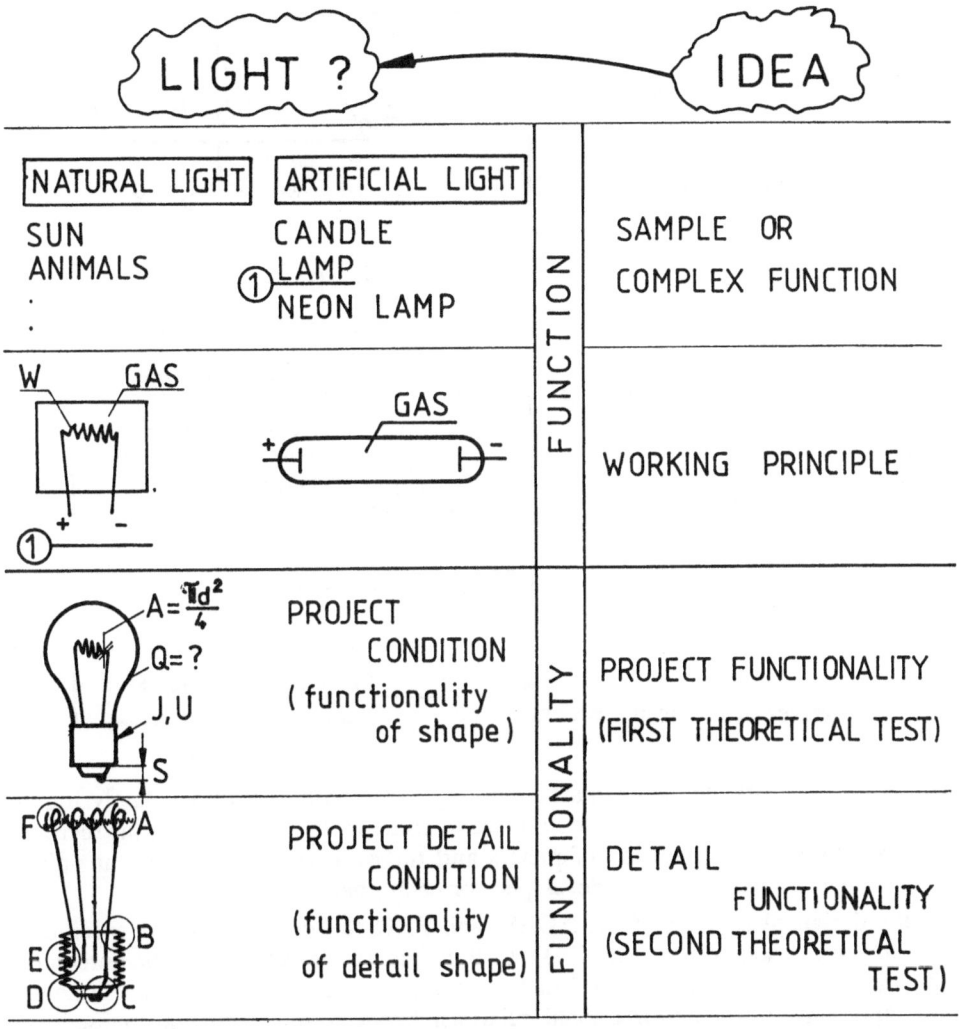

NATURAL LIGHT	**ARTIFICIAL LIGHT**			
SUN ANIMALS	CANDLE ① LAMP NEON LAMP	F U N C T I O N	SAMPLE OR COMPLEX FUNCTION	

Fig. 13.1. Correlation between function and functionality from natural phenomena to manufacturing drawing

These connections can be made intuitively or by means of an expert system for a definite phase of dimensioning. The design process runs according to a certain sequence of activities (Duhovnik, Kimura, and Sata 1983; Koller 1985; VDI 1979). This sequence of activities enables that the idea about a new design is developed until the phase of the ready-made technical documentation and the prototype. According to Duhovnik et. al. (Duhovnik, Kimura, and Sata 1983), the characteristic sequence of activities can be grouped into the following main phases TASK, RESEARCH and DEVELOPMENT (RESDEV - phase) and project (PRO - phase) as shown in Fig. 13.2. According to our definition of the activity, it is important that each phase should

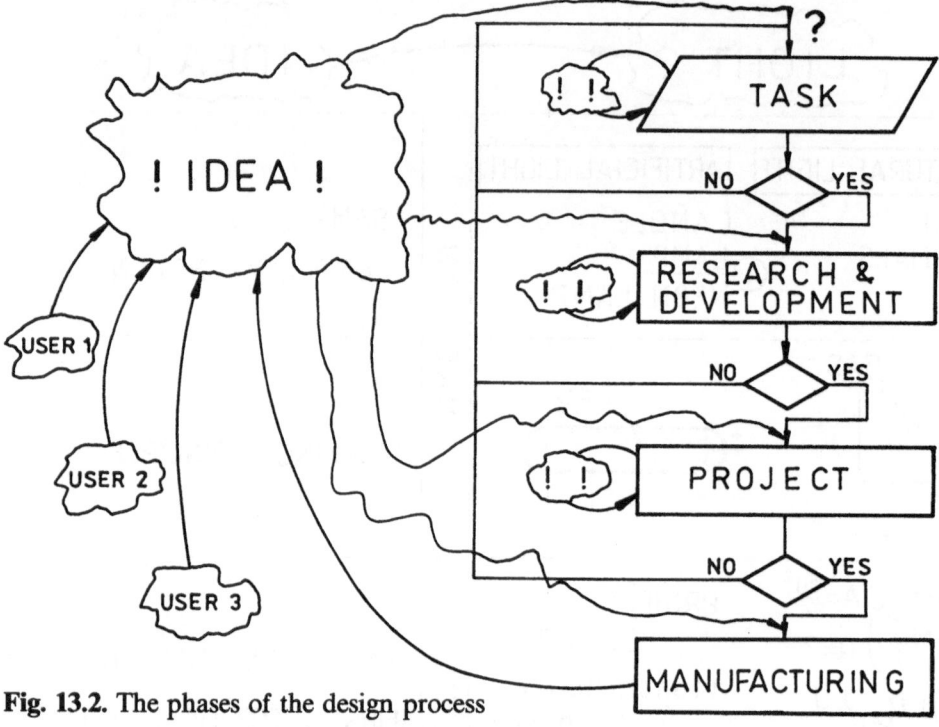

Fig. 13.2. The phases of the design process

enable a full creativity which is marked in Fig. 13.2 with the idea. Analyzing the possible variations of the expert system, we have come to the conclusion that there is a connection between the separation of phases and the estimation; therefore our opinion is that the expert system has to be developed for each phase separately.

In the design phases TASK, RESDEV, and PRO the jobs have to be specified in detail. However, each job in the CAD system has to be related to the interaction between the function and the shape. In the analysis of the design process (Duhovnik, Kimura, and Sata 1983) unidirectional relations between the particular phases were found out (Fig. 13.3).

The connection between the particular phases can be very intensive (Duhovnik and Matičič 1985). It is very important that the graphical presentation takes place always after the determination of the function or functionality. The conditions defining the shape (rough or detailed) are obtained from the criteria of function and functionality. For each criterion either the allowable or maximum efficiency value (stress, pressure, etc.) is known. On a mechanical model the designer finds out the actual efficiency values and compares them with the allowable ones. In determination of the shape the criterion for each parameter is equal to a unit and has the value 1.0. In general we have to take into account that the shape is presented in 3D-space. For the presentation of the shape in detail, cross sections of full bodies are worked out. The lower is the level of the working phase in the design process the more precise is the shape. On the lowest level in adaptation design the graphics presentation is carried out in 2D space

natural phenomenon having a certain material effect can be explained or repeated by a mechanical assembly. A natural phenomenon is presented in a concrete shape of a mechanical assembly in order to recognize the process. Any process consists of material, energy, and information. With a mechanical assembly a designer gives a concrete shape to a function whose aim is to describe a natural phenomenon.

In the description of a function by means of a mechanical assembly a certain working principle has to be followed. This working principle defines the functional relationship between phenomena which are presented mechanically, giving the function a material shape. For the same function one or several working principles can be applied. The notion of a function is defined by the natural phenomenon and the working principle specially worked out for it.

The working principle enables the recognition of physical phenomena (mechanics, optics, electricity, nuclear physics, etc.). For any physical phenomenon the engineering methods for the determination of its rough shape are known. It is the determination of the rough shape by engineering methods which represents the first step towards the assessment of the functionality of a product.

The described procedure of the determination of shape is usually called product projecting, but in order to be able to define the design process more precisely this activity was called functionality projecting. The projecting of functionality is in fact the first theoretical testing of the functionality of a product. A rough shape obtained at the first functionality test is already so clear that it can be considered as a classical geometric CAD model. The shape projected cannot be realized yet because the manufacturing details (i.e., the manufacturing functionality) are not yet known.

The manufacturing functionality contains the conditions of the technical shape for manufacturing, assembly, control, maintenance, application, etc. (Fig. 13.1). For both function and functionality we have to design a model, i.e. a product, which is graphically presented on the screen.

3. PHASES AND THEIR SIGNIFICATIONS IN CAD

The function of a product is defined on the grounds of a goal and a working principle (Koller 1985). An objective working principle gives a schematic model which represents a rough technical shape. This has to be tested for its functionality which is defined by computations. The shapes of a functional product which is tested are the technical shape and architectural one. The technical shape is defined by manufacturing, assembly, exploitation, etc. According to the present conception, the architectural form represents the industrial design. The result of a design process is a shape that is defined in all details and tested for its functionality (Duhovnik, Kimura, and Sata 1983; Duhovnik and Matičič 1985; VDI 1979).

. Based on such a conception of the design process, we can distinguish the working principle of a function and its functionality. A similar classification exists for shapes. We distinguish a schematic model and a product with a technical shape and an architectural form. By means of the sequence dimensioning of the product we can find out that there are more working principles corresponding to a function, and more schematic models for a working principle, due to the technical conditions of presentation.

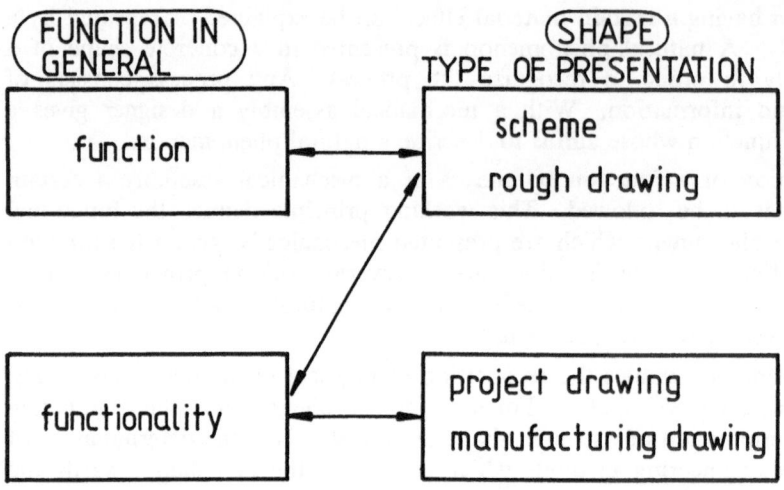

Fig. 13.3. Correlation between function and shape

3.1. Connection of the Function with Scheme or Rough Drawing

For each function we have to define the working principle which is not necessarily a singular one. In the conventional design process this connection is not defined algorithmically but intuitively. With an expert system on this level having a suitable mechanism of logics and a larger database for the working principles, we will obtain the solution of function and their possibility as a result of logical operations of expert systems. Each working principle can be described by a scheme. A scheme represents the first graphic solution of the task. However its shape as an object is undefined. The relationships of the basic dimensions can be different and thus also the function of the element (Fig. 13.4). Taking a simple example, we can see that for certain conditions of a shape, the working can have only one function and for certain other conditions of a shape, the working can be multi-functional (multi-purpose).

In the analysis of the geometric parameters defining a three-membered lever we get a smaller number of changeable parameters. With the reduction of the first function (angle) only functions of a and b remain. In this case two parameters defining the characteristics of the lever remain. Analyzing the functional relationship, we can find out that, under the consideration of the actual values of the function parameters (Duhovnik et al. 1986), the main parameter of the geometry of the shape can be defined. In this phase we thus obtain a rough drawing where the connection is defined by a functional matrix. The shape matrix represents the individual characteristic parameters of a certain shape which can be expressed by a vector. The characteristic parameters are defined by means of the shape matrix. A geometric modeller usually operates with rough primitives which are interesting especially for the assessment of the shape but are of little help in the determination of the shape to be machined. Rough primitives are cylinder, cube, etc.

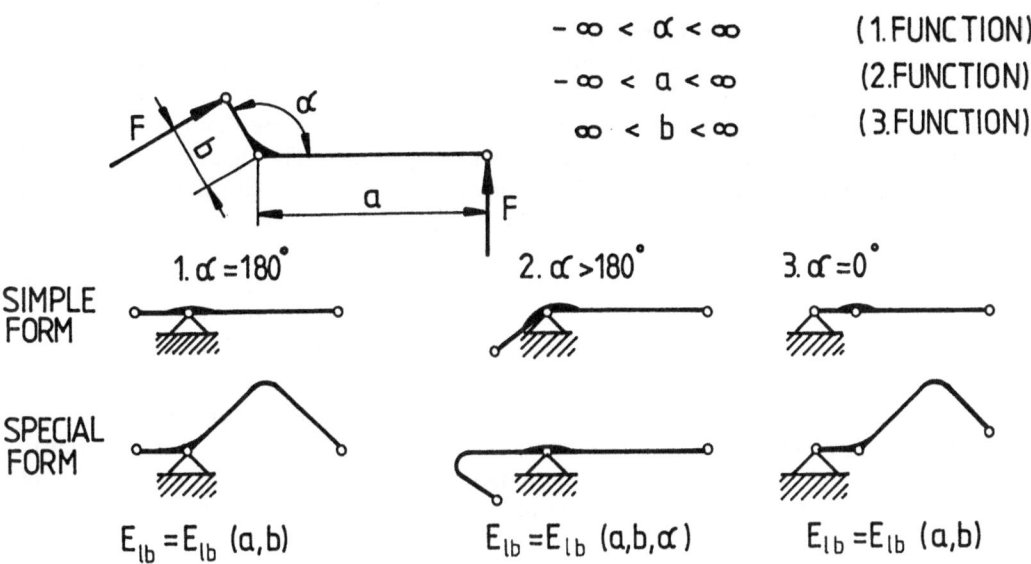

Fig. 13.4. Example with changeable number of parameters by the three-membered lever

The description of an element is simple. Considering the local space and the direction of the abscissa of the coordinate system for a parameter with the main axis of the element, we obtain:

$$E_{lb} = E_{lb}(b_1, b_2, \cdots b_n) \tag{13.1}$$

where b_i [meter] is the parameter defining a characteristic e.g. b_i for the diameter and b_{i+1} for the length of a cylinder in a local coordinate system.

Trying to transfer the element into a world coordinate system, a shape of parametric expression is obtained as

$$E_b = E_b(x, y, z, \alpha, \beta, \gamma, b_1, b_2, \cdots b_n) \tag{13.2}$$

where x, y, z [meter] are coordinate points for the positioning of an element in the world coordinate system, and α, β, γ [rad] are angles of rotation of an element in the world coordinate system.

The expression of the parameters in such a shape is worked out in all geometric modellers. When we have matrices of shape it is possible that the vector of shape is transferred into expression (13.1). This connection between the matrix of function and expression parameter of shape is worked out automatically.

3.2. Relationship of an Element Functionality with Drawing or Manufacturing Drawing

A machine element satisfies one or several functions. As an example we can take a gear pair transferring energy (function 1) from one location to another (function 2) in the conditions of equal circumferential speed rates (function 3). We can determine the schematic relation of these functions and also the full shape. In the RESDEV phase we determine the function whereas in the PRO phase we determine the functionality which includes all the details. For a gear the functionality conditions are defined by module, number of teeth of the pinion and the gear, helix angle, etc. The details of the special shape are e.g. exaggerated profile modification, bombing tooth width lines, etc., which are also important in terms of stress and strain relationship of the tooth body and the gear. The requirements for the realization of these details enter into the conventional design process accidentally depending on the designer's knowledge. However, in the computer aided design process these relations have to be uniformly defined, and this very fact then makes possible that the geometry arising from the functionality requirements can be written in a special shape.

The conditions of function and functionality are different and in general do not have any physical similarities. The different conditions can be sorted out into a matrix $n \times m$ where n means the number of functionality criteria and m means the number of control value. Thus, in general we can write

$$[\Sigma] = [S] \times [V] \qquad (13.3)$$

where Σ is actually reached values for the criteria such as strength, pressure, etc., and S is the load dependent on the criteria, and V is the functionality criteria.

The actually reached values for the particular functionality criteria can be defined for each part of an element separately or on the whole. For each functionality criterion we also have to define a corresponding load which however does not have to be mechanical (it can be e.g. thermal). A functionality criterion defines the characteristics of an element including also its geometry parameters.

A functionality criterion can be expressed by a product of matrix

$$[V] = [G] \times [F] \qquad (13.4)$$

where G [meter] is detail shape matrix and $[F]$ is matrix of functionality. The detail shape of an element can be written by the elements of a matrix

$$G = G(b_1, b_2, \cdots b_n) \qquad (13.5)$$

where the individual elements in the matrix b_i represent the parameters of the characteristics shape of an element. The more detailed is the presentation of the functionality, the more precise is the determination of the shape parameters of an element. Examining the matrices, we can notice very solid diagonal matrices with the elements lying but seldom outside the diagonal. As a result, this enables the determination of the values of particular elements.

The consideration of the geometry with all its special details affecting the functionality of an element is proved. A problem arises only in how these criteria should be considered in the modelling of an element. What is meant here is especially the transfer of the geometric details defining an element to the level of manufacturing documentation.

In the same way, as for the function, we try to work out for functionality. A great difficulty usually arises when we try to transfer the functionality criteria automatically into shape generators (geometric modeller). As there does not exist any worked out parametric description for the generators of detail shape, after the functionality test, all the data have to be put together into a detail shape and the shape of the element had to be created going from one detail to another. The ready-made macros help to solve the task faster, yet they do not enable a complex treatment of the problem. Macros do make possible to combine special details on an element, but the global shape generation of an element with special details is not possible.

Considering that the present determination of all the details is insufficient, we introduced into the parametric determination of a shape all the special details which are relevant in defining the global shape of a certain machine element.

For a local coordinate system, we can write

$$E_{la} = E_{la}(b_1, b_2, \cdots b_n, a_1, a_2, \cdots a_n) \tag{13.6}$$

where a_i [meter] is the parameter of the detail shape and b_i [meter] is the parameter of the global shape. Furthermore, for the world coordinate system we get

$$E_a = E_a(x, y, z, \alpha, \beta, \gamma, b_1, b_2, \cdots b_n, a_1, a_2, \cdots a_n). \tag{13.7}$$

The determination of a shape following this method requires a very precise treatment of each element.

We found out that for the evaluation of the function and functionality a direct connection can be reached by a geometric modeller. Here it has to be emphasized that the matrix expression was used only for sorting out the mathematical expressions. The individual parts of the matrix can be interrelated physically (e.g. the vector of stress and surface pressure on an element, lost frictional work and the thermal load on the element), yet in making the particular assessments it would be difficult to unite them into one element of the matrix. Likewise, this method also requires that the loads with all the characteristics are known in advance. Therefore the software has to be worked out so that it leads the designer towards a more integral treatment of an element.

For this purpose, an expert system for machine elements was worked out (Duhovnik 1984) where the machine elements for driving units were sorted out. An application of this system in industry has shown that by using the software the designer learned all the conditions of functionality and made decision about the detailed shape. In case that his requirements for functionality were to low, he would still get a solution which would be however accompanied by descriptions of possible better solutions. The experts system is used in each phase of the design and it is also worked out for each phase separately. The diversities of its use in particular phases can be seen especially if different types of structures are treated.

4. PRESENTATION OF DESIGN MODES

In the design process there are several design modes. Very often only a partial change of a product is needed. The change is perhaps conditioned by specific manufacturing conditions which have to be entered into the documentation. Manufacturing conditions are manufacturing technology, assembly, transport technology, etc. It is exactly the manufacturing conditions that very often dictate a change in documentation and it seems that this has led to a belief that because of a change of a detail we must necessarily restart the whole design process. In the theory (Duhovnik, Kimura, and Sata 1983; VDI 1979) such a design mode is supposed to be organized so that the functionality remains unchanged while the details can be adapted to a specific manufacturing technology. Thus in the CAD process the metafile of a product remains unchanged while the work files are subject to change.

In general, four design modes are known (Fig. 13.5):

* New design.
* Innovative design.
* Variation design.
* Adaptation design.

New design	New working principle A complex or individual function	Scheme
Innovative design	Variation in working principle A complex or individual function	Rough drawing
Variation design	Variation in functionality Individual function	Project drawing
Adaptation design	Variation of detail Individual function	Manufacturing drawing

Fig. 13.5. Correlation of design modes with graphics presentation

A new design is defined by a quite new working principle which has to be determined for each function separately. For a new design either a complex or an individual function can be studied, and on this basis a new working principle can then be defined. The development of a new principle is in this case carried out only for a completely new and exceptional idea. The developed working principle is then entered into the knowledge base of an expert system.

For the physical presentation of the working principle mathematical algorithms were used while for the graphical presentation a scheme was used. The research and development of the working principle can be either experimental or theoretical. The solutions of the principle cannot be foreseen in advance and thus a computer can only be used for specific analyses, e.g. numerical calculation, geometrical modelling, or as a part of an experimental work. The rest of the design modes follow in the order presented above.

Innovative design represents the selection of the most suitable working principle to meet a certain function. The selection of the working principle is carried out by means of an expert system. The innovation which is obtained with this design mode represents a part of the development tasks which are determined during the solution. The selection is assessed also by a variety of mathematical models. The assessment can be done only by simple values or a field of real values. The results of the variation of the working principles are presented by schemes which have found their materialization in the rough drawing version containing the physical parameters defined by the task.

In general, innovative designs are the most numerous. The essential reason for this is that the research of the working principles is much more expensive and thus small firms cannot afford them. Innovative design is the most frequent in the production of industrial equipment.

Variant design enables a change of the carrying capacity of the design. This means that the working principle is defined, and the structure of the elements and assemblies is the same. The shape changes only due to the variation of load. In the case that the functionality conditions change (e.g. additional lubrication of gears due to the size of the gear train) we talk of variant design. Lubrication is of course necessary also for smaller gear trains but the principle of pouring on the lubricant is different. The main function of a gear train is energy transfer and an auxiliary function is lubrication of gears. Since the main function does not change (neither does the working principle), the design belongs to the variant type. Variant design is mostly present in specialized firms in Yugoslavia.

Adaptation design was already described at the beginning of this section.

Depending on the design modes, one has to define the design tasks and also the more important CAD software. This has to be joined into a whole enabling a systematic approach to the solving of design problems. The connection between expert systems and geometric modellers has to be automated. In our work we do not dispose of a complete software for the design process, so in parallel with the development of the software new and innovative designs are worked out using the conventional design methods. Individual sections of the design process are performed on computer, and in this way the systematics of the design is also checked.

5. SYSTEMATIC DESIGN

The design process can be divided into two characteristic phases; the research and development phase where the function of the new or innovative design is defined and the second phase called the project phase dealing with the functionality of a variant or adaptation design. The relation between the particular phases and the graphic presentation can be illustrated by a tree structure of the design process (Fig. 13.6).

However, in this tree structure some limitations emerge. For example the function cannot be directly expressed in the manufacturing drawing. The limitation is built into the pre- and postprocessors. In a similar way, the limitation can be carried out upwards with regard to the level of difficulty of machining. Thus, e.g. functionality is

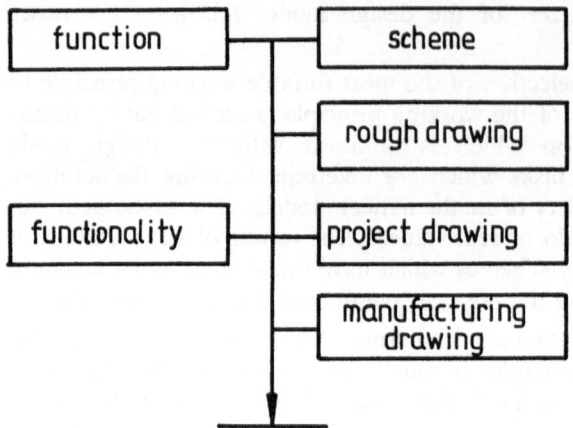

Fig. 13.6. The tree structure of the design process

not presented by a scheme as the first graphic presentation but only later on a higher level (manufacturing drawing, project drawing, and rough drawing). In case that these higher levels of graphic presentations do not exist the function and functionality cannot be defined either. With this condition, i.e. that the functionality can be worked out only after all graphic presentations are enumerated, a certain systematics of the design process is possible. The presentation of design modes related to the tree structure of the design process is given in Fig. 13.7.

From these interrelations we can see that after the definition of the task when defining the type of design, it is determined in advance which parts of the design procedure will be used. On this basis, a corresponding software can be prepared. A general presentation of it is shown in Fig. 13.8. Every software contains program packages that can process a given level in a given phase of the design process. As an example we can mention a geometric modeller for the graphic presentation in the design process which has to have the following functions of drawing:

- scheme 2-D:
  - preprocessor with parametrics of function.
  - processor with function scheme.
  - postprocessor with parameter of body (shape).
- rough drawing 3-D:
  - preprocessor with parameter of body.
  - processor with a generator of primitive bodies.
  - postprocessor with rough shape of elements.
- project drawing 3-D:
  - preprocessor with rough shape of elements.
  - processor with a generator of primitive elements.
  - postprocessor with shape of elements.

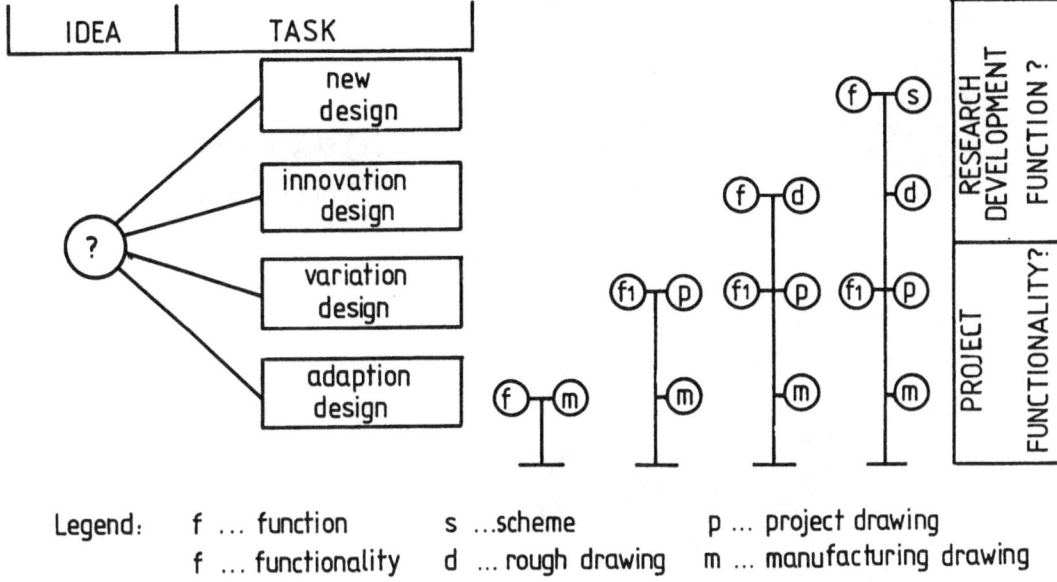

| Legend: | f ... function | s ...scheme | p ... project drawing |
| | f ... functionality | d ...rough drawing | m ... manufacturing drawing |

Fig. 13.7. Design modes related to the tree structure of design process

- manufacturing drawing 2-D:
  - preprocessor with shape of elements.
  - processor with a generator of primitive details.
  - postprocessor with detail shape of elements.

In a similar way, we can sort out any part of the software built into intelligent CAD systems.

6. INTELLIGENT CAD SYSTEM AND SYSTEMATIC DESIGN

Intelligent CAD systems will have to contain the concepts of systematic design. The management program which activates the individual parts of the program will include an analysis of the state in the process and will present those possible further steps which would enable further treatment of the model (Fig. 13.9). In an integrated production system intelligent CAD systems will be connected with the management and marketing on the one hand, and production (CAPP[1], CAM[2], CAQ[3]) and commercial

---

[1] *Computer Aided Process Planning*
[2] *Computer Aided Manufacturing*
[3] *Computer Aided Quality Control*

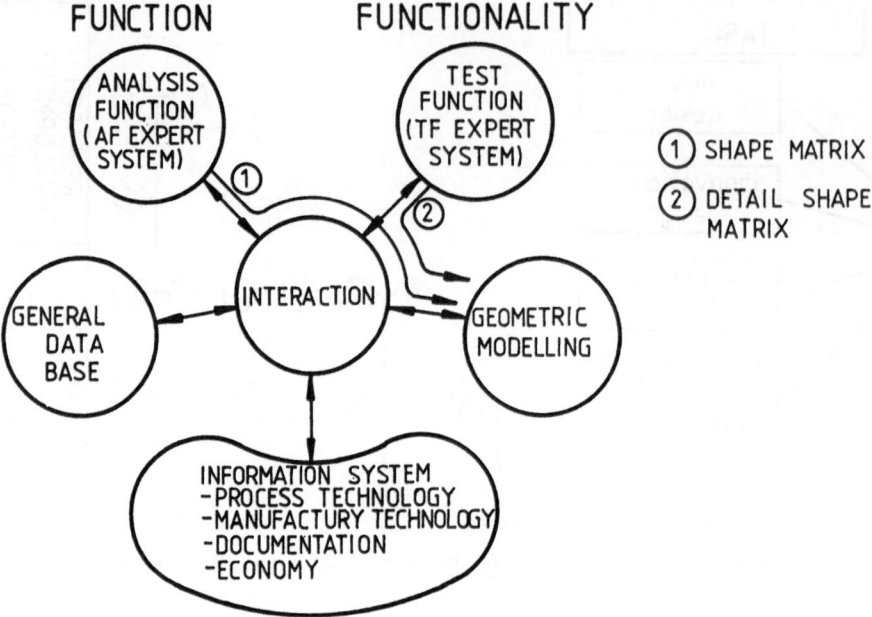

Fig. 13.8. General scheme of software in design process

department on the other hand. CAD management accepts the information and via interactive interface transfers it to the experts systems relating the function and functionality with the shape. Analyses of working principles, stress analysis, and testing in the model and its details built into the program are used in connection with the expert systems if needed. The design database contains all those data the designer needs for the definition of the technical realization. The shape of the model depends on the type of presentation. With respect to the phase of design process, the geometric database has therefore separate parts for each particular phase. The connection of the scheme to the 3D-modeller is realized via an expert system and not directly.

The database contains data of momentary models. In this way the optimization of space for the registration of the necessary data describing the model (geometrical and topological data) which would have to be done for each case separately is avoided. The input data can be performed directly or indirectly through an interactive interface. At the output the data are made ready by the translators of standards (GKS (ISO 1985), IGES (NBS 1983), etc.). In the formation of the software for expert systems the function and functionality characteristic is built in separately for each element. From the data thus compiled, those data which appear several times are gathered. The data obtained in this way are built into a common database which is used also by the software for the analysis and testing of the function and functionality. The software for the analysis of the function and functionality enable the determination of rough dimensions of the model, while the software for the testing of the function and functionality enables a theoretical testing of the definite dimensions of the model.

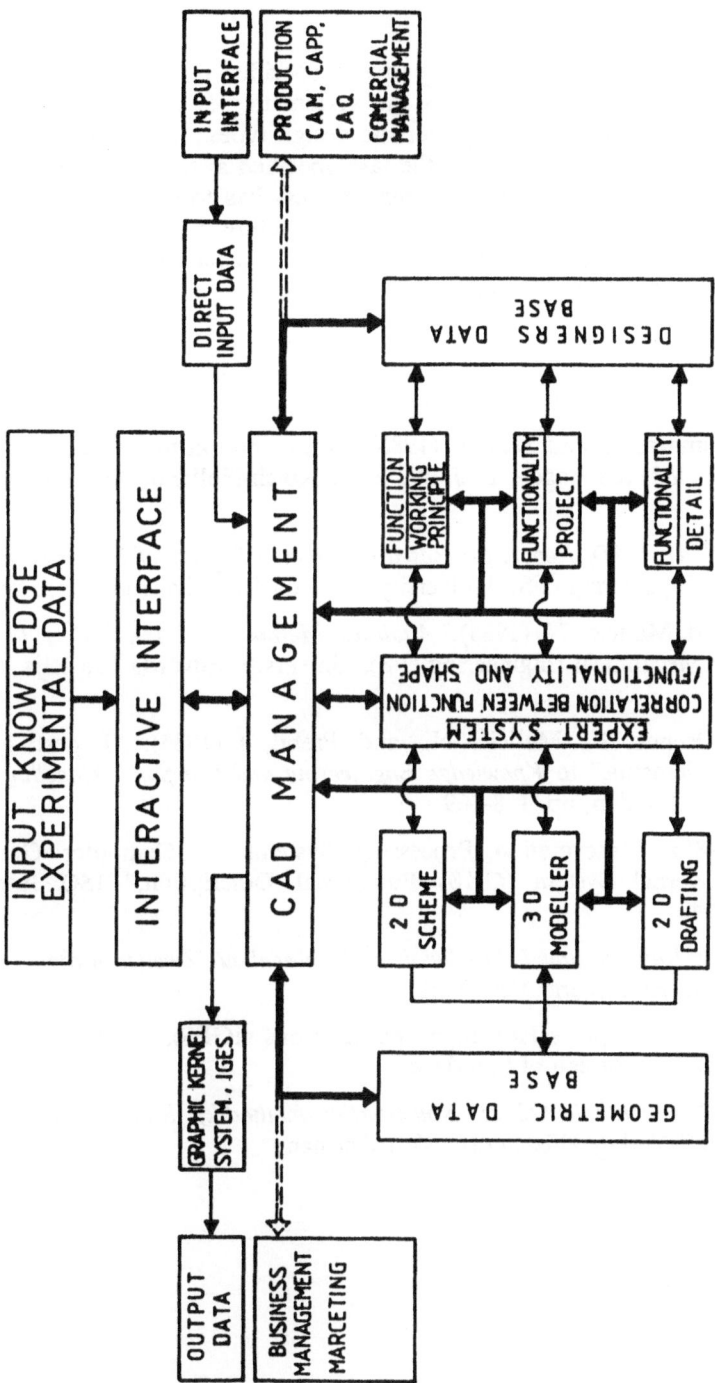

Fig. 13.9. Information flow in intelligent CAD system which have principle of systematic design

7. CONCLUSION

The design process is executed according to well specified phases which have to follow each other in a systematic order. It is the systematics of the design process which enables the categorization of the characteristic design types. The type of design has to be determined within the framework of the task specified at the beginning of the design process. The software supporting the design process has to contain the characteristic elements of the process for each phase. The tree structure of the interrelations between the operative tasks in the particular phases enables in general only the transfer in one direction — from function towards shape.

REFERENCES

Duhovnik, J., Kimura, F., and Sata, T. (1983): "Contribution to methodic in CAD," in *Eurographics '83*, ten Hagen, P. J. W. (ed.), North-Holland, Amsterdam, pp. 113-132.

Duhovnik, J. (1984): *CAD, Report for Research Council Society of Slovenia*, Faculty for Mechanical Engineering (FS), University Edvarda Kardelja, Ljubljana, Yugoslavia.

Duhovnik, J. and Matičič, N. (1985): *Analitics Method in CAD, Lakos 11 Seminar*, Faculty for Mecahnical Engineering (FS), University Edvarda Kardelja, Ljubljana, Yugoslavia.

Duhovnik, J., Ocepek, P., Matičič, N., and Prebil, I. (1986): "Expert system for mechanical elements," in *Knowledge Engineering and Computer Modelling in CAD*, Butterworths, London, pp. 158-169.

ISO (August 1985): "Information Processing Systems — Computer Graphics — Graphical Kernel System (GKS) Functional Description," ISO International Standard 7942.

Koller, R. (1985): *Konstruktion Lehre für den Machinenbau (Zweiter Auflage)*, Springer-Verlag, Berlin, Heidelberg, New York, Tokyo.

NBS (1983): "Initial Graphics Exchange Specifications (IGES), Version 2.0.," NBSIR-82-2631, National Bureau of Standard.

VDI (1979): *VDI Richtlinie 2222: Konstruktionsmethodik, Koncepieren Technischer Produkte*, VDI-Verlag, Düsseldorf, West Germany.

Report on Session 6

Chair: T. Kjellberg †
Cochair: J. Rogier ‡
Edited by: V. Akman ‡

Koegel discussed the support a designer should have when working with an advanced CAD system based on integration. Having outlined the need for working on a higher conceptual level well-supported by graphics, he presented an overview of a "paper" system. He proposed that his system should be realized through the use of object oriented programming in logic, object oriented databases, and molecular object data modelling. He frequently referred to a number of research papers in AI, databases, and logic programming.

Duhovnik discussed a narrow view of design for mechanical engineering, using small examples from the real world. He stressed the importance of a separation between "function" and "functionality," where the latter stands for detailed design solution of form and technical specifications to realize a function. His paper had very few references to other work in AI and intelligent CAD.

The two papers are complementary. Koegel outlines his theoretical model of an intelligent CAD system with a modular approach. Duhovnik talked about systematic design and its importance for intelligent CAD. Both papers should have tried to answer more questions within the area of intelligent CAD. The papers also touched little on the background of intelligent CAD such as theory of knowledge, theory of design, intelligence in design, etc. Koegel's paper is interesting from the viewpoint of implementing intelligent CAD systems. Duhovnik's paper is striking from the angle of practical experience with industrial design. It seems that his ideas are very pragmatic and comprise a rather myopic view of design. In any case, both papers present a curious mix of science and art in engineering.

During the Q&A period, Duffy inquired how far Koegel got with an implementation of his theory. The answer was that a parallel logic programming

---

† Department of Manufacturing Systems, Royal Institute of Technology, Brinellvägen 81, 100 44 Stockholm, Sweden.
‡ Centre for Mathematics and Computer Science, Kruislaan 413, 1098 SJ Amsterdam, The Netherlands.

language (POOL), an interface to a relational database, and a graphics editor do already exist. Mac an Airchinnigh wanted to know how it is possible in Koegel's system to reuse design specifications. His reply used an analogy with software reusability where one approach is to store descriptions of software modules in a database automatically searched by a library assistant. In his case, the data model would be the basis of identifying reusable designs.

Ten Hagen suggested that Koegel might consider placing intelligence of the system into the user interface in order to extend its means in communicating with the user. Then the question naturally arises: What part of the communication between the user and the system may be intelligently supported by interpreting the designer's strategy? Koegel noted that the system is meant to be reconfigurable both by the system designer and the user. This permits the system designer to tailor the system for each domain and provide the intelligence in the interface if he chooses to do so.

Arbab's questions were as follows: Koegel is advocating an object oriented data model. Why does he use the entity relation (ER) model for descriptions? Is he advocating the use of ER model? Is he able to provide a justification for his choice of an object oriented data model? Koegel emphasized that the molecular object model has been described in more detail in the reports provided in his bibliography. The ER diagram he is using is actually a modified ER diagram to represent the model and to demonstrate its mapping to the relational data model. He studied other object oriented models and the particular features unique to his model are version control and mapping to the relational data model.

Akman criticized Duhovnik for complicating the issues by using rather imprecise terms in his paper. Novak wondered, considering the distinction Duhovnik made between function and functionality, how the system could assist in each of the design types Duhovnik distinguishes (e.g. new, innovative, variational, adaptive designs). Hoeltzel remarked that, regarding Duhovnik's distinction between function and functionality, his description of the latter seems to be a static one. He then asked: How will you encounter the fact that description of functionality may change during the design process, e.g. how do you take dynamic forces into account? Duhovnik's reply was that functionality means that one defines dimensions via numerical calculations and one changes shape for some specific domain of manufacturing technology. This means that description of functionality is not static but dynamic.

Session 7

14. A Multiparadigm User Interface for Intelligent CAD Systems

Z. Ruttkay, R.H. Allen †, and B. Laczik ‡

Computer and Automation Institute, Hungarian Academy of Sciences
POB 63, Budapest, H-1052 HUNGARY

Abstract: *A generic, multiparadigm user interface for intelligent CAD systems for mechanical design is presented. The needs for an interface are examined from the engineer's standpoint as well as the standpoint of integration with CAD systems. A general architecture for such an interface is discussed. As an example, a frame-based implementation for a mechanical component design is presented.*

Keywords: user interface, natural language processing, interactive graphics, knowledge representation, CAD.

1. INTRODUCTION

In recent years, many efforts have been made by both the top-down and bottom-up methods to meet the challenge of CAD systems. The theoretical approaches make it clear that design is a highly complex process involving sequences of decisions in a comprehensive, often conflicting, and ill-formed decision space.

Simultaneously with the theoretical research, *practical CAD applications* can be seen, each of them either developed for the given domain, or adopting one of the CAD systems commercially available. These so-called CAD systems — with a few exceptions — are highly restricted in scope: they can be used only at certain stages of the design process to perform special tasks, e.g. to give a visual representation of the object being designed, or to evaluate some critical values for the design (Hatvany 1984). These systems can partially aid the engineer in some stages of his design activity. The crucially important decisions need to be worked out by the engineer himself and the gaps have to be bridged without hints (Lansdown 1986).

---

† Department of Mechanical Engineering, University of Houston, Texas, USA
‡ Department of Mechanical Engineering, Technical University, Budapest, HUNGARY

The "real" CAD systems are those which are meant not only to perform certain — albeit rather tiresome or inhuman — tasks, but to aid the user during the entire design process, relying on comprehensive and complex domain knowledge at each stage, considering the design in context. It is these systems that have been called intelligent CAD (ICAD) systems (Melkanoff and Kamvar 1985) or CAD expert systems (Sriram and Rychener 1986). In spite of numerous efforts, this challenge has not been met, as shown by the fact that only a few of the experimental applications have gained ground in industry (Fox 1987).

While much work has been done concerning the "content" of CAD, little activity appears in the field of user interfaces for CAD systems. It has been often stated that enlargement of the scope and intelligence of CAD systems poses new challenges concerning the interface, providing efficient and convenient usage for the interacting human (Tomiyama and Yoshikawa 1985), and that the question of an adequate interface is of first importance (Popplestone et al. 1986). All the same, these questions has not been dealt with theoretically, and in experimental ICAD systems there is not much attention paid to the interface. At the same time, considerable progress has been achieved in related fields, such as user interface management systems, natural language processing, interactive graphics, geometric modelling. There are projects on multi-media access to encyclopedical knowledge bases (Bolt 1979; Herot et al. 1980), experiments with integrating natural language and interactive graphics (Adorni and Massone 1985), and more engineer-friendly graphical packages are coming into the market.

Here we give the outlines of a user interface which supports communication via different media in a mixed and flexible way. The paper is organized as follows. In Section 2 the state-of-the-art in engineering communication is critically evaluated with an outlook to interface some of those methods with ICAD systems. In Section 3, the architecture and the theoretical discussion of a multiparadigm user interface is dealt with. It is shown how a frame-based product model can serve as a basis for such an interface in Section 4. Finally, a summary is presented, conclusions are drawn, and further issues are discussed.

2. HOW SHOULD THE USER COMMUNICATE WITH A CAD SYSTEM?

In establishing mechanisms by which a user — who is considered to be an engineer, with more or less convictions about how engineering processes are to occur — is easily able to communicate with a CAD system, those of the conventional tools of design have to be selected, which are inherent to the design of a special product, or to the human's activity in the design process. In addition, from the rich choice of the devices and techniques for man-machine communication the ones which could aid and extend the designer's activity should be included. There is no consensus about the most "natural" media to be used: graphics, text and speech processing are favoured exclusively, or assigned to output and input. There are experiments to interweave these media in a more sophisticated way.

In the case of interfaces for CAD expert systems the crucial point is the exchange of information concerning the object under design. In the traditional engineering design process, the initial step after goal specifications are set is typically the creation of a series of sketches.

These *sketches*, often with symbolic elements, determine the functional structure and the vocabulary of the object being designed. The initial design sketches evolve into *detailed two-dimensional engineering drawings* with dimensions indicated to define the object and its parts, often with further information concerning manufacturing (such as tolerances and raw materials).

There is a *conventional language for detailed engineering drawings*: different types of lines, symbols for shapes and distances, numbers indicating dimensions, character strings referring to standards, and often words and sentence fragments are the elements to describe the designed entity.

When a two-dimensional representation is insufficient to either check the object from the aesthetical point of view, or to indicate the assembly of the parts and make the structure clearer (often for manufacturing purposes), *three-dimensional views* are provided. At times, a *clay model* or other scaled *prototype* is produced to serve as a representative model for the manufacturing process (e.g. casting). To date, drawings have been considered as both the technical and legal *documentation* for design.

One natural question arises: what should be used from all these in communicating with a CAD expert system? It is clear that engineering drawings are highly efficient tools for defining design entities and referring to specific portions of such objects, to get an impression about the shape of the object at a glance. However, the static nature of a design drawing is a disadvantage. The domain-dependent conventions in engineering drawings are not sufficient for the representation of conceptual parts, even by adding symbols or textual references. In addition, a precise drawing — which is required for manufacturing purposes — involves formidable effort. Moreover, a detailed drawing invariably misrepresents information that is flexible as being fixed. For example, the shape and size of a corner in a drawing may actually have a wide range of equally acceptable alternatives. However, the drawing does not reflect that and the manufacturing process is more restricted than required. Recent results in variational design and parametrical design are intended to fulfil this aspect of CAD (Light and Gossard 1982; Lin, Gossard, and Light 1981).

Noting the advantages and disadvantages of the primary method of conveying engineering information, the characteristics of the engineering drawing desirable for a user interface are:

- icons, to symbolize different objects and parts (Fig. 14.1).

- two-dimensional drafts for the assignment and visualization of dimensions (Fig. 14.2).

- two-dimensional or three-dimensional drawings to indicate the dimensions and tolerances, topology, structure, and assembly of the object (Fig. 14.3).

- two-dimensional free-hand drawings to define shapes, if they cannot be defined by geometrical concepts or functions.

- three-dimensional, isometric graphic of the object or a two-dimensional projection of it, in cases when the complete and faithful representation of the object is necessary for the evaluation of the design.

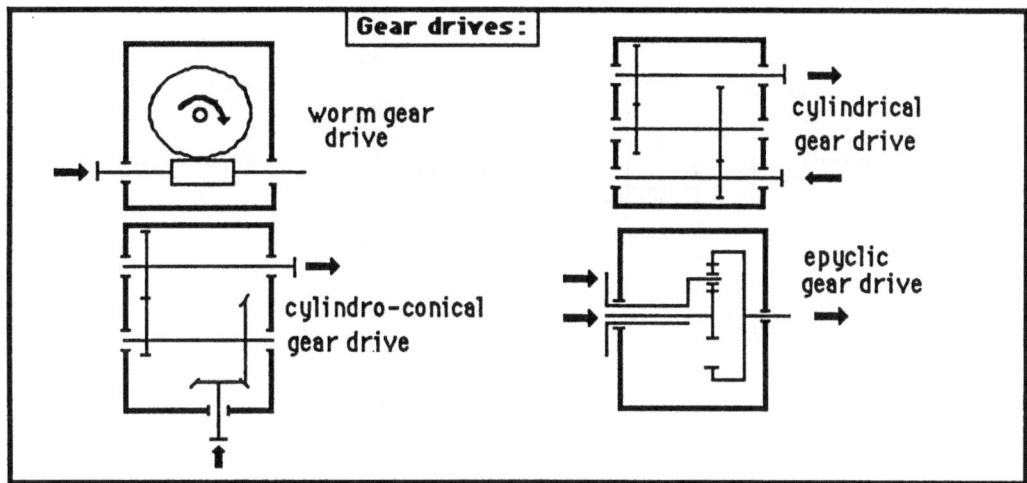

Within the figure:

Gear drives:

worm gear
drive

cylindrical
gear drive

cylindro-conical
gear drive

epyclic
gear drive

Task: To transmit rotary motion of 20 kW. Parallel shafts. Driving speed: 1440/min, driven speed 150 /min. Non-stop operation, span of life 30 000 h. Modest dynamic load. As cheap and simple as possible. Humid, slightly corrosive atmosphere. No noise.

Fig. 14.1. Icons for gear-drives and functional specification

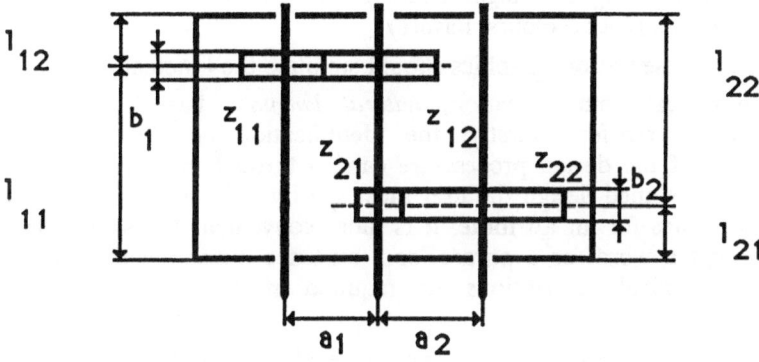

Fig. 14.2. Draft of a cylindrical gear-drive with 2 transmissions

245

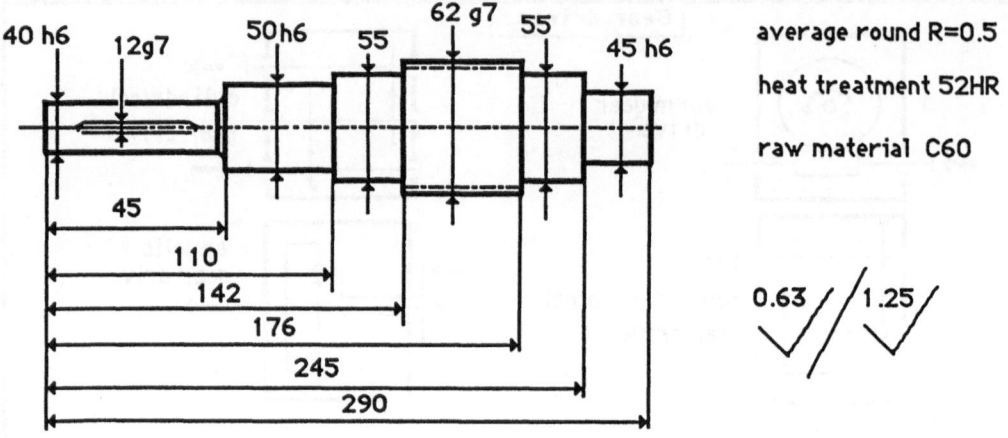

Fig. 14.3. Drawing for production

However,

- drawings should not be the tools to completely define the designed entity.
- instead of a few, overcrowded drawings, a choice — varying on the application domain — of different drawings should be offered.
- according to the role of the different drawings, interactive manipulations should be supported.
- the conventions of engineering drawings should be enhanced by the additional techniques of computer graphics (colours, flashes).
- in addition to static representation, graphics should simulate the kinematics

In addition to graphical communication, *natural language* has fundamental importance in information transfer. Firstly, the identification of objects, parts, functions at the beginning of the design process are carried through by using natural language. In course of the detailed design the user must specify different attributes. If there are many possible values for an attribute, it is more convenient to ask the user about the value and accept a word, or a phrase, or a sentence indicating the chosen value. In many cases, technical descriptions are required in the form of natural language.

During the entire design process, the user should be allowed to have a look at different chunks of data, concerning the object and the design environment, i.e. the standards, technical prescriptions in large databases. At a given stage, the user's questions cannot be predicted. Different types of questions referring to varying data may arise, according to the user's characteristics such as experience, designing style, and according to the specific design task. Such flexibility cannot be achieved by menu techniques; the user should be allowed to ask questions in natural language. Of course, the grammar and vocabulary are only a subset of the entire natural language, but the limits of the vocabulary should be easily extensible to allow the user to use synonyms for built-in words and to define new grammatical structures.

Besides the graphical and textual communication, the *numerical data sheets* have their distinct role. The part of information controlling the design specification is given in databases, such as tolerance descriptions, raw material characteristics, and data on standardized component parts. Although the CAD system handles the information in databases hidden from the user, the user may need to peruse parts of some databases, either to guide the choice or to inform himself about the design environment.

In CAD expert systems a part of the knowledge is given as procedures. For explanation purposes, the conventional *mathematical formulas* should be used rather than the source or a metacode to describe the procedure. Users should be allowed to use some of the basic mathematical symbols (e.g. $=$, $\leqslant$, $<$) in graphical and textual context too.

In many cases a precise or approximate *graph of functions* and inequalities should promote a more clear and concise representation of relations originally given as mathematical formulas or rules.

3. ARCHITECTURE OF THE INTERFACE

The above discussion of the formal activities of engineers gives the firm belief that:

- different types of information (graphical, numerical and textual) should be processed in a mixed, sometimes overlapping way.

- the semantics of the different representation media should be mapped to a common base.

- the interpretation of the user's commands should be highly based on the semantics which may change dynamically according to the state of the current design.

- the medium actually used can be determined automatically by the "nature" of the information to be expressed, or explicitly by the user, or implicitly inferred from the history of the communication.

- the interface should act in a mixed-initiative way to aid the user in the rather intricate context of design and to help him in correcting errors.

All these requirements bring into focus the knowledge representation technique used to represent the objects of design. In many cases of user interfaces it was proven that the surface representation, e.g. natural language cannot be used · successfully without regarding the deep semantics e.g. conceptual model. Also in the case of graphical tools the graphical representation of concepts has become primarily important (Szalapaj and Bijl 1984; Tomiyama and Yoshikawa 1985). These trends are in accordance with the new way to describe the designed objects by "product modelling" (Sata et al. 1986).

The base for the interface proposed in this paper is the *frame-based representation* of the objects to be designed, as well as the entities and concepts used the design process. The frame as a tool for conceptual representation has roots in natural language processing (Charniak and Wilks 1976; Simmons 1984). Minsky (1975) invented the frame concept and discussed the adequacy of this technique to such diverse human activities as vision and language understanding. This representation technique has been widely used, and is an established tool (Adorni and Massone 1985; Fox and Baykan 1985; Sriram and Maher 1986).

A frame-based representation language is rich and flexible enough to express:

- specific entities, with attributes and attribute values
- conceptual taxonomy of the entities, with attributes and attached meta-information on attributes: restrictions on attribute values, default and inheritance rules for the value, procedures to evaluate the attribute value, side effects of manipulating the attribute value, etc.
- relations, with consequences on the inheritance of attributes.
- different points of view to filter out the germane attributes for the given context.
- hooks to the support language and other software.

The advantages using such a language for representation purposes are the following:

- The different representations of an object can be considered as the different views of the same object.
- The special attribute values of an object can be given in a flexible way: they can have zero, one or more values, inherited or specified, predefined or interactively given or modified.
- The frames representing entities, concepts, and relations together with the frame-manipulation language serve as a common semantical representation for the different communication interfaces and the design modules.
- The "updating" of natural language utterances, drafts, and detailed drawings happens automatically by activating procedures attached to the attribute with altered value.
- The "semantical correctness" of a command expressed in terms of any of the communicational tools can be derived from the local test procedures and constraints attached to the attributes of objects or concepts, forming a part of the domain knowledge.
- The relations describing the structure and classification of objects can be used to inherit the representatives for it in natural language, drafts and detailed drawings.
- Special software tools — such as standardized graphical packages — can be reached via the supporting language.

The main architecture of the proposed interface is given in Fig. 14.4. We briefly discuss each of the modules.

The kernel of the interface is the *communication supervisor* which guides the communication in the following way:

- It keeps track of the changes in the communication media. The user's command is sent to the corresponding interface and the frames describing objects and concepts become available in the given context.

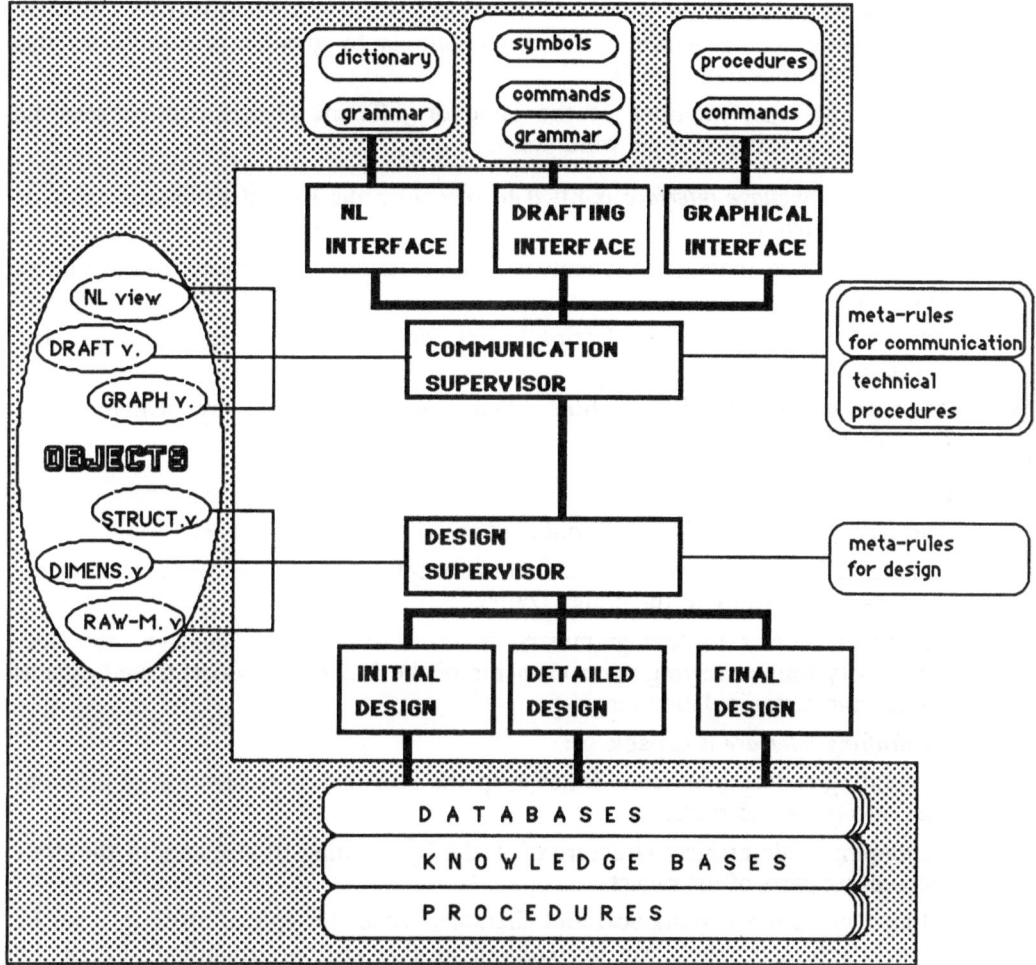

Fig. 14.4. Architecture of the interface

- If a command is interpreted with success by a chosen interface, the meta-representation of the command is passed to the design supervisor. After the successful performance or the failure of a command by the user, the answer is given "in the most reasonable way"; e.g. the communication supervisor — using the metarules for communication and the type and quantity of information involved in the answer — determines which interface is the most adequate to express the answer. It is difficult to establish the rules to choose the most convenient tools for different types of information. There can be given rules reasonable for most of the users, e.g. to manipulate a dimension of an object, a draft indicating that dimension is more convenient than a lengthy description, or sequence of choices to identify the dimension in question. However, often it is not clear-cut whether a natural language explanation, a detailed data-sheet, a draft, or a smart three-dimensional picture of the object is the most informative. In such cases, the user should decide, either one-by-one, or by giving a preference for the different communicational tools.

- The different representations of the same or different objects can be seen in parallel too, by using multiple windows.

The *natural language interface* is fitted to recognize the limited structures used for special purposes such as:

- noun phrases and sentence fragments to specify design goals, objects and attributes of objects.
- what, which, why questions referring to data on the object being designed or the constraints on attributes.
- non-processed textual information which can be browsed by searching for keywords.

This interface uses:

- a dictionary for morphological variants of the words to be processed.
- the natural language view of objects and concepts, defining the meaning of words and phrases according to the design context.
- a grammar, similar to case grammars, to transform natural language utterances into activity frames, relying upon elements of the frames for concepts and objects and the frame manipulation language.

The *drafting interface* is capable of:

- representing concepts by icons which help the user in fast recognition of different types of objects and parts.
- maintaining a hierarchical structure of drafts by pointing to elements of the draft representing parts of the object.
- editing a given draft: numerical and character values for special attributes can be edited, graphical symbols can be replaced to alter the corresponding part of the object.
- asking about attribute values.
- drawing and modifying elementary icons.

The drafting interface should use tools similar to the drawing facilities of personal computers (e.g. MacDraw[1]): the same drawing functions could be chosen on each draft level, and the design context and the evaluation of the entire draft can be guided by activating one of the currently available commands.

The drafting interface uses:

- a collection of elementary icons and graphical symbols.
- the draft view of objects, to define the draft of an object by using the draft of its parts.

---

[1] MacDraw is a trademark of Apple Computer Inc.

- a collection of draft commands such as delete, replace, edit.
- a draft grammar to map the sequence of commands on the frame manipulation language.

The *graphical interface* produces direct graphics which can be considered as the refinement of drafts. They can serve two purposes:

- to give a true picture of the object being designed. In this case the user uses the picture in a passive way, as a factor in evaluating a finished design and in possibly directing the redesign.
- to specify and edit some parts of the object (those which cannot be defined conceptually).

4. EXAMPLE

We give an example from the field of gear-reducer design to highlight the above ideas. For simplicity's sake we concentrate only on a few aspects of the design which are illustrative of what has been said. We ignore a number of particulars which otherwise play an important part in the design, e.g. efficiency, heat and noise generation, cost of production.

All concepts and entities are represented in the form of frames. The taxonomy of objects and parts is represented by the conceptual description of the objects and parts, by listing their attributes. The value of an attribute can be — in addition to numerical values and character strings — another frame representing an object or a procedure to perform a specific task. Meta-information concerning the value of a given attribute is also given by using slots attached to the attribute. The natural language expressions — nouns, noun-phrases or abbreviations — used for a concept in natural language context, and the icon and drafts used to represent it are also given.

    gear_drive
        synonyms: 'gear drive' 'gear reducer'
        effect:
            synonyms: 'effect'
            type: **effect**
        driving speed:
            synonyms: 'driving speed'
            type: **angular velocity**
        driven speed:
            synonyms: 'driven speed'
            type: **angular velocity**
        operational factor:
            synonyms: 'operational factor' 'dynamic load'
            type: **operational factor**
        setting_of_shafts:
            symbol: 'Σ'
            synonyms: 'setting of shafts' 'position of shafts' 'angle of shafts'

            type: **angle**
IT_quality:
            synonyms: 'quality' 'ITquality'
            range: **IT_standard**
            demon: **check_noise**(noise, IT_quality)
noise:
            synonyms: 'noise'
            range: **ranks**
            default: **unrestricted**
            demon: **check_noise**(noise, IT_quality)
duration_of_service:
            synonyms: 'service life' 'life-time' 'duration of service'
            type: **time**
complexity:
            type: **scale**
box:
            synonyms: 'box' 'house'
            type: **gear_box**
lubrication:
            synonyms: 'lubrication'
            range: **oil_standard**

cylindrical_gear_drive
            synonyms: 'cylindrical gear drive' 'helical gear drive'
            draft: **draft_of_gear_drive()**
            is_a: **gear_drive**
                        complexity: **simple**
            number_of_transmissions:
                        synonyms: ' transmissions'
                        range: 2 or 3 or 4
            list_of_gears:
                        type: list **gear_pair**
                        demon: **compare**(list_of_gears, number_of_transmissions)
            list_of_shafts:
                        type: list **shafts**
                        demon: **compare**(list_of_shafts, number_of_transmissions)

gear_pair
            synonyms: 'gear pair' 'transmission'
            draft: **draft_of_gear_pair()**
            driving_gear:
                        synonyms: 'driving' 'frist' 'small'
                        is_a: **gear**
            driven_gear:
                        is_a: **gear**
                        synonyms: ' driven' 'second' 'big'
            ratio_of_teeth:
                        demon: **check**(driving_gear, driven_gear, ratio_of_tooth)
            distance_of_shafts:
                        range: **standard_distance**

The draft of an object or a partially determined entity is given by draft frames. Some attributes from the concept of the object to be drafted are used to draw the draft. If any of them is not given a value, the drafting procedure uses default values for them. The shared attributes are the ones which can be modified referring to the draft, while the inherited ones cannot be modified. Some parts of the draft — corresponding to parts of the drafted object — can be clicked and more detailed draft of that part can be accessed and edited.

draft_of_gear_drive()
    inherited: number_of_transmissions, setting_of_shafts
    shared: distance_of_shafts, number_of_teeth, box
    click_to: gear_pairs, shafts
    by: gear_draft_proc()

The domain-dependent as well as the universal engineering concepts can also be defined in terms of frames:

dimension
    nom_val:
    tolerance:
        type: **tolerance**
    unit:
    char_symbol:

IT_standard
    from:
    till:

angular_velocity
    is_a: **dimension**
        unit: 1/s
        symbol: 'n'

operational_factor
    is_a: **dimension**
        nom_val:
            range: [1,3]

5. CONCLUSIONS AND FURTHER ISSUES

In this paper outline of a multiparadigm user interface was given to integrate the techniques and concepts of distinct fields of interfaces. This interface is to meet the demands of the user from both the conceptual and technical points of view and seems to be an adequate tool for CAD systems relying on product models.

We pointed out that a rich conceptual representation can be the theoretical basis for such an interface. The existing frame-based knowledge representation languages can serve as tools of implementation.

The further work focuses on two fields. The practical task is to develop and implement such an interface for an existing frame-based CAD system. This formidable task can be achieved through modules. The frame-representation is an advantage from the points of view of development and modification: the refinement of concepts and views can de done in parallel to development.

The research can be continued to extend the interface to such areas as:

- drawings representing assembly.
- animated drawings to represent kinematics.
- hooks to standard graphical systems.
- charts and graphs.

The authors would like to thank Joe Hatvany, Andras Markus and Jozsef Vancza for inspiring discussions and reviewing the manuscript.

REFERENCES

Adorni, G. and Massone, L. (1985): "Graphics and natural language: an integrated interface for man-machine communication," *IFAC Conference on Analysis, Design and Evaluation of Man-Machine Systems*, Varese, Italy.

Bolt, R. A. (1979): "Filing and retrieving in the future — Spatial data-management?," in *Man/Computer Communication, State of the Art Report*, Infortech, Maidenhead, UK, pp. 19-36.

Charniak, E. and Wilks, Y. (1976): *Computational Semantics*, North-Holland, Amsterdam.

Fox, M. S. and Baykan, C. A. (1985): "WRIGHT: An intelligent CAD system," *SIGART Newsletter*, **92** , pp. 61.

Fox, M. S. (to appear, 1987): "Industrial applications of AI," in *AI in Manufacturing*, Mowforth, P. (ed.), Springer-Verlag, Berlin, Heidelberg, New York, Tokyo.

Hatvany, J. (1984): "CAD — State of the art and a tentative forecast," *Robotics and Computer-Integrated Manufacturing*, **1**(1), pp. 61-64.

Herot, C. F. and et. al (1980): "A Prototype Spatial Data Base Management System," *Computer Graphics*, **14**(3), pp. 63-70

Lansdown, J. (1986): "Do CAD systems fulfill designers' needs?," *Computer Aided Design*, **18**(6), pp. 299-300.

Light, R. A. and Gossard, D. C. (1982): "Modification of geometric models through variational geometry," *Computer Aided Design*, **14**(4), pp. 209-214.

Lin, V. C., Gossard, D. C., and Light, R. A. (1981): "Variational geometry in computer-aided design," *Computer Graphics*, **15**(3), pp. 171-178.

Melkanoff, M. A. and Kamvar, E. (1985): "A manufacturing-oriented intelligent CAD-system," *Annals of the CIRP*, **34**(1), pp. 159-162.

Minsky, M. (1975): *The Psychology of Computer Vision*, McGraw-Hill, New York.

Popplestone, R., Smithers, T., Corney, J., Koutsou, A., Millington, K., and Sahar, G. (1986): "Engineering Design Support Systems," Report No. DTOP/EXT/EDAI/09/1, Edinburgh University, Department of Artificial Intelligence, Edinburgh, UK.

Sata, T., Kimura, F., Hiraoka, H., Suzuki, H., and Fujita, T. (1986): "Comprehensive modelling of a machine assembly for off-line programming of industrial robots," *IFIP Working Conference on Off-line-programming of Industrial Robots*, Stuttgart, West Germany.

Simmons, R. F. (1984): *Computations from the English*, Prentice-Hall.

Sriram, D. and Rychener, M. D. (1986): "Expert systems for engineering applications," *IEEE Software*, 3(2), pp. 3-5.

Sriram, D. S. and Maher, M. L. (April 1986): "The representation and use of constraints in structural design," in *Applications of Artificial Intelligence in Engineering Problems, Proceedings of 1st International Conference (1986), Southampton University, UK*, Sriram, D. and Adey, R. (eds.), Springer-Verlag, Berlin, Heidelberg, New York, Tokyo, pp. 355-368.

Szalapaj, P. J. and Bijl, A. (1984): "Knowing where to draw the line," in *Knowledge Engineering in Computer-Aided Design, Proceedings of the IFIP WG 5.2 Working Conference 1984 (Budapest)*, Gero, J. S. and Gero, J. S. (eds.), North-Holland, Amsterdam, pp. 149-169.

Tomiyama, T. and Yoshikawa, H. (1985): "Requirements and principles for intelligent CAD systems," in *Knowledge Engineering in Computer-Aided Design, Proceedings of the IFIP Working Group 5.2 Working Conference 1984 (Budapest)*, Gero, J. S. (ed.), North-Holland, Amsterdam, pp. 1-23.

15. SIDESMAN: A CAD System for VLSI Design

H.J. Kahn

Department of Computer Science, University of Manchester
Oxford Road, Manchester M13 9PL, UK

Abstract: *This paper examines the criteria for 'intelligent' CAD systems for VLSI design in the light of practical requirements. It identifies the importance of the total support environment in creating a system which is sufficiently integrated and flexible to provide the design engineer with an intelligent assistant. The role of knowledge-based tools in CAD for VLSI is also examined, considering their use as individual 'expert' tools and as rule-driven intelligent adjuncts or controllers of conventional CAD applications. The SIDESMAN Design System is then used to illustrate some of the techniques which can be employed to produce an intelligent CAD system.*

Keywords: CAD, data modelling, knowledge-based systems

1. ELECTRONIC CAD: THE CURRENT STATE

In the field of electronic CAD, as far as many designers are concerned, the CAD system is an obstacle that must be faced on the route from design concept to design realisation. The reasons for this attitude are many; for example, the CAD system may have a poor user interface, it may enforce too strict a design regime on the engineer, or it may be populated with CAD tools that are difficult for the engineer to understand. In this last case, the use of a CAD tool becomes a matter of following ritual and hearsay rather than a task performed with understanding and control. Sometimes, the system is only really understood by users with years of hard-won experience, and the less experienced designer is left to do the best he can. Consequently, resistance to change among experienced designers is great and they frequently prefer to stay with the outdated system that they have eventually learned to conquer, rather than face a new system with added mysteries and unexplained phenomena.

Given this gloomy diagnosis of the state of some, if not all, current CAD systems, what can be done to raise the status of CAD tools in the eyes of the designer? One obvious answer is to make the CAD system easier to use and more suited to the real requirements of a designer. A useful approach involves turning the CAD system into an intelligent design assistant, capable of carrying out mundane tasks unaided, but intelligent enough to realise when it needs to invoke the expertise of the designer. It should also be capable of offering advice to designers whose own expertise may be limited. The designer wants the system to ensure that his designs are correct, but he does not want to lose the freedom to be creative.

2. CHARACTERISTICS OF INTELLIGENT CAD SYSTEMS

The CAD environment required to support the VLSI design process is large and complex. The amount of data needed to represent a completed chip design for even a modest sized chip (say, 100,000 transistors) is very large indeed. The design process itself involves many stages from initial design capture, via synthesis and design elaboration, to physical design representation and eventually manufacture. Different tools are appropriate at different stages such as simulation (from the functional level, via the gate level down to circuit level), placement (the relative and absolute arrangements of devices), routing (achieving design connectivity), and testing.

The provision of the full range of design tools is rarely possible from scratch. In practice, previously developed tools are grafted into the evolving system. This has in the past been achieved by simply developing appropriate bridging software which allows data to be passed from one stage to the next via *ad hoc* transformations. The effect is a system which is far from intelligent. One way of avoiding the problem of inelegant bridging software is to refuse to incorporate externally developed applications. The only realistic solution, however, to the problem of producing an appropriate CAD system is to define an 'open' support environment which explicitly caters for a wide range of applications and for rapidly changing user demands.

The important characteristics of any intelligent CAD system must be those features that will allow the system to perform in a manner that satisfies the *user's* own expectations of expertise. Such features cannot be provided simply by the development of a few one-off 'expert' applications because the size of the CAD problem for VLSI design is far too great. What is required is an integrated environment which supports the system-wide need for a sound underlying data model and yet is 'open' enough to allow new classes of applications, including knowledge-based processes, to be included when needed.

2.1. The Support Environment

The overriding requirement for any CAD system for electronic design is that it should correctly support the VLSI design process. It is therefore essential to consider the main stages in that process and to identify the environment that is most likely to provide satisfactory support at all stages.

One significant factor about the design process is that it is *not* a steady progression from high level concept to detailed implementation. The process is by its very nature iterative and at any one stage in the process, other than when the design is fully implemented, the data is incomplete, inconsistent or perhaps wrong. It is therefore important when designing a support environment for CAD that the data model upon which all applications rely and from which they typically draw their design data should be capable of maintaining structural consistency. This must be achieved even though the data may be in a transitional phase and will therefore be incomplete. In practice, the problem can be limited by the fact that designers usually design in a reasonably modular manner, so that while some parts of a design may be rapidly changing, other parts remain relatively stable.

Designers may adopt different design strategies or methodologies at different stages of the design process. In particular, it is reasonable to consider a 'top-down' or successive refinement approach at the early stages of the design process. However, as the design is elaborated, it may be useful to switch, possibly temporarily, into 'bottom-up' or constructive mode in order to create the primitives into which the more abstract levels of design will map. Associated with variations in the design methodology used, is the possibility that part of a design may be purely speculative and the designer may reject various solutions before a satisfactory one is found. It is therefore important that the CAD system should have the ability to manage design subsets and to maintain the necessary relationships between parts of a design.

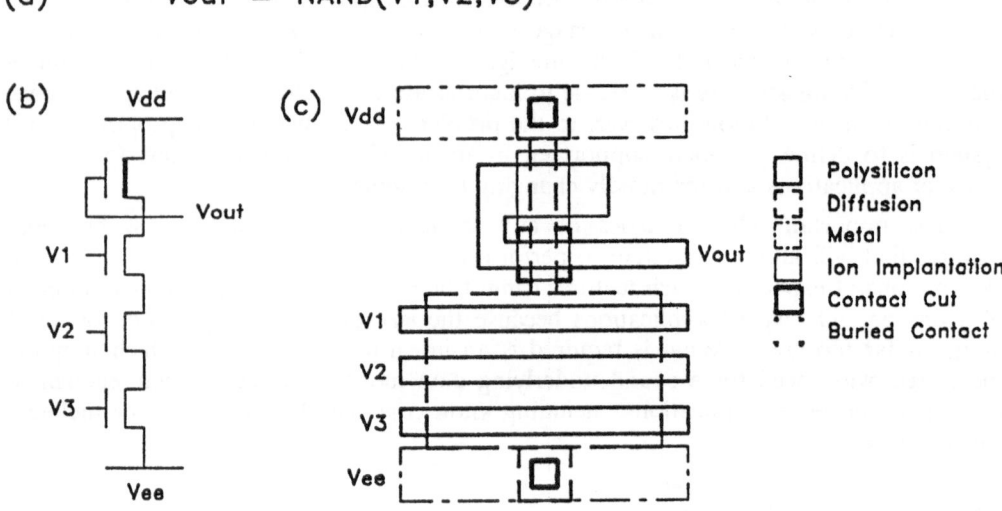

(a)   Vout = NAND(V1,V2,V3)

Fig. 15.1. Representations of a NAND gate

Another important characteristic of the electronic CAD environment, and of CAD systems for VLSI design in particular, is that multiple representations coexist for the same conceptual entity. These representations may be used to model an object at different stages in the design process or may simply be needed to match the data requirements of particular application tools. For example, Figure 15.1 illustrates three different representations of a 3-input NAND gate; Figure 15.1(a) shows the functional representation, Figure 15.1(b) might be suitable for circuit level schematics, and Figure 15.1(c) represents an NMOS mask level form of the gate.

There is clearly a requirement on the CAD system that multiple representations should be kept 'in step' and self-consistent as far as possible. Figure 15.1, however, illustrates a problem that may occur. Note that the representations have a small but significant difference in that, in Fig. 15.1(a), only four ports (input/output points) of the NAND gate are specified (V_1, V_2, V_3, and V_{out}), whereas in the other two representations additional ports, V_{dd} and V_{ee}, have been included. It is necessary, therefore, that the CAD system should not only support multiple representations of the same object, but that it should also understand *when* and *how* it is appropriate for the representations to diverge.

The type of support required from the CAD environment differs depending on the type of object that the designer may be developing at a particular time. For example, if the designer expects to implement in terms of gate arrays or cell arrays, then he may naturally tailor his design requirements to use cell designs, from a predefined library, as the building blocks. On the other hand, if the designer has freedom to design a full-custom chip, he may be less likely to see his task as the skillful assembly of predefined blocks and may handcraft his own cells and structures. Different types of application will be involved in these different design styles and a different task agenda will be followed by the designer. It is important to ensure that the freedom to select a particular design option remains with the designer.

2.2. Technological Accuracy

When a designer is concerned with a serious design project it is of paramount importance that his design should be checked for technological correctness. In many systems, this requirement means only that very detailed implementation and technology checks are carried out at the lowest level when the design nears completion. For example, physical design rule checking validates all geometric spacing requirements at the point at which detailed masks are about to be sent off to the VLSI chip manufacturing process. This interpretation is, however, insufficient. In practice, throughout the design process, an experienced designer uses his knowledge of the likely implementation technology both to guide him in defining the structures that will eventually make the design and to allow him to estimate the performance characteristics to be achieved.

It must be a fundamental requirement of an intelligent CAD system that it should support the experienced designer as well as guide the novice to ensure that technology violations do not occur. Therefore, once a designer has stated which technology is to be used or what target implementation is intended, he should be *prevented* from violating technology rules or selecting inappropriate design structures. Furthermore, he should be *advised* on how best to use his selected technology. Current CAD systems are often weak in this area because generality is achieved by avoiding technology issues altogether or by modelling design objects as 'neutral' or 'technology-free' devices. Attempts are then made to map these universal models on to technology-dependent implementations (Krekelberg et al. 1986; Royal, Hunter, and Buchanan 1985). Technological neutrality may be usable at an early design stage when it can support an abstract representation of the design. It is also sometimes used at the lowest physical level to permit a fully specified design to be mapped between variations of the same technology (Bergmann 1985). In general, however, a neutral form is insufficient to allow applications throughout the design process to correctly validate design aspects such as functionality, structure or timing. What must, therefore, be provided is an environment where technology related rules and constraints are incorporated into applications — but only when the user has identified his chosen technology. This means that the data held in the CAD database is likely to be closely tied to the user's selected technology, but the applications are neutral until specifically tailored to allow technology checks to be performed. This topic is developed further in Section 3.3.

2.3. Knowledge-Based Techniques

The use of knowledge-based techniques offers a potentially powerful means for improving the 'intelligence' or the 'expertise' of CAD tools. There are a number of ways that knowledge-based approaches can be fitted into a CAD system. Autonomous 'expert' applications may be developed to perform appropriately constrained application tasks. Such tools are likely to operate on specific data, drawing design information from the CAD database and supporting the application with a domain-specific rule set. The use in CAD applications of this kind of rule-based tool often provides a fast tool prototyping method. In practice, however, an application operating entirely as a knowledge-based system may prove slow or unwieldy to handle the large amounts of data found in VLSI CAD systems.

A common application of knowledge-based techniques in the Artificial Intelligence environment is associated with diagnostic processes. This use of knowledge-based tools is appropriate to CAD too. It is often the case that a conventional CAD application, such as a tool to assist in the layout of a design, fails to achieve complete success. When this occurs, the designer needs to examine the reasons for the failure so that he may modify his design in the hope of achieving better results next time. There is inevitably a large number of hypotheses, drawn from a variety of sources, which will need to be considered. Possible sources include the design itself, the ordering of the tasks carried out by the application, the technology used, and the functionality of the application tool. An intelligent advisor is particularly useful in this context.

Another area in which knowledge-based tools are of particular interest is concerned with the ordering of tasks in the design process. Deciding *which* tools to use and *when* to use them requires experience; guidance on what to do at a particular point in the design can be invaluable. Once again, decisions are based on an assessment of the design state obtained from various data sources. Examples of this type of control application might be deciding whether a placement is sufficiently good to be worth passing the data on to the next stage, i.e. routing, or whether the likelihood of achieving satisfactory interconnections with the current layout is such that the placement should be modified. An even more constrained example might be one in which the knowledge-based system decides which algorithm to use to tackle a specific task.

2.4. Combined Conventional-Rule Driven Applications

In current CAD systems, the application software is usually conventional procedural code written in languages such as C and Pascal — or even Fortran. These applications may represent significant investments as far as CAD developers are concerned. There is, therefore, an understandable reluctance on the part of CAD vendors to displace this software in favour of knowledge-based applications. Sometimes, this reluctance is further justified by the fact that the application in question is a reliable product, well-understood and accepted by the design engineers. In these circumstances, it is often a retrograde step to replace this software with a 'modern' rule-driven equivalent. Instead, it may be appropriate to improve the 'intelligence' of the procedural application by embedding it within a more supportive environment. This ability to combine rule-driven and conventional techniques is a particularly useful way of allowing the level of intelligence of a system to be raised without imposing unacceptable overhead on the design processing time.

The use of such combined conventional/rule-driven techniques is particularly relevant to the VLSI CAD area because of the size and complexity of the tasks required. Many CAD systems include specialist hardware (Carpenter, Gosling, and Kahn 1986; van Brunt 1983) which accelerates time-consuming processes such as layout and simulation. It is important that access to such tools should be maintained within more intelligent environments.

3. CAD FOR VLSI AT THE UNIVERSITY OF MANCHESTER

The preceding discussion has stressed the importance of the design of a *complete* environment for an "intelligent" CAD system so that both rule-driven and conventional applications may be used to support designers. The process of designing and developing a CAD system with this broad characterisation is complex. This section considers how some initial steps on the route to producing such a system are being taken during the design and development of SIDESMAN (SIlicon DEsign System at MANchester). It is acknowledged that the CAD System development process involves the use of advanced techniques in a wide range of areas — notably, user interface design and interaction management, data modelling and database management, knowledge-based systems and, of course, suitable applications and algorithms. Here, only a subset of the relevant methodologies and techniques will be considered. Figure 15.2 illustrates the overall structure of the SIDESMAN System.

3.1. Multi-View Data Model

As SIDESMAN is required to be capable of acting as an intelligent assistant, it is an important feature of its design that the system should be fully integrated. Only in this way can consistent operations be maintained. The main support for this integration comes in the form of a multi-view data model (Kahn 1986) which incorporates a Design Database in which *instance* data for a given design is stored and Component Libraries which the design database and application programs reference in order to obtain *generic* component data. This permits the design to be modelled as a (possibly hierarchical) collection of object instances (e.g. devices and their

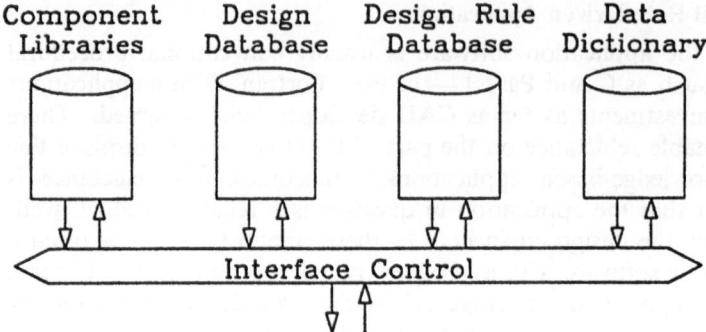

Component   Design   Design Rule   Data
Libraries   Database   Database   Dictionary

Interface Control

Expert and Conventional CAD Applications

Fig. 15.2. Overall diagram of the SIDESMAN system

interconnections) whose detailed generic characteristics (such as schematic representation or behaviour) are referenced from a library. In addition, there is a system wide <u>Data Dictionary</u> which is used to define the terminology and notation used throughout the system.

In order to ensure that the user may select the design style rather than have it imposed by the system, the data model offers support for different design methodologies. In particular, both constructive ('bottom-up') and successive refinement ('top down') approaches are supported. The design database is highly structured, with information held in multi-level hierarchies, and modelled in terms of a number of 'views'. These views are loosely correlated with the main design steps (e.g. logic description, layout and physical manufacture) and their supporting applications, but are not specific to any given application. As a result, views provide a way of ensuring that data relevant at a given stage is identifiable and can be kept consistent. It should be noted that, within a given view, the design hierarchy may differ from that in other views. For example, as illustrated in Fig. 15.3, one view (b) uses a multi-level hierarchy to define a given design object, whereas, in another corresponding view (a), the same object is represented as atomic.

Structuring data into views provides control of complexity and size over and above the control provided by a hierarchically organised data model. A consequence of adopting this structured approach is that it must be possible to move between views and for data across different views to be correlated. In the case of the design database, this is achieved in two ways: firstly, by the use of certain 'global' views which control the definition of objects and, secondly, by requiring that any transitions from one state to another in the design database have formally defined pre- and post-conditions. The use of formal specification methods has assisted the specification of this database.

The Component Libraries also make use of a multi-view approach to controlling complexity and organising data access. Within a given component library, data is related by membership of the same logic family (e.g. TTL or CMOS). Components are

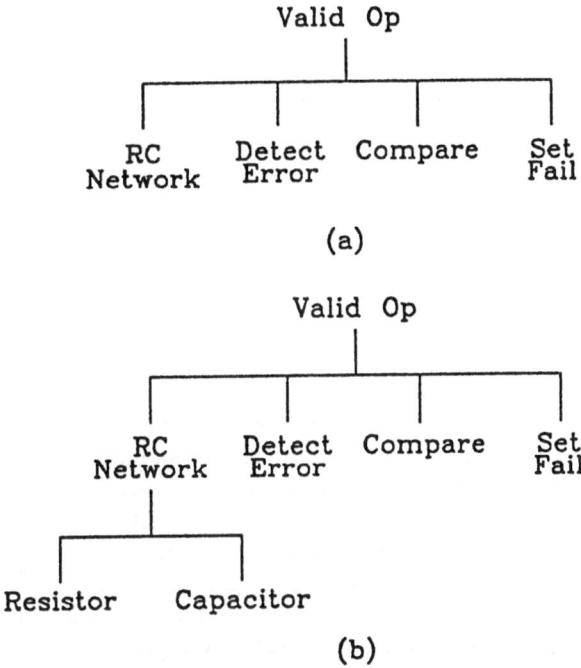

(a)

(b)

Fig. 15.3. View hierarchies

then described in terms of an extensible range of views. For example, there may be a view defining the schematic appearance of an object, another view defining its physical manufacture and a third view defining the function of the component. Furthermore, any view may support an unlimited number of variants. For example, there may be many different ways of drawing a component on a schematic, as illustrated in Fig. 15.4 or there may be multiple ways of modelling the component's functionality, each one targeted at a different design level or a different CAD application. As a result of this flexibility, support can be offered when new applications, requiring component representations different from those currently supported, are incorporated into the system.

3.2. Data-Driven Software

In considering the design of the SIDESMAN system, an important goal has been to develop as flexible an environment as possible. This flexibility is felt to be important both because SIDESMAN is intended to provide an experimental environment for CAD system design, and because preconceived restrictions and conventions are incompatible with a general approach to system design. Only if this degree of flexibility is maintained throughout the system will it be possible to move towards an intelligent design assistant operating without imposing limitations on the designer's ability to innovate.

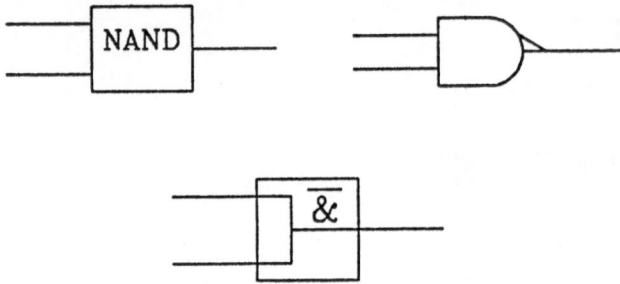

Fig. 15.4. Variant component representations

In order to achieve this flexibility, data-driven techniques are used throughout the SIDESMAN system in various guises. As a result, the system has the potential to be a suitable support environment for electronic CAD for a wide range of architectures, hardware implementation strategies and technologies. This approach is different from many commercial CAD systems (ULACAD[1] 1986) which are specific to given technologies and implementation processes. Other systems (SL-2000[2] 1985) offer generality to the user by imposing most of the responsibility for validation of technology and implementation constraints directly on the individual designer.

3.3. Technology Adaptability

The most important example of the use of 'data-driven' techniques in SIDESMAN is associated with the methodology adopted to support technology adaptability. Unlike systems which use 'neutral' models to achieve some degree of independence from predefined technologies, the approach adopted in the SIDESMAN system is to specifically ensure that the flexibility is incorporated into the CAD applications themselves. CAD programs are, therefore, written without technology related rules embedded in their code, but those same CAD applications become fully technology dependent during execution. This is achieved by dynamically passing technology rules to the application program from a Technology Design Rule Database (Aude and Kahn 1986). These design rules are then executed within the scope of the data structures of the application program. In other words, the rules examine and manipulate the data structures of the application. Hence, technology rules, which in other systems are totally embedded within application program code, are explicitly excluded from program code and are defined in an independent manner.

The design rules that may be defined within the Technology Design Rule Database have a wide range of application, including architectural, electrical, geometric and testability rules. In current CAD systems, technology constraints are mainly limited to geometric issues; in SIDESMAN, the Technology Design Rule Database makes it

---

[1] ULACAD is a trademark of Ferranti Electronics Ltd.

[2] SL2000 is a trademark of the Silvar-Lisco Corporation.

possible to offer more intelligent support. For example, rules to indicate which hardware structures are appropriate for certain technologies can be described in the database. As a result, an intelligent data capture application can be tailored so that it only suggests structures which are permissible for the user's selected technology or a (possibly) less intelligent application can be tailored to validate a user defined design in any technology.

The design rules are defined in a language, DRDL (Design Rule Description Language), which uses the entities, relationships, and functions defined in the system-wide data dictionary. A DRDL rule takes an English-like form and includes constructs such as the *there is* sentence, which may be made conditional by an *if..then..else* clause or repeated using a *for every* clause. Both system-defined and user-defined functions are available. For example *current(wire)*, *width(wire)*, *spacing(shape* 1, *shape* 2). Relational expressions and clauses define relationships between entities, for example

a shape1 is around a shape2 or pin1 is in a net with topology = 'bus'.

In order to control the focus of attention and to allow rules to be categorised in a supportive manner, special attributes may be associated with a rule. These attributes are classified as *tool* related, *design area* related and *design level* related. The following example illustrates a rule and its associated attributes. The attribute settings for this example allow the controlling environment to ensure that any *layout* application dealing with the *routing* at a relatively high (i.e. *system design*) level has access to this rule.

rule ecl01

> *for every path representing a net with environment = 'transmission line',*
> *topology(path) = 'chain link';*

Attributes:

> *Tool related:Routing*
> *Design Area:Layout*
> *Design Level:System Design*

In the Technology Design Rule Database, the classification of rules is by technology, sub-technology and fabrication process (e.g. MOS, CMOS, CMOS P-WELL, etc.). For any particular technology, say CMOS P-WELL, the rules that apply are all those defined for its antecedents (MOS-CMOS) excluding any rules explicitly stated as inapplicable.

The process of dynamically tailoring an application with these externally defined rules, involves four stages:

a. *Entity Mapping* which establishes the relationship between the SIDESMAN entities and the individual data structures within the application program. This stage is needed so that the executing rules associated with the design can run within the scope of the application data structures.

b. *Function Evaluation* which performs appropriate evaluations using the data structures mapped in the entity mapping process.

c. *Operator Execution* which is similar to *function evaluation*, but can modify the application data structures.

d. *Rule Execution* which coordinates the execution of the rules.

This process has been implemented in a prototype version to allow design rules to be applied to a geometric cell generation and manipulation program called DEMAND (Spink and Kahn 1986). Currently, use is made of the design rules to ensure that the spacings between geometric objects is correct, and to ensure that the system recognises semantic objects such as transistors which the user may have deliberately or inadvertently created out of basic geometric figures.

The following examples illustrate the DRDL form of rules passed to the DEMAND program to enable it to carry out width and spacing checks. The entities, such as *layer* and *shape*, referred to in the rules are mapped into appropriate objects in the data structures of DEMAND prior to operation and rule execution according to the process indicated above.

rule nmos01:

> *{This is the minimum width rule for NMOS Polysilicon}*
> *for every shape, in a layer, in a polysilicon,*
> *width(shape) >= 2*lambda;*

rule cmos01:

> *{This is a CMOS diffusion minimum spacing rule}*
> *for every {shape1, in a layer, in a diffusion} and*
> > *{shape2, in a layer, in a diffusion}*
> *such that shape1 is different from shape2,*
> *spacing (shape1, shape2) >= 4*lambda;*

These rules are precisely those that might have been hand-coded in Pascal into the source of the DEMAND program. However, if they had been coded in directly, DEMAND would have been limited to a specific technology class; because the rules are defined externally, DEMAND is capable of checking geometric correctness for any technology for which rules are defined.

Another way of using the rules from the design database is to execute them directly within a Prolog environment (Aude 1986). This usage is limited to performing tasks such as Design Rule Checking of a complete design at the detailed physical level. This facility offers a relatively simple way of prototyping a design rule checking program, but in practice is too slow to be used in a production environment.

3.4. Knowledge-Based Techniques

The aspect of SIDESMAN which is most obviously relevant to the development of an 'intelligent' CAD system is concerned with the use of knowledge-based techniques. A number of prototype applications have been developed using the knowledge-based environment created for the SIDESMAN system. These are noted here in order to indicate the diversity of applications to which the techniques are applied and to provide a focus for comments on the requirements for knowledge-based approaches.

All three applications described are implemented in PESL (Prolog Expert System Language). PESL stores facts and rules as pseudo-natural language sentences (Steele 1981), for example, *X is fact* or *X if A and B*. The use of pseudo natural language is convenient as it allows the rule base to be examined easily and supports constrained explanations. Rules may be qualified by certainty factors where appropriate. This facility ensures that rules derived from 'secure' sources (such as the component library) can be sensibly combined with other less secure, possibly heuristic, knowledge.

Ideas concerning the combination of conventional and knowledge-based techniques have resulted in PESL containing facilities, called *actions*, for access to procedural knowledge. This could, of course, have been limited to the use of Prolog in procedural rather than declarative mode. However, that solution would not have satisfied the criterion that access to CAD applications written in conventional languages (Pascal, Fortran) should be supported. PESL *actions* can be activated when a rule succeeds. This allows the circumstances under which a procedure is used to be fully controlled by the knowledge-based system. The following example illustrates an action firing a conventional routing application by calling *track( )*:

route point P1 to point P2 for net NETNAME and action track(NETNAME, P1, P2, WD)

pre-conditions not variable P1 and not variable P2 and not variable NETNAME

if

    initialised router and
    wrong direction movement for net NETNAME connecting point P1 to point P2 is WD

explanation

    This rule attempts to route point P1 to point P2 for the net NETNAME provided the router has been initialised and that a wrong direction movement WD for this connection has been derived.

justification

    A net can be routed provided the router has been initialised successfully and a reasonable wrong direction movement has been derived for this connection. The threshold of 1.0 controls the uncertainty mechanism in order to ensure that the PESL system is completely sure that both conditions are met before the track( ) procedure is called.

The flow of reasoning in PESL can be strictly scheduled using contexts. These allow focus of attention to be controlled and also support the structuring of rules into 'modules'. Scheduling is achieved by specifying the temporal relationship between contexts. Examples here include:

Context A before Context B and Context C again after Context B.

An important requirement of PESL is that it should provide some explanation capability. Currently this is supported partly by allowing free text to be associated with rules and partly by information carried in the system-wide Data Dictionary. The text is output in response to *why?* queries. If variable values are available, these may be embedded in the text. This facility is seen as a basic minimum for explanation. In the CAD context, more is needed. In particular, the ability to output graphical explanations would greatly enhance the naturalness of the system from a design engineer's point of view. Other constrained explanations in response to *how* and *why not* queries are felt to be of limited usefulness so browsing systems are being considered as a way of allowing the user to wander about the knowledge base.

Three quite different applications have been developed using PESL. The first of these, NOP, was a knowledge-based process which could control the behaviour of a conventional routing program (Kahn and Filer 1985). This initial work showed how a rule-driven environment could be used to assess the current state of a design and as a result influence the actions of a conventional application. The result of adopting this approach is that a previously unintelligent application program can be made to appear considerably more intelligent.

An entirely different process which is currently under development is a cell abutment 'expert', EASY (Expert Abutment SYstem). This is activated from the interactive cell geometry manipulation application, DEMAND. The role of the knowledge-based system in this case is to assist the conventional application in making decisions in an uncertain environment. Within VLSI chips, cells may abut under a variety of circumstances. Although abutment is often a desirable strategy, it is not easy to identify which cells should abut and in what configuration. Figure 15.5 illustrates one of the simpler conditions which control abutment. In Fig. 15.5(a) the desired interconnections align precisely, so abutment is a possible strategy; in Fig. 15.5(b), the crossing of interconnections rules out abutment. The strength of the rule is indicated (0.8) in order to show how strongly the evidence supports the current hypothesis. Part of the EASY rules to assist in this decision making is:

cell CELLA is a candidate for abutment with CELLB

strength 0.8

preconditions not variable CELLA and not variable CELLB

if

> *ports on cell CELLA which are connected to cell CELLB can be aligned with the ports on cell CELLB which are connected to cell CELLA such that only connected ports will abut.*

justification

> *In order to abut two cells the connected ports on both cells must be aligned so that connected ports touch in such a way that all relevant design rules for connected objects are satisfied.*

The third application, called HOPE (Laithwaite 1986), was developed as part of a commercial CAD system rather than SIDESMAN, but made use of the PESL support environment. The problem to be solved by HOPE is twofold. Firstly, a knowledge-based tool was required to advise designers on how best to use a conventional language design placement program. The algorithm used by this program is particularly sensitive to the hierarchy used in the design, but there were no tools or metrics available to designers to tell them how the design hierarchy should be manipulated to yield a satisfactory layout. The second requirement followed on from the first in that the aim of HOPE was to advise the designer, not replace him. The system therefore included a

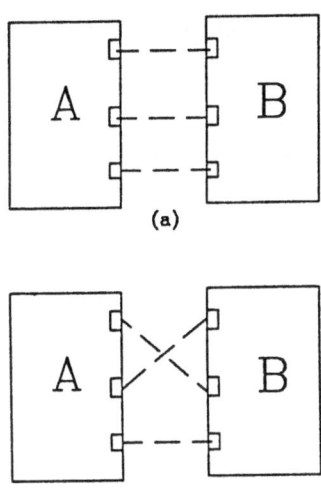

Fig. 15.5. Simple abutment

conventional interactive 'front end' from which the knowledge-based process could be activated if required. HOPE, therefore, combines features of both NOP and EASY in that it controls a conventional application (such as NOP) but is itself callable from a conventional application (such as EASY).

4. SUMMARY

The process of designing and developing any CAD system for VLSI Design is a complex task. The support provided for such a system must include an integrated environment with sound interfaces between the user and the system and within the system itself. This can only be achieved if a well-structured data model exists to support the user environment. When the additional requirement to include 'intelligence' in the CAD system is postulated, it is sensible to ensure that the conventional CAD system provides a support environment for emerging 'expert' applications.

The addition of knowledge-based techniques offers opportunities to improve the behaviour of the CAD system, but it is not realistic to assume that such support can come from knowledge-based tools alone. Therefore, the knowledge-based tools must be interfaced to conventional CAD applications, either to provide processes which can operate within uncertain environments or to provide intelligent interfaces between the user and conventional applications. The work carried out on the development of the three prototype 'experts', NOP, EASY and HOPE, indicates how successful such a combination of knowledge-based and conventional applications can be.

There is, however, an additional requirement imposed by CAD for VLSI. This arises from the complexity of the design process when various technologies, implementation strategies and target architectures are used. These technology-related characteristics change very frequently as advances in device physics and chip implementation techniques occur. An experienced designer is aware of the effects of varying these characteristics and he uses this knowledge of technology issues to assist him in tasks such as the selection of hardware structures, evaluating design performance, and achieving area-efficient implementations. If a CAD system is to behave intelligently, it must be able to assist with many of these technology-related decisions and it must ensure that the knowledge it uses is always up-to-date. The techniques used in the Technology Design Rule database system developed for SIDESMAN offer a powerful way of handling this problem. By providing a means of expressing technology-related rules explicitly in a manner independent of an individual application, the system can share rules between applications, support browsing through the rules to learn about design strategies, and dynamically readjust application software.

ACKNOWLEDGEMENTS

Part of the work described in this paper is supported by the Science and Engineering Research Council as part of the Alvey Initiative (Project CAD011). This support is gratefully acknowledged. Thanks are, of course, also due to all the members of the CAD group who have worked with enthusiasm and interest on the various aspects of SIDESMAN. Particular thanks are due to Hilary Bannister and Nick Filer, whose help during the preparation of this paper was invaluable.

REFERENCES

Aude, J. S. and Kahn, H. J. (1986): "A design rule database system to support technology adaptable applications," *Proceedings of 23rd Design Automation Conference*, Las Vegas, USA, pp. 510-516.

Aude, J. S. (1986): "Design Rule Representation within a Hardware Design System," Technical Report (Ph.D. thesis) No. UMCS-86-11-4, University of Manchester, Department of Computer Science, Manchester, UK.

Bergmann, N. (1985): "Generalised CMOS — A technology independent CMOS IC design style," *Proceedings of 22nd Design Automation Conference*, Las Vegas, USA, pp. 273-278.

Carpenter, A. F., Gosling, J. B., and Kahn, H. J. (1986): "An accurate hierarchical simulation engine," *Proceedings of 3rd Silicon Design Conference*, London, pp. 257-266.

Kahn, H. J. and Filer, N. P. (1985): "An application of knowledge-based techniques to VLSI design," *Proceedings of Expert Systems 85*, Warwick, UK, pp. 307-321.

Kahn, H. J. (1986): "A multi-view data model for VLSI," *IEE Colloquium on Design Databases, Digest No. 1986/29*, London, pp. 8/1-8/4.

Krekelberg, D. G., Shragowitz, E., Sobelman, G. E., and Lin, L. (1986): "Automated layout synthesis in the YASC silicon compiler," *Proceedings of 23rd Design Automation Conference*, Las Vegas, USA, pp. 447-453.

Laithwaite, R. N. W. (1986): "An expert system to aid placement on gate arrays," *Proceedings of 3rd Silicon Design Conference*, London, pp. 305-313.

Royal, N., Hunter, J., and Buchanan, I. (1985): "A case study in process independence," *Proceedings of 22nd Design Automation Conference*, Las Vegas, USA, pp. 591-596.

SL-2000 (1985): *SL-2000*, Silvar Lisco Corporation, Menlo Park, CA, USA.

Spink, M. A. and Kahn, H. J. (1986): "Hierarchical design manipulation within a general silicon compiler," *Proceedings of 3rd Silicon Design Conference*, London, pp. 97-104.

Steele, B. D. (1981): "EXPERT — The Implementation of Data-Independent Expert Systems with Quasi Natural Language Information Input," M.Sc. thesis, Imperial College, Department of Computing Science, London.

ULACAD (1986): *ULACAD Design System*, Ferranti Electronics Ltd., Manchester, UK.

van Brunt, N. (1983): "The ZYCAD logic evaluator and its application to modern system design," *Proceedings of IEEE ICCD 83*, Port Chester, NY, USA, pp. 232-233.

16. SYNERGIST: A Schematic Capture and Fault Diagnosis System

R. Milne

Intelligent Applications Ltd., Kirkton Business Centre
Livingston Village, West Lothian EH54 7AY, UK

Abstract: *In the past few years, the use of computer aided design for electronics has become widespread. However, automated testing support from design is only now beginning to take place. In this paper, we describe a system, Synergist, which derives automatic testing procedures from the computer aided design description of a circuit. Synergist uses the Artificial Intelligence technique of structural isolation in order to automatically derive testing procedures for printed circuit boards.*

Keywords: artificial intelligence, CAD, schematic capture, fault diagnosis, structural isolation.

1. INTRODUCTION

The process of electronic circuit design fabrication, layout and testing is gradually becoming automated. Currently, there are a wide range of tools available to automate the design process of circuits. Many of these tools provide checking to assist the designer and make sure he has not made mistakes. The more powerful CAD tools also interface to a wide range of printed circuit board layout and fabrication tools. Today, however, only the most advanced tools provide any support for testing. Our research is focused on how to develop testing procedures from the design of a circuit. Our current work is focused at printed circuit boards from the high level description of board down to component level. The goal is to develop effective testing procedures from the design information of a circuit and simulation data with little or no human intervention or additional assistance.

In current CAD practice, the designer is usually left to develop a circuit without any thought or concern for testing. The test engineer is generally not involved until after the design phase has finished. In our work, we want to bring those two aspects, design and test, together. If one can automatically decide how a circuit would be tested from the design then it is possible to give better testing feedback to the designer as well as designing more effective tests by capturing design information. As with other systems, such as software development, testing should be considered an integral part of the design system. Better tools are needed so that testing is automatically done as the design is produced, resulting in both better and more effective testing, but also better and more effective designs.

This work relates to computer aided design in two ways. Although strictly not a design tool, schematic capture systems are an important part of any CAD system. In the second half of this paper, we discuss our intelligent schematic capture system based upon object oriented programming. Although not part of the design process, using computer aided design databases to derive test results is an important use of CAD systems. In this paper, we build upon the capabilities of CAD systems and tools to drive and derive testing. We assume that the design process has been completed or is at an intermediate stage where testing now needs to be evaluated.

Once a circuit is designed, testing occurs in two major phases. First, the circuit when it is first fabricated must be tested to verify that the fabrication was accurate and the circuit is working. For digital circuits, many modern computer aided design systems provide assistance in developing test vectors and then generating patterns for digital test equipment from these test vectors. That work tackles a very important bottleneck, namely, developing automatic test programs from the basic circuit design. It suffers currently from several limitations. First, it only covers digital circuits. Second, there is much additional work needed to develop automatic test equipment programs, such as board identification, thorough checking for shorts and opens, and understanding the limitations of the test fixture. Finally, it still requires extensive manual intervention.

The second major phase of testing occurs after the board has been proven and initially used and then a fault somehow develops. Normally, one can assume that the board has passed the design and fabrication tests and been in use for one or two years. It is now returned to a repair centre which must isolate the fault and make an effective repair. Usually this repair centre does not have access to the full design information and many times the repair is done almost exclusively in a manual fashion.

Our work is focused on this second phase of testing, that of manual repair for printed circuit boards after they have been in use for a period of time. We have not specifically addressed testing with automatic test equipment, although many of the techniques discussed here will be applicable in developing more powerful connections between design and automatic test.

The *Synergist* system has two primary goals. The first one is to derive an effective testing procedure using as little human intervention as possible. It is desirable to use the schematic and wirelist of a circuit and simulation data to automatically derive an effective test procedure to assist with manual testing. The second goal is to provide a primary tool for a test engineer performing manual or benchtop troubleshooting. The output of the system is not directed towards automatic test equipment but towards the items a manual test engineer needs to know. These two requirements generate a variety of specific functions that a system such as *Synergist* must perform in order to be useful in tracing signals and signal paths through a schematic. These techniques are very important since the main bottleneck in manual troubleshooting is the length of time a test engineer requires. A large portion of that time is spent tracing signal paths on a paper schematic. Automating this step alone provides a significant increase in the effectiveness of manual testing. The other class of testing procedures are designed to take the schematic, wirelist, and simulation data and automatically develop testing procedures. The key technology for these is structural isolation which will be described in more detail later in the paper.

This paper is organized as follows. In the next section we examine the high level strategies which *Synergist* is able to derive from a circuit description as well as other features which assist with testing. We then explore the fundamental techniques — structural isolation — upon which *Synergist* is built. We finish with a discussion of the intelligent schematic entry portion of the system used to introduce circuits into *Synergist*.

2. AUTOMATED FAULT DIAGNOSIS

Synergist is designed to assist the test engineer who manually troubleshoots printed circuit boards using only a schematic, a test manual, and some basic test equipment. *Synergist* offers automated fault diagnosis for faster and more accurate testing. Once a schematic has been captured and annotated with test information, *Synergist* can diagnose the circuit based purely on the wirelist computed from the schematic.

The types of test information that can be annotated on a schematic include voltage information, the waveform patterns that should exist on a data bus, the sine wave or square waves that should exist at various points in the circuit, or the images that should be seen when using logic probes and other instruments. Additionally, the schematic can be annotated with testing procedures indicating where to test first and, in each instance, what to do if the test fails.

Synergist, designed by Intelligent Applications Ltd., is a fault diagnosis and schematic capture system for board-level circuits. Typically, a senior test engineer captures the schematic electronically into *Synergist* and then test engineers can use *Synergist*'s diagnostic methods to diagnose faulty boards.

Synergist offers five different methods of fault diagnosis. Test engineers can use one or any combination of the methods to completely test a printed circuit board:

Connectivity — Using connectivity (the connections between components), the test engineer can work backwards from an output on a diagram to its inputs, testing each connection to compare the value shown on the schematic with the voltage on the board. When the faulty component is pinpointed, the user can quickly switch to a diagram of that component and test it.

Test points — The test engineer can use test points on the schematic to have *Synergist* guide him through tests described in the circuit's test manual. For each test, *Synergist* prompts him to set up conditions and compare the values on the schematic with the values for the board. If a test fails, *Synergist* informs the engineer what to test next.

Automatic path searching — Synergist provides automatic path searching between two points on a diagram to pinpoint bad data paths. *Synergist* finds the middle of a path and asks whether the value at that point is good or bad. If it is good, *Synergist* then makes a new path of the second half and searches for its middle. If the value is bad, *Synergist* then makes a new path of the first half and searches for its middle. It then asks whether the value at this midpoint is good or bad. *Synergist* continues in this manner, until it has narrowed down the number of components to the faulty one.

Automatic switching between diagrams — Synergist can automatically switch to diagrams that connect with the inputs and outputs of the current diagram, which makes it easier to move around the design while testing it. The automatic path searching is implemented through a variety of traditional search approaches. Either within an individual sub-diagram or through the entire circuit, a depth-first or breadth-first search can be employed to discover where connections lead.

Manipulating the schematic — Synergist's schematic capture tools allow the user to describe board-level circuits comprised of a mixture of analogue and digital components through a hierarchical system of black box and detailed views. A schematic can be annotated with as much information as desired, such as the names of signals and various signal traces, the voltage values that should be found at each point, and the connections and signal paths in the circuits.

Synergist also outputs a knowledge representation schema, suitable for loading into other expert systems. Currently, *Synergist* is coupled with three expert systems: KEE[1], OPS5 and YAPS. (YAPS is a common forward-chaining production system). Once a design has been captured, *Synergist* can output a wirelist suitable to wire boards and to diagnose faulty boards.

In *Synergist* the schematic is represented in an hierarchical form where each black box has a detailed view containing other basic components or other black boxes. For example, an entire transmitter is one black box, but may be made up of four or five printed circuit boards. Each of these may contain several logical sub-divisions and logical sub-circuits. When a test engineer would like to find out where a particular electronic path leads from one diagram to the next, he merely indicates this to the system and the system automatically moves from one diagram to the next, always trying to go to the highest levels of the hierarchy and then working its way down to the lowest level.

3. DIAGNOSTIC STRATEGIES

Let us now turn and study in more detail the diagnostic reasoning that is used in this approach. This is normally classified as a *structure and function expert system* (Milne and Chandrasekaran 1985).

---

[1] KEE is a trademark of IntelliCorp.

In early expert systems (Barr and Feigenbaum 1981), knowledge was represented in the form of production rules. The inference engine was a very simple mechanism. It merely found the set of rules that could conclude the current hypothesis, and tried each one in turn. On occasion, in order to satisfy a rule, other rules would be needed. This led to recursion and backward chaining.

This approach worked for simple problems. In the early days of expert systems development, the systems could only solve simple problems. Expert systems could be developed because the inference engine gave a very simple paradigm for solving problems. After a time, however, problems became too complex for simple rule bases. Especially in diagnostic reasoning, *structure and function* (Milne and Chandrasekaran 1985) has begun to be used to guide the effort to isolate the fault. *Structure* is first used to isolate the region which contains the fault. *Function* is used two ways: first to assist with isolating the fault in the structure, and then to further reason about the possible faults. In this paper, we will only discuss reasoning from structure.

3.1. Structural Reasoning

One of the simplest types of knowledge that one can use in a diagnostic system is structural or connectivity information. In many systems, it can be difficult to understand or describe the behaviour and function of the system, but it is simple to describe the connectivity. Diagnosticians are also good at using structural information to guide a diagnosis, even when they do not understand the function of part of the device. Reasoning about structure can be very efficient, hence simple and fast. As a result, it is often desirable to use structural information to isolate a faulty region and then employ more complex reasoning to identify the exact problem.

Structural information also helps to provide *closure* in building systems and guarantee *completeness*. Closure is gained by using a simple, uniform inference mechanism to derive a large number of possible faults directly from the description of the system. Instead of writing hundreds of rules, the system builder merely describes the connectivity and the system generates the rules. Completeness is derived from examining all structural connections and paths, guaranteeing that nothing is forgotten. Once the structure has been used to isolate the fault to a region, functional information can be used to further reason about the possible faults.

To use structural isolation, one first identifies the paths of information flow from the inputs of the device of interest, to the outputs. The paths contain all the sub-components involved in that information flow. If there is a problem with a particular path, structural information can only say that any element of that path may be faulty. Many troubleshooting systems work by using this simple step alone. If there is no output at all from that path, then structural information can only declare that the path is broken at some point. Again all elements of the path are possible candidates.

One normally uses this approach by collecting two sets of known good and bad paths. By intersecting these sets, faults in some portions of the system can be ruled out. That is, the sub-paths of a bad path, that also occur in a good path, can be assumed to be correct. The candidate set is then reduced to only those elements which occur in bad paths. This decision is based solely on *GO/NOGO* information and can be very efficient.

Given a path of several possibly faulty elements, it is desirable to isolate one bad element. This is done by testing for *GO/NOGO* between each element. Each test should generate a new sub-path. In order to minimise the number of tests, the simplest algorithm is to split the possible fault paths in half. In order to pick the optimal test to perform next, information about the cost of each test, the reliability of each element and the possible information gain of each test can be considered.

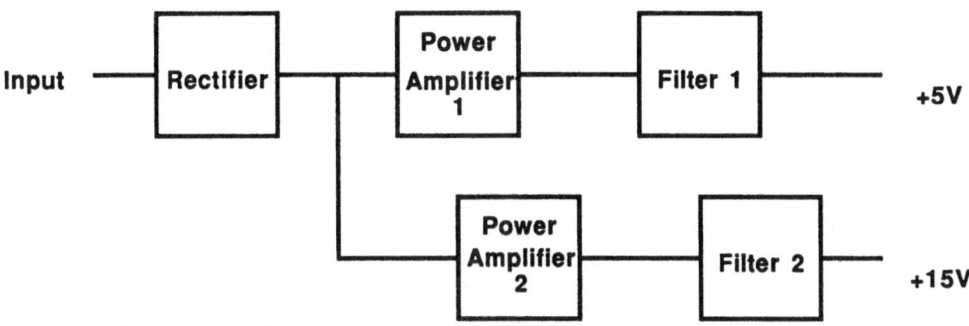

Fig. 16.1. Structural isolation

Figure 1 illustrates structural isolation. There are two outputs of the system, +5 Volts and +15 Volts. There are also two primary paths. From the AC Input through the Rectifier, Power Amplifier-1, and Filter-1 to the +5V output, and through the Rectifier, Power Amplifier-2, and Filter-2 to the +15V output.

Assume that the +15V output is good. Then we can assume that the Rectifier, Power Amplifier-1, and Filter-1 are good. If we then measure the +5V output and discover that it is incorrect, we can conclude that that part of the path leading to the +5V that does not intersect with a known good path (Power Amplifier-2 and Filter-2) are possibly faulty.

We now have a possibly faulty path of Power Amplifier-2 and Filter-2. The best measurement to make would be to split this path equally, that is to measure between the two components. If the signal is correct at this point, then we can conclude that Filter-2 is faulty. If the signal is incorrect, then we can conclude that the problem is in Power Amplifier-2.

Eventually, the presence of faults becomes insufficient to reduce the number of possible candidates any further. At this time, the expected outputs of the devices which make up the system can be used to further identify the fault. By examining how the function of each device should alter its inputs, candidates can be eliminated. Davis and Shrobe (1985) use the propagation of values and the function of digital devices to rule out possible faults. Genesereth (1984) further reduces the possible faults by reasoning which inputs could not produce the wrong output. For example, if an AND gate has a 0 on *input1* and a 1 on *input2* and produces 1 on the output, then we conclude that *input1* is at fault since its 0 value was responsible to cause the output to be 0. This approach reduces the search considerably.

If there is no output at all, then no information is present on which to base reasoning about the function. In this case Milne (1985b) uses the ways devices can behave such that they produce no output and propose faults. For example no current will flow through a resistor if it is faulted in the open state. In a typical series circuit, there is only one path for the current, so each resistor could be open.

When the structure cannot be used to further isolate the fault, then function alone must be used. Often in electronic circuits, it is not possible to test between groups of components. In this case function can be used to deduce the possible fault. The cited works of Davis and Shrobe, and Genesereth illustrate this in the digital domain, where the function of the devices is invertible and values can be propagated backwards. The cited work of Milne explores what may be done when the functions cannot be inverted and hence only forward reasoning is possible.

Scarl et al. (Scarl, Jamieson, and Delaune 1985) describe a more complicated reasoning mechanism to declare whether a device could not have caused the fault by inferring that their function could not have caused the possible fault. This work illustrates the reasoning and issues involved in using function in this way. White and Fredericksen (1985) use a simple view of function in electrical circuits to rapidly guide a binary search for the fault. Their approach is at the other end of the spectrum from Scarl's work. Whereas Scarl et al. use a complex combination of reasoning, White's simple method is just as effective although not as general.

4. INTERFACING WITH OTHER EXPERT SYSTEMS

The previous discussion has focused on how to use the schematic as the primary troubleshooting vehicle. Another way to look at computer aided design systems is that the design system is merely a graphics front end to a larger expert system. From this viewpoint, a test engineer would like to develop an expert system using a variety of toolkits to troubleshoot the circuit. However, traditional means of developing expert systems require extensive work to build the description of the circuit as part of the knowledge base. For a simple circuit, it is necessary to create a knowledge base entry for every component and all its values. It is also necessary to create a knowledge base entry for every node and set of connections. All of these items must automatically keep track of what they point and are connected to. One of the defects of such an approach is that if a minor change is made to the diagram, extensive knowledge base editing is required.

The ideal is to be able to change the diagram and automatically have the knowledge base description updated. Synergist provides this capability. Its output can be sent directly to either the KEE expert system shell from IntelliCorp or expert system shells in the family of OPS5 using simple data structures and very fast forward chaining. In Synergist, this is accomplished by a reasonable mapping between the object oriented representation of the circuit and the types of data structures the rule-based expert systems use.

5. SCHEMATIC CAPTURE

All previous discussions in this paper have focused on how to do testing once the schematic is actually entered. In the *Synergist* system, we have developed our own object oriented schematic capture system to facilitate this. Although the functionality of *Synergist*'s schematic capture will be familiar to those knowledgeable in the computer aided design area, the key difference is the way it is implemented through object oriented programming. Many of the key features, such as automatic alignment are taken care of simply by sending messages to the objects to instruct them to be at the same vertical or horizontal alignment.

Typically, the senior test engineer uses the schematic capture portion of *Synergist* to draw the schematics of a circuit on the computer screen. The schematics are entered as modular diagrams in a hierarchical system so that diagrams can be arranged and abstracted as appropriate. Typically, a schematic design has one top-level view which contains many abstract modules. A module itself may contain many other modules and so on. The entire design is then represented as a hierarchy of modules. The *Synergist* drawing routines are fast and flexible. Also, once captured, a component or sub-circuit can be reused any number of times in any designs in *Synergist*.

5.1. Schematic Capture Features

Synergist provides many features which make it easy to use:

Automatic alignment — *Synergist* automatically maintains the connections between components in diagrams during moves. For example, if you move one component that already has connections to other components, *Synergist* adjusts the lines so that they continue to connect with both components as before.

Annotation of diagrams — *Synergist* allows the user to annotate each view with different kinds of text. Signal text matches the signals from one module to another. Port text matches the connections between the views in a module. Comment text allows the user to write comments directly on the view. Different sizes and typefaces of fonts are available to help the user visually distinguish types of text.

History of views — *Synergist* keeps a history of views that have been displayed during a session. The user can list this history at any time and rapidly redisplay any previous view.

Multiple screen configurations — Many different screen configurations are available at all times, allowing the user to choose the one most appropriate for the current task. For example, the user can display two views at the same time, either side-by-side, or one above the other. A chip menu can be displayed, from which chips can easily be copied into the design. The user can choose between different sizes of drawing and mouse prompt windows.

Types of commands — Most drawing commands are called with a combination of mouse buttons and key strokes. Each type of view has a set of commands for editing it. The entire set of commands can be displayed while drawing so that the user can choose a command. Other commands are available as extended commands or on menus and can be called while displaying a view.

Online help — *Synergist* provides extensive online help at any time in all parts of the system.

Customising the environment — The user can customise the environment by choosing different values for a set of user variables. These values can be saved to the user's directory and reloaded.

Scaling — A range of scale sizes is available for changing the display and actual sizes of drawn objects. Two grids with variable sizes help the user place and select objects on the screen.

Undo facility — An undo facility can sometimes reverse the last action performed.

Synergist provides other features which make it easier for the user to draw schematics, such as:

- automatic black box-creation, automatic port creation, auto-increment of component or port names, and automatic bus connector creation.
- text centering and multiple fonts.
- picture centering and scrolling facilities.
- rotation of placed components.
- vertical and horizontal alignment of placed components.
- copy facilities.

The schematic capture system is built upon object oriented programming. In this programming paradigm each element of the schematic, that is every actual component such as a capacitor, a resistor, a transformer, is an object. Each line is also an object as well as each port and each point and each group of text. Items can be moved in a uniform way by sending them the move-message. If the user attempts to move a line versus a bit of text or a capacitor, radically different things happen. Yet, from the software viewpoint, they all receive the move-message and through information hiding, the details of how each move is taking place is concealed from the programmer.

Object oriented programming makes many useful features easy to develop. For example, maintaining vertical alignment of a series component is accomplished in the following manner. When an object is moved, it tells each of its ports to send a message to anything connected to it to move as well and stay vertical. Each of those connections then receive a move message which pass the move message on to any of their connections. As a result, by automatically sending move messages, a variety of diverse components can be moved in a simple way. Because a diagram is fundamentally composed of many separate objects, each with high level uniform commands that can be asked of them, but fundamentally different low level ways they should be performed, schematic capture is a natural for object oriented programming. The uniform meetings of knowledge representation and message passing provide an efficient and rapid way to program and develop the system as well as providing cleaner and readable code.

6. CONCLUSION

In this paper, we have examined *Synergist*, a system designed for intelligent schematic capture and fault diagnosis. *Synergist* is designed to be used at an intermediate or final stage of design to give help for area of design testing. In this paper, we have examined ways in which *Synergist* automatically derives testing procedures from the design using the Artificial Intelligence technique of structural isolation. We have also briefly discussed the schematic capture front end to the system, built upon object oriented programming.

REFERENCES

Barr, A. and Feigenbaum, E. (eds.) (1981): *The Handbook of Artificial Intelligence, 3 Volumes*, William Kaufmann Press, Palo Alto, CA, USA.

Cantone, R., Lander, W. B., Marrone, M. P., and Gaynor, W. (July 1985): "Automated knowledge acquisition in IN-ATE using component information and connectivity," *ACM SIGART Newsletter*, **93** , pp. 32-34.

Clocksin, W. and Morgan, A. (July 1986): "Qualitative control," *Proceedings of ECAI-86*, Brighton, UK, pp. 350-356.

Davis, R. and Shrobe, H. (July 1985): "The hardware troubleshooting group," *ACM SIGART Newsletter*, **93** , pp. 17-20.

de Kleer, J. (September 1979): "Causal and teleological reasoning in circuit recognition," Technical report No. TR-529, MIT AI Lab., Cambridge, MA, USA.

Erman, L. D., Hayes-Roth, F., Lesser, V. R., and Reddy, D. R. (June 1980): "The HEARSAY-II speech understanding system: Integrating knowledge to resolve uncertainty," *Computing Surveys*, **12**(2), pp. 213-253.

Forbus, K. (1984): "Qualitative process theory," *Artificial Intelligence*, **24** , pp. 85-168.

Fox, M. S. (August 1983): "Techniques for sensor-based diagnosis," *Proceedings of the Eighth International Joint Conference on Artificial Intelligence*, Karlsruhe, Germany, pp. 158-163.

Genesereth, M. (January 1984): "The use of design descriptions in automated diagnosis," Stanford Heuristic Programming Project Memo No. HPP-81-20, Stanford University, Stanford, CA, USA.

Harmon, P. (August 1986): *Expert Systems Strategies*, **2**(8), Cutter Information Corp.

Milne, R. (October 1985): "Diagnosing faults through responsibility," *1985 ACM Annual Conference*, Denver, CO, USA, pp. 88-91.

Milne, R. (August 1985): "Fault diagnosis through responsibility," *Proceedings of the Ninth International Joint Conference on Artificial Intelligence*, Los Angeles, CA, USA, pp. 424-427.

Milne, R. and Chandrasekaran, B. (eds.) (July 1985): "Reasoning about structure, behavior and function," *ACM SIGART Newsletter*, **93** , pp. 4-59.

Milne, R. (July 1985): "The theory of responsibilities," *ACM SIGART Newsletter*, **93** , pp. 25-30.

Milne, R. (May 1986): "Artificial intelligence applied to condition health monitoring," *Chartered Mechanical Engineer*, **33**(5), pp. 45-46, Mechanical Engineering Publications Ltd.

Milne, R. (April 1986): "Fault diagnosis & expert systems," *The 6th International Workshop on Expert Systems & Their Applications*, Avignon, France, pp. 603-612.

Milne, R. (April 1986): *Applications of Artificial Intelligence in Engineering Problems, Proceedings of 1st International Conference (1986), Southampton University, UK*, Springer-Verlag, Berlin, Heidelberg, New York, Tokyo.

Milne, R. (to appear, May 1987): "Strategies for Diagnosis," *IEEE Systems, Man and Cybernetics*, **17**(3).

Moore, R. (April 1986): "Expert systems in process control: Applications experience," in *Applications of Artificial Intelligence in Engineering Problems, Proceedings of 1st International Conference (1986), Southampton University, UK*, Sriram, D. and Adey, R. (eds.), Springer-Verlag, Berlin, Heidelberg, New York, Tokyo, pp. 21-30.

Morris, P. and Nado, R. (August 1986): "Representing actions with an assumption-based truth maintenance system," *Proceedings of the AAAI-86*, Philadelphia, PA, USA, pp. 13-17.

Scarl, E. A., Jamieson, J. R., and Delaune, C. I. (July 1985): "Process monitoring and fault location at the Kennedy Space Center," *ACM SIGART Newsletter*, **93** , pp. 38-44.

White, B. Y. and Fredericksen, J. R. (July 1985): "QUEST: Qualitative understanding of electrical system troubleshooting," *ACM SIGART Newsletter*, **93** , pp. 34-37.

Winston, P. H. (1984): *Artificial Intelligence*, Addison-Wesley, Reading, MA, USA. 2nd ed.

Report on Session 7

Chair: F.J. Schramel †
Cochair: P. Veerkamp ‡
Edited by: V. Akman ‡

In order to use an intelligent CAD system efficiently, attention needs to be paid to the user interface. Ruttkay presented a generic, multiparadigm user interface to be used with an intelligent CAD system for mechanical design. The demands for such an interface were given from the viewpoint of the engineer as well as integration with existing CAD systems. A general architecture for such an interface was proposed and a frame-based implementation was presented as an example.

Kahn presented an intelligent CAD system for VLSI design. Her SIDESMAN Design System is based upon the apprentice-like of approach. It provides the user with a total support environment which is sufficiently integrated and flexible to be an intelligent assistant. The system combines existing and new tools such as knowledge based design aids for VLSI.

Milne started by stating his belief that although a lot of attention has been paid to automate the design process, automated testing support has not been provided so far. He then presented a system which can be an assistant for automatic testing of design objects, especially for printed circuit boards. His Synergist system uses AI techniques developed for fault diagnosis to construct such a system.

During the Q&A period, ten Hagen asked Ruttkay what kind of strategy she uses to implement a multi-media interface and how she combines/splits the various semantic domains. Ruttkay's reply was that she would like to make some experiments on mapping the syntactically different languages on a common semantic representation. But that would be slow and expensive. Bernus asked if she thinks that the design supervisor and the communication supervisor have to share the same knowledge base. Ruttkay asserted that this depends on the amount of the semantics expected from the user interface. But even for the natural language grammar, she thought that concepts of the application domain should be used.

---

† Philips International B.V., Corporation ISA, Building VN 305, PO Box 218, 5600 MD Eindhoven, The Netherlands.
‡ Centre for Mathematics and Computer Science, Kruislaan 413, 1098 SJ Amsterdam, The Netherlands.

Ten Hagen directed a question to Kahn after pointing out that she adapted an existing system by simply adding rule based methods in an *ad hoc* manner: Does not this give problems with the architecture? Kahn said that the system is designed to include both conventional and rule based parts. She only needs a sophisticated architecture to improve performance and the new rule based methods are not *ad hoc*. Hoelzel commented that Kahn uses views to describe components and asked: How are they structured and how do you describe the different stages of the design process? Kahn replied that the views are loosely correlated with the design process. One has different types of views at different stages. A check is made whether one has the right pre- and post-conditions for the views.

Rogier asked Milne if he reuses the previously developed strategies, i.e. whether Synergist has a learning capability. Milne's answer was that the ordering of the tests can be changed since history data may be used. But some isolated problems cannot be handled.

Session 8

17. Intelligence beyond Expert Systems: A Physiological Model with Applications in Design

T. Takala

Information Processing Laboratory, Helsinki University of Technology
Otakaari 1, SF-02150 Espoo, FINLAND

Abstract: *Differences between human intelligence and knowledge engineering are observed. Based on a brain model developed in physiology, a structure for learning and creative expert systems is proposed. The model is related to general design theory. Generative grammars are pointed out as a tool for applications.*

Keywords: knowledge engineering, human reasoning, design theory, production systems, CAD.

1. KNOWLEDGE ENGINEERING AND HUMAN THOUGHT

Most CAD systems currently in use are intended for modelling a design object, in order to analyse it visually or computationally. They can mechanically perform algorithmic operations upon user command. Knowledge engineering (KE) techniques are recently used to increase the systems' reasoning ability to be more flexible and intelligent (Gero 1985; Yoshikawa and Warman 1987).

An apparent problem with expert systems is knowledge acquisition, which usually is a separate process done by knowledge engineers. Also the expert knowledge, like the knowledge in most artificial intelligence (AI) systems, is shallow, and specialised, and rather rigid. The systems do not easily adapt into changing situations during their use. This also means that design solutions tend to be bound to the predefined knowledge, and that unexpected, creative solutions are seldom found.

In order to envision the next step in the future, let us imagine a fictitious design assistant, who behaves like a human apprentice except for actually being an automatic robot. It would perform well-defined tasks exactly as it is told to do — just like a computer, but it would also have the ability to learn and to relate a particular situation to its wider experience. Left for a while without exact orders, it would soon play with something, or it might actively make suggestions on how to solve a current design problem. It would make mistakes (*errare humanum est*), but that is expected to happen sometimes anyway, and is thus acceptable.

Such fiction enlightens some fundamental differences between machine reasoning and human intelligence. One of them is that the latter is not strictly logical. Often human reasoning is based on fuzzy "maybe" rules or "it-reminds-me" associations. Another particular difference is that human mind is continuously active, not only answering questions but also making suggestions. Such properties should be added to expert systems in order to make them adaptive and learning.

2. A PHYSIOLOGICAL MODEL OF CONSCIOUSNESS

A model where both logical and free-associative aspects are combined is presented by the physiologist Bergstrom (1986) The model is depicted in Fig. 17.1. It consists of two parts one of which is the active, noisy generator (located in the brain stem), and the other is the logical, well-organised reasoning system (located in the cerebral cortex). Both are tightly interconnected to each other and to the sensory organs.

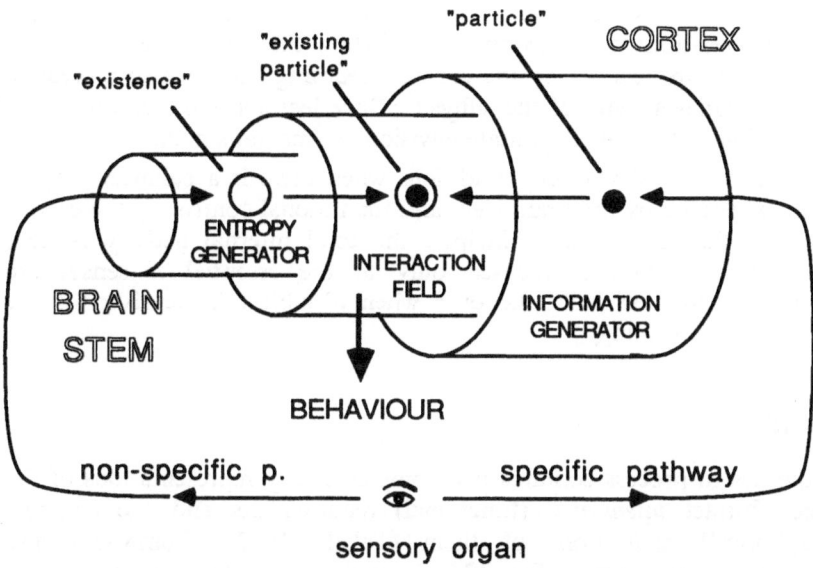

Fig. 17.1. The two-way model of sensory experience (Bergstrom 1986)

The model explains that the "understanding" of a sensory observation is two-fold. The specific neural pathways will bring the sensory signals from an organ to the corresponding areas on the *cortex* where they are organised and analysed into separated features together with their temporal and spatial order relations. On their way to the cortex, the specific pathways pass the *brain stem* which will also be activated by the sensory signals but in a non-specific way. Rather, its activation is a general arousal about the existence of something observed. The *consciousness* comes from the combination of signals from both channels at the *interaction field* (not yet located exactly, but probably in the columns of the cerebral cortex). "Knowing" a particle seen by the eye, for example, is the sum of its syntactic features (the neural microstate) and the holistic arousal that something is seen (the neural macrostate).

The *behavioral activity* comes out from the interaction field. It is the response partly to sensory experiences (reactions to external stimuli), but also partly to imaginary experiences (self-activation), which from the brain's point of view are as real as any. Imagination originates from the brain stem as a continuous stream of more or less random signals. They activate the cortex by starting chains of associations in it. When these chains return to the interaction field, they will be well-organised and are (together with the original activation) experienced as observations. Thus there are two signal sources competing for dominance in the interaction field: the chaotic "noise generator" (or entropy generator) of the brain stem, and the organising "active filter" (or information generator) of the cortex.

Creativity appears as the system's ability to produce new and unpredictable events and combinations. It is not possible without motivation, i.e. without certain initial activity from the brain stem. Rational *sensibility*, on the other hand, develops when consistent input/output relations of the external world (how the environment responds to the subject's excitation) are learned by the cortex. Learning itself is also a creative process, for it requires initial activity of the subject. Once learned, these relations can participate in imagination too and will facilitate envisioning (mental simulation).

The brain will have optimal working conditions when there is a balance between the "anarchic" noise generated by the brain stem and the rational control by the cortex. This is managed with the *homeostatic* principle: the total internal activity is kept roughly constant. The attention is focused only to the essential if senses are overloaded, and free imagination will take over when a subject is deprived of the external stimuli (e.g. while dreaming).

3. DESIGN THEORY

A model of the design process contains two complementary representations of the design object: the abstract *intensional* (functional) requirements and the concrete *extensional* (metaphorical) realisations of them (Takala 1987a; Tomiyama and Yoshikawa 1987). The design process (Fig. 17.2) is aiming at a situation where these are consistent with each other. The consistency can be checked with *analysis* functions which can recognise relevant properties of extensional representations, and can then verify if they agree with the requirements. The inverse of these analysis functions is design *synthesis*. It is a problem-solving process to find a solution for the given specifications and implicit constraints.

The more the analysis functions can be formalised, the more formal their inverse can be. If analysis can be done with backward reasoning by inference rules, the corresponding forward reasoning performs a synthesis process. Expert systems can be utilised in such design situations. However, often formal algorithmic synthesis is not

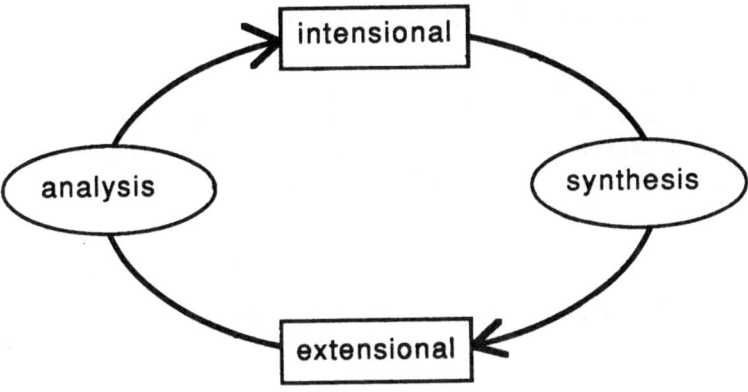

Fig. 17.2. The design process

possible. Designing is an iterative trial-and-error process. It is usually the designer's task to create potential solution proposals which can be checked by automated analysis functions. If there is no logical explanation of how such a proposal is derived, then it can be considered to be selected at random. *Randomness* in this sense includes heuristics and pseudorandom numbers — we neither know nor care how a selection is done.

The generate-and-test paradigm (Mitchell 1977; Winston 1984) applies generally to iterative solution methods of design problems. It can be depicted as a data flow diagram (Fig. 17.3), where the output of an initial generator process (G) is reduced by successive filters (F), which together form the complete specification (set of constraints) of the product. In order to be efficient, the distribution of the generator should be as close as possible to the set passed by the filters. Thus the generator is normally controlled by feedback from the filters.

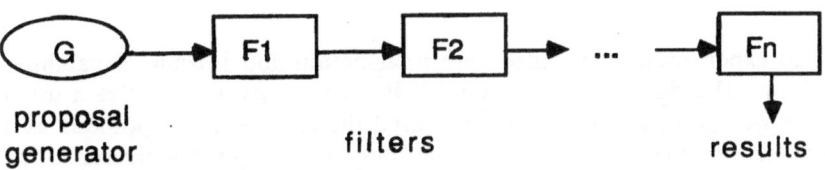

Fig. 17.3. The generator-filter paradigm

This paradigm resembles the physiological model described above. We can interpret the brain stem as a random generator, whose output is modified and filtered by the associative network of the cortex. Practical implementation aspects of this are discussed next.

4. HOW TO ADAPT THE BRAIN MODEL TO COMPUTERS

Although mechanisms for performing adaptive filtering and self-organisation are known and could in principle be used to model the cortical activities (Kohonen 1984), their implementation in full scale is still far in the future. Today we are to restrict ourselves to logical inference networks instead of those with analog signals.

However, the physiological model might be partially applied to AI systems in the following ways. First, the inference rules of an expert system are used for bidirectional reasoning — both backwards to check assertion of required facts, and forwards as a generative grammar. Together these correspond to the filtering and associative operations of the cortex. In the context of design the former performs analysis and the latter makes synthesis.

Second, the analogue of the brain stem as a random generator would be a system sometimes making random variations in the knowledge, like if there were software errors. Facts or rules can be added to a knowledge base in order to get a richer set of consequences in forward reasoning. The other possibility is to delete something at random in order to avoid conflicts in backward reasoning, and thus obtaining a higher probability to assert a hypothesis. For example, if the set of predicates appearing in the rules is

$$\{X_1, X_2, \cdots X_n\},$$

we can set any of them to a constant (true or false), or we can add new rules like

$$\text{IF } X_i \text{ AND } X_j \text{ THEN } X_k,$$

(or any other combination in the left side) with indices i, j, and k selected at random from $\{1, \cdots, n\}$.

Seeding bugs into a knowledge base (random modifications) may first seem to be dangerous. However, they can be controlled with the filtering method; only a restricted part of the knowledge is varied in order to generate something new. The other part acts as a safe organising process which does not let nonsense to come out from the system.

A sketch of an expert system extended with imagination and learning capabilities is shown in Fig. 17.4. The right part of the knowledge base (KB) behaves like a usual expert system; only well-grounded knowledge is stored there. Given a hypothesis as a query, it will prove or disprove it by backward reasoning and supply the answer to the user through the user interface management system (UIMS). The left part behaves more actively: by forward reasoning or random variation it will generate its own hypotheses (guesses) which can then be checked either by the right part or by the user. If confirmed facts are produced, they can be moved on to the right part. This way the system can gradually learn by experiment: its own original suggestions are found to be consistent with the external world. Generally, the right part is reliable but mechanical and the left part is creative but inclined to inconsistency and to make mistakes. Depending on the purpose the user can either confidently rely on the right part or use the left part as a source of vague ideas.

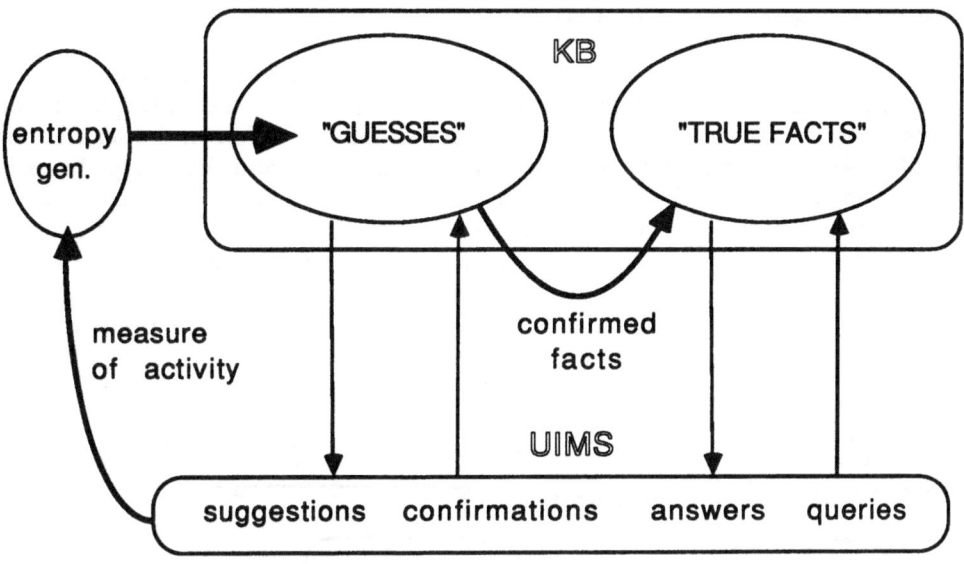

Fig. 17.4. Expert system with partitioned knowledge base

The activity of the left part is not left uncontrolled. It can be used in two modes: either switched on/off directly by the user, or gradually adapting to the general average activity of the other parts (it will stay still during heavy logical concentration, but will gradually wake up when nothing else happens for some time). The latter mode resembles the homeostasis in brain. Actually the parts of the system should be implemented as a larger set of continuously running concurrent processes (demons) should we want to simulate the subconscious brain processes.

The partitioning corresponds to the attention focusing mechanism in the brain. More effective search within a partition is possible, fore irrelevant knowledge is temporarily filtered out. Also the partitioning makes it possible to handle contradictory facts in the same system. This is especially important during a design process where the modification of one part may require changes in others but the changes cannot be done immediately. Each partition can quite easily be made internally consistent, and the interconsistency is managed later on a higher level of hierarchical partitioning, corresponding to the logical structure of the design object or to the transaction structure of the design process (Takala 1987a).

5. TOWARDS ARTIFICIAL CREATIVITY

There is a striking similarity between inference systems and formal grammars. Both can be represented as a set of rules, the rules can be used both backwards (parsing) or forwards (generating), and pattern matching plays an important role in the selection of applicable rules. The main difference seems to be that the logic in knowledge bases is usually considered monotonic (the application of rules can only add knowledge), but in generative grammars the pattern matched to the condition part of a production rule is replaced with the consequence part. Thus the evolution of a knowledge base of an expert system tends to the transitive closure of the original facts, but that of a production system will be a potentially infinite sequence of state generations some of which may be totally different from each other.

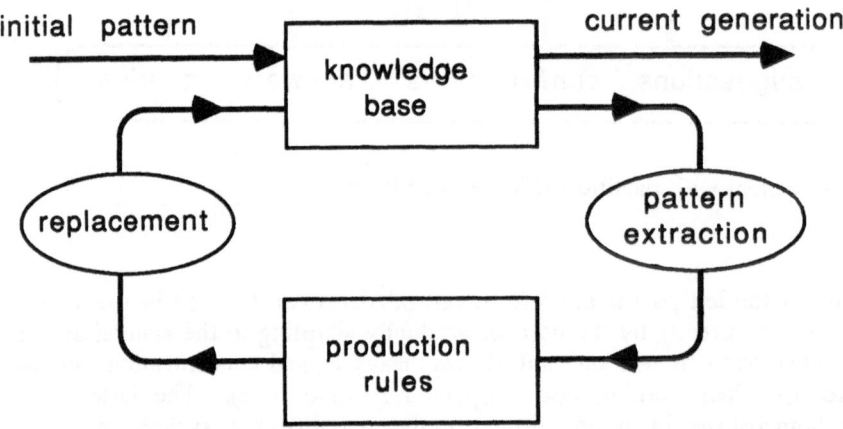

Fig. 17.5. Production system as an evolution process

Perhaps the most creative and original products of computer graphics are the fractals. Though most often generated by recursive procedures, they can also be described in terms of grammar rules controlling a production system (Fig. 17.5). Interesting and aesthetically pleasing images are generated, because either statistical variations (random numbers) are utilised, or the system of deterministic rules is complex enough to appear like a pseudorandom, unexplainable process (Smith 1984). Such grammar systems have also been experimented in architectural design (Gero 1987). Essentially the same idea on metalevel is an evolution model of rule-based design generators (Takala 1987b). Research and development is going on by the author, aiming at learning generative tools, which will be used in practice for industrial design of geometric shapes and ornaments.

Surprisingly, the fractal idea applies successfully not only to geometric design, but also to other creative arts like music (Dodge and Bahn 1986). A system generating Markov chains can be understood as a sequence of random noise generator followed by a rule-based filter (cf. Fig. 17.3). With proper selection of the statistical distribution

("1/f-noise") and the state transformation rules (a small and harmonic set of allowable voices), endless and non-structural but still non-boring pieces have been produced.

Creativity is a balanced combination of chaos and order. It can only be achieved by adding randomness (in some form) to otherwise strictly deterministic, logical computer systems. The advantage of the methods described here is that the balance of chaos and order can be fully controlled. Only randomness that is needed to find enough variation for design solutions will be allowed. The design assistant will be a quiet and reliable servant, if told so, but it may also make discoveries by using simple mechanisms when the control is relaxed for a while.

ACKNOWLEDGEMENTS

This work has been supported by the Academy of Finland and the Technology Development Centre. Parts of it have been done while I was a visiting researcher at the Centre for Mathematics and Computer Science (CWI), Amsterdam.

REFERENCES

Bergstrom, M. (1986): "Mind-brain interaction: Consciousness as a neural macrostate," in *Finnish Artificial Intelligence Symposium (September, 1986), Vol. 1, AI and Philosophy*, Karjalainen et al. (ed.), Finnish Society of Information Processing Science.

Dodge, C. and Bahn, C. (June 1986): "Musical fractals," *BYTE*.

Gero, J. S. (ed.) (1985): *Knowledge Engineering in Computer-Aided Design, Proceedings of the IFIP WG 5.2 Working Conference 1984 (Budapest)*, North-Holland, Amsterdam.

Gero, J. S. (1987): "Knowledge-based planning as a design paradigm," in *Design Theory for CAD, Proceedings of the IFIP W.G. 5.2 Working Conference 1985 (Tokyo)*, Yoshikawa, H. and Warman, E. A. (eds.), North-Holland, Amsterdam, pp. 339-379.

Kohonen, T. (1984): *Self-Organization and Associative Memory*, Springer-Verlag, Berlin, Heidelberg, New York, Tokyo.

Mitchell, W. (1977): *Computer Aided Architectural Design*, Petrocelli Charter.

Smith (1984): "Plants, fractals and formal languages," *ACM Computer Graphics (SIGGRAPH'84 Proceedings)*, **18**(3).

Takala, T. (in preparation, 1987): "Essays on Design Theory," CWI report, Centre for Mathematics and Computer Science, Amsterdam.

Takala, T. (1987): "Theoretical framework for computer aided innovative design," in *Design Theory for CAD, Proceedings of the IFIP W.G. 5.2 Working Conference 1985 (Tokyo)*, Yoshikawa, H. and Warman, E. A. (eds.), North-Holland, Amsterdam, pp. 323-338.

Tomiyama, T. and Yoshikawa, H. (1987): "Extended general design theory," in *Design Theory for CAD, Proceedings of the IFIP Working Group 5.2 Working Conference 1985 (Tokyo)*, Yoshikawa, H. and Warman, E. A. (eds.), North-Holland, Amsterdam, pp. 95-130.

Winston, P. H. (1984): *Artificial Intelligence*, Addison-Wesley, Reading, MA, USA. 2nd ed.

Yoshikawa, H. and Warman, E. A. (eds.) (1987): *Design Theory for CAD, Proceedings of the IFIP WG 5.2 Working Conference 1985 (Tokyo)*, North-Holland, Amsterdam.

18. An Integrated Data Description Language for Coding Design Knowledge

B. Veth †

Centre for Mathematics and Computer Science
Kruislaan 413, 1098 SJ Amsterdam, THE NETHERLANDS

Abstract: *We present in a unifying framework the basic notions of IDDL (Integrated Data Description Language) to code design knowledge in the IIICAD system. IIICAD is an intelligent, integrated, and interactive computer-aided design environment we are currently developing at the Centre for Mathematics and Computer Science.*

Keywords: CAD, design theory, logic, theory of knowledge, theory of design objects, qualitative reasoning, object oriented programming, logic programming, knowledge engineering, software engineering, prototyping.

1. INTRODUCTION

In this paper we deal exclusively with the following issue: How to code design knowledge? We shall start in Section 2 with the IIICAD (Intelligent Integrated Interactive CAD) concepts and present our methodology to develop a useful representation language — i.e. a theoretical approach. Accordingly, we first concentrate on the theory of CAD and then derive requirements and specifications for IDDL. A full account of the implementation details will be left to an upcoming paper although we touch on this subject briefly.

The theory of CAD consists of three parts: theory of design, theory of knowledge, and theory of design objects. In Section 3.1 we introduce a design process model which is derived from a design theory, and in Section 3.2 we show its logical notation. In Section 4 we deal with the theory of knowledge. In Section 5.1 we describe the theory of machine design as an example of the theory of design objects. Designing is a process where we materialize our imagination. Any design process, therefore, cannot escape

---

† Group Bart Veth consists of (in alphabetical order): Varol Akman, Peter Bernus, Paul ten Hagen, Jan Rogier, Tetsuo Tomiyama, and Paul Veerkamp.

from the real world restrictions. In Section 5.2 we illustrate naive physics which treats this aspect. In Section 6 we count the general requirements for IDDL from CAD and software engineering viewpoints. Section 7 closes the paper by citing design policies for IDDL and showing our prototype implementation with an example of bridge design. A word about presentation: throughout the paper we prefix with **DM** the so-called design maxims (Yeomans, Choudry, and ten Hagen 1985) which will be collected in Section 7 and converted into specifications for IDDL.

2. OVERVIEW OF IIICAD

2.1. The Concept of IIICAD

CAD systems are vital elements of almost every facet of the technology but it is also admitted that they are plagued by inflexibility. It is not unjust to claim that the majority of the existing systems are but sophisticated workbenches for engineering drawing. As the application domain becomes serious, designing becomes unmanageable with only this type of support. Since design is essentially an intellectual activity, we need, not surprisingly, more *intelligence* in a system — hence the first I of IIICAD.

Borrowing an analogy from (Bobrow, Mittal, and Stefik 1986), until now CAD systems were built using the *low road* and *middle road* approaches. The low road approach involves *ad hoc* programming (mostly in prehistoric languages like Fortran) and is biased towards geometric information. Middle road systems are more interesting in that they are aware of the fact that they have to incorporate intelligence. They focus on a well-defined domain and collect specialized knowledge coded as say, *if-then* rules. In other words, they become expert systems (e.g. PRIDE (Bobrow, Mittal, and Stefik 1986)). An annoying problem with expert systems is that genuinely expert performance can only rest on knowledge of a model in which an underlying mechanism *understands* what is going on (Kuipers 1986).

Finally, one distinguishes the *high road* systems which IIICAD is aiming at. High road systems are deep systems (as opposed to low and middle road systems which are shallow) in that their knowledge represents the principles and theories underlying the subject "design." In the case of IIICAD, the fundamentals of General Design Theory which is based on axiomatic set theory can be found in (Tomiyama and Yoshikawa 1987).

We do not deny the fact that there are several domain-specific sides to design. For instance, VLSI design is two-dimensional (although this is changing) while mechanical design is inherently three-dimensional. IIICAD incorporates similarities in design, leaving the application-dependent issues to further consideration as side requirements and using intelligence based on a clean and robust design theory. Thus here we are not working on yet another geometric modeler or expert system.

The other two I's of IIICAD correspond to *integration* and *interactivity*. Design systems should support integration because human designers have a unified view of design objects. Interaction requires almost no validation. Good design systems cannot be obtained without using the best man-machine communication techniques.

To summarize:

- In IIICAD we pursue a top-down theoretical approach incorporating more intelligence than expert systems, more integration than geometric databases, and high-level interaction using advanced computer graphics.

- We want IIICAD to be a system based on expandable ideas and a framework where designers can exercise their faculties at large. We believe that the essential thing in a designer is that he builds us his world and IIICAD must give him the freedom to do so.

2.2. Elements of IIICAD

The Supervisor (SPV) is at the core of IIICAD and controls all the information flow. It adds intelligence to the system by comparing user actions with *scenarios* which describe standard design procedures, and by performing error handling when necessary. Since SPV is the central authority for control the following becomes relevant.

DM 1. *IDDL should be able to describe status and control information of the system with origin, destination, and time stamp of the control information.*

While SPV corrects the obvious user errors, it does not have the initiative for the design process itself because IIICAD is envisaged to be a designer's apprentice, not an automatic design environment.

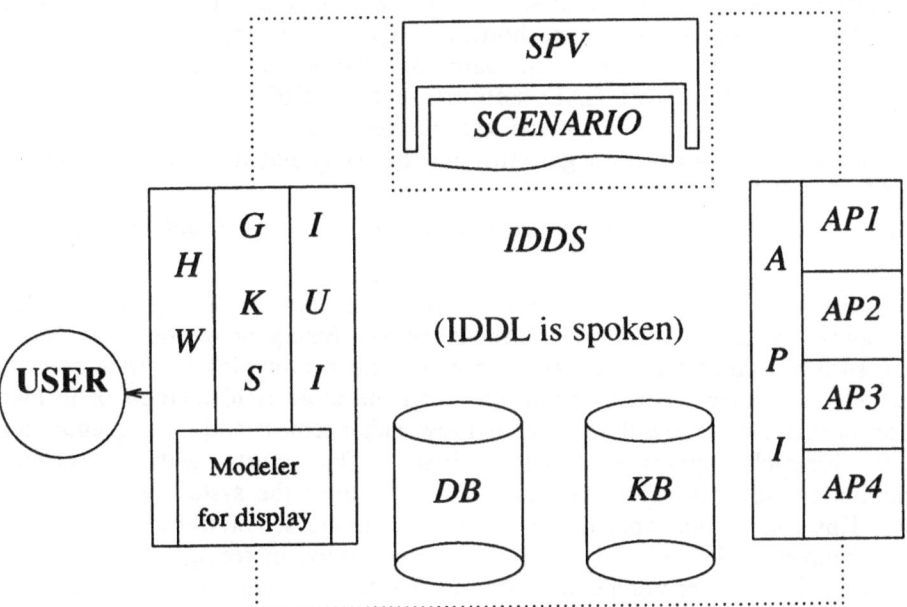

Fig. 18.1. IIICAD architecture

The Integrated Data Description Schema (IDDS) regiments the data and knowledge bases relieving the user from the burden of specifying where and how to store/retrieve data. IDDS has a language called Integrated Data Description Language (IDDL) spoken by all system elements. IDDL is the means to code the design knowledge and the design object to guarantee integrated descriptions system-wide. Like most modern programming languages, IDDL differentiates between what is commonly known as the *external* and the *internal* contents. The former is essentially nonmathematical information such as input-output behavior and diagnostics. The latter consists of mathematical operations which do the job. More on IDDL will be said in Section 6 and Section 7 which show how IDDL codifies design knowledge. Here it should suffice to remark that internally IDDL will be based on logic and accordingly knowledge engineering is the key factor in building the IIICAD system. IDDL is an essential step in developing IIICAD.

In addition to the above principal elements, IIICAD has a high-level interface called Intelligent User Interface (IUI) which is also driven by scenarios written in IDDL, and the Application Interface (API) which secures the mappings between the central model descriptions about the design object and individual models used by application programs such as geometric modelers, finite element analyzers, etc. Figure 18.1 shows the preceding elements in block diagram level.

2.3. Software Engineering Viewpoint

Maintainable software systems should be modular both "in the small" to allow alteration of minor components in specific applications and "in the large" to allow changes in major components based on say, the advances in technology. They should be designed for evolution for long time horizons. They should be sturdy and open-ended (Wegner 1984). In this regard, software engineering will always be a leading concern in developing intelligent CAD software such as IIICAD because even the conventional CAD systems are large and complicated. In other words, knowledge engineering is more than software engineering but probably not much more (Bobrow, Mittal, and Stefik 1986).

DM 2. *IDDL, as a language to construct a knowledge base, should support easy maintenance.*

In the development of intelligent CAD systems the underlying strategy is "Plan to throw one away. You will anyhow." (Brooks 1975). Emerging trends of software engineering such as exploratory programming and rapid prototyping are thus crucial. These methods are somewhat more permissive than the more rigid method of formal specification in that they follow the idea of iterative enhancement and consequently, an evolutionary life-cycle approach (Wegner 1984). One starts with a skeletal implementation (rapid prototype) and adds new parts until the system is reasonably completed. This incremental approach is fruitful when the set of tasks and the end result are incompletely defined. Also one is more interested in seeing a glimpse of a future system built as a prototype in order to assess its strengths and weaknesses globally. Exploratory programming using powerful workstations and modern languages (e.g. Smalltalk-80[1]) makes this process very effective (Ramamoorthy, Shekhar, and Garg 1987).

DM 3. *IDDL must support incremental programming.*

---

[1] Smalltalk-80 is a trademark of Xerox Corporation.

3. DESIGN THEORY

3.1. Modeling of Design Processes

Design theory (Yoshikawa 1981) provides a strong basis for formalizing design processes and design knowledge. For this purpose, we use General Design Theory (Tomiyama and Yoshikawa 1987; Yoshikawa 1981) which is based on axiomatic set theory and models designing as a mapping from the function space where the specifications are described in terms of functions, onto the attribute space where the design solutions are described in terms of attributes.

There are many interesting results derived from General Design Theory; we emphasize in Section 3.2 the possibility of a logical formalization of design processes. Figure 18.2 shows a design process model derived from General Design Theory. The basic idea is as follows (Tomiyama and ten Hagen 1987a; Tomiyama and ten Hagen 1987b):

- A designer, given the specifications, may try to select a candidate and refine it in a stepwise manner, rather than trying to get the solution directly from the specifications.

- Therefore, a design process can be regarded as an evolutionary process of such intermediate descriptions of the design objects rather than just a mapping. The collection of these intermediate descriptions can be used as the central model about the design solution and we call it a *metamodel*.

- The designer will evaluate the candidate to see whether it satisfies the specifications or not. To do so, he derives various kinds of models of the design object from one central model (i.e. the metamodel).

Our discussion leads to the following design maxims:

DM 4. *IDDL should be able to describe not only design objects but also design processes.*

DM 5. *IDDL should be able to describe metamodels and models (for evaluation) derived from the metamodel.*

DM 6. *IDDL should be able to describe the stepwise nature of the design process.*

DM 7. *IDDL should be able to describe knowledge to detail the metamodel, to check its feasibility, and to control the detailing process.*

DM 8. *IDDL should be able to describe knowledge to derive models for evaluation from the metamodel and knowledge to evaluate models.*

DM 9. *IDDL should allow multiple views of a design object, which are possibly independent but still correlated.*

To illustrate a design process, we need to recognize three major components: *entities*, *attributes* of entities, and *relationships* among entities. A design process is thus a collection of small steps to obtain complete information about these three. We can also observe components which do not change during design. For instance, when we design a board we simply use a VLSI chip as a building block that cannot be changed. Let us call such objects *invariants* in a design process. Descriptions about the board at this level are, oppositely, dynamically changed during the design process. Let us call

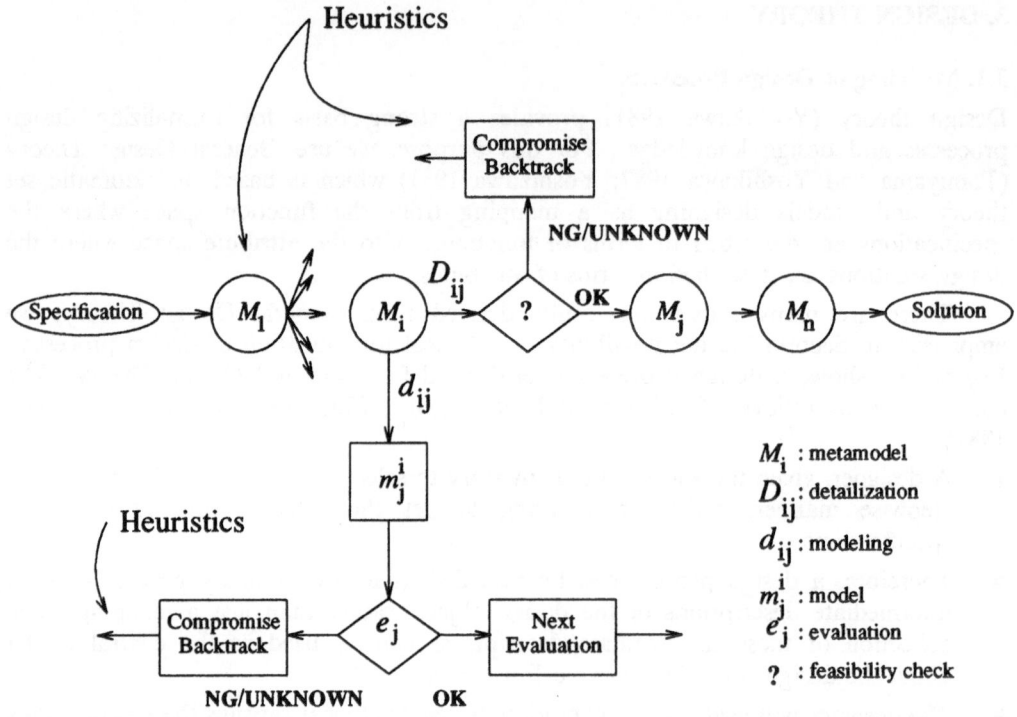

Fig. 18.2. Design process model due to General Design Theory

them *variants* in a design process. Clearly, the concept of e.g. a point should not change while designing a geometric entity; however attributes such as coordinates of a point may be frequently changed. Let us call such changeable concepts associated with variants *covariants*.

DM 10. *IDDL should be able to describe invariants, variants, and covariants in design processes.*

Another important thing about the design process model is that eventually we need to check physical constraints (i.e. feasibility check). This implies further that we need to distinguish success and failure, known and unknown, necessity and possibility, etc.

DM 11. *IDDL should be able to describe positive and negative information, known and unknown, and modalities such as necessity and possibility.*

3.2. Design Processes: Logical Formalization

Let us assume predicate logic as the basis of our discussion. To control the stepwise refinement of the design process there is a need to express unknown, uncertain (default), and temporal information about the design object. For this reason we equip our logic language with three-valued logic, modal logic, inheritance, and situational calculus.

Considering the metamodel evolution model in Fig. 18.2, the system starts from the specification S of the design object and it continues the design process until the goal G is reached:

$$S \to M^0 \to \cdots \to M^i \to M^{i+1} \to \cdots \to M^n \to G$$

We define $\{q^i\}$ as the set of propositions at the metamodel state M^i. In other words $\{q^i\}$ is the current state of knowledge about the design object. There are two possibilities: either the current state of knowledge is complete and consistent or there is some incompleteness or inconsistency. In the first case the goal is reached and we finished the design process. In the latter case we need to proceed to a next metamodel in order to solve the incompleteness or inconsistency.

DM 12. *IDDL should let the inconsistency of a certain metamodel be represented, but this inconsistency needs to be resolved when transferring to a next metamodel.*

We need language constructs to evaluate a metamodel and to derive new properties or to update uncertain or unknown properties in order to get more detailed knowledge about the design object. The decisive point is how to proceed from M^i to M^{i+1}; i.e. given $\{q^i\}$ how do we find $\{q^{i+1}\}$? We shall adopt the following strategy:

- From $\{q^i\}$ we can derive p, so the next state $M^{i+1} = \{q^i\} \cup p$. This means that the knowledge base is extended by asserting property p. In practice p might come from design procedures, default assumptions, results of engineering analyses, and so on. Before the acceptance of state M^{i+1} the consistency of the new metamodel has to be checked. For this purpose we can use the appropriate set of logical inference rules.

- We use the modal operator $\Box$ to express default values. Thus $\Box p$ means it is possible, but probably not the case that p; $\blacksquare p$ states it is necessary that p. Note that $\Box p$ is equivalent to $\neg \blacksquare \neg p$ (McDermott 1982).

DM 13. *IDDL should incorporate modal logic.*

- If we have $\Box q_j$, can derive $\neg q_j$, and it is not based upon default properties then we assume $\neg q_j$. In other words $M^{i+1} = \{q^i\} - \{\Box q_j\} \cup \{\neg q_j\}$. To see this, imagine that the designer wants to design a bridge and the system needs the length which is unknown. In this case the system knows that bridges normally have a length and it can conclude by an inheritance mechanism that the length of the bridge is L. This will be asserted to the knowledge base as $\Box\, equal(length(bridge), L)$. If in a next state of the design process the real length RL of the bridge is determined, $\Box\, equal(length(bridge), L)$ can be changed into $\blacksquare\, equal(length(bridge), RL)$.

- The Skolem constant ω is used to denote unknown values; so if we have $q_j = p(\omega) \wedge \cdots$ and we can derive q_j' without unknown values, then we have the next metamodel $M^{i+1} = \{q^i\} - \{q_j\} \cup \{q_j'\}$. Referring to the previous example about the length of the bridge, the system can assume an unknown value instead of a default value in $equal(length(bridge), \omega)$. When we want to express the fact that a certain property is unknown, we use $\perp p$.

DM 14. *IDDL should have both the Skolem constant and the unknown operator.*

- If a severe inconsistency is encountered which cannot be resolved by the system in terms of deriving more knowledge or stepping back to a previous metamodel, apparently an inconsistency in the specification provided by the designer has been found. The designer should be notified with this inconsistency together with exact transactions so that he can fix it and restart the design.

DM 15. *IDDL should have facilities for error handling, when it encounters inconsistent or incomplete states, with the help of the designer.*

To add a certain proposition an assertion operator $assert(p(x))$ is needed. To modify propositions we need a $change(p(x))$ operator. After these operations the knowledge base must still be consistent (cf. **DM 12**).

DM 16. *To control the behavior of the system IDDL should have metaknowledge that chooses which rule to apply at a certain time.*

In other words, IDDL must be able to describe a design process so that during designing we can "design" designing procedures using metaknowledge.

4. THEORY OF KNOWLEDGE

Theory of knowledge is necessary especially to put our knowledge into a particular framework and to utilize it in the IIICAD architecture. The following discussion is based on our result (Tomiyama and ten Hagen 1987c) and its most significant contribution is the distinction between two opposing knowledge representation methods, i.e. *extensional* vs. *intensional* descriptions.

In order to discuss design, we need to describe entities, their properties, and relationships among entities as mentioned in Section 3.1. In an extensional description method, the fact that an entity e has property p is described by $p(e)$ and the fact that entities e_1 and e_2 are in a relationship r is described by $r(e_1, e_2)$. In an intensional description method these two facts can be represented by $e(p)$ and $relation(e_1, r, e_2)$, respectively. The extensional descriptions do not assume any preconceptions while the intensional descriptions assume preconceptions such as that e's property is limited to one particular p. This means that an intensional description is equivalent to an extensional description with some assumptions, such as the number of arguments, the order of arguments, the type of arguments, etc.

These two description methods are basically equivalent except for assumptions. Since an intensional description assumes something predefined, when it has to be changed this results in changing those predefined (and perhaps implicit) conditions. For instance, a mechanical part, say a shaft, might be represented by

$$shaft(diameter, length, bearing_1, bearing_2).$$

If we now want to add new attributes, such as transferring power, this results in a redefinition of this *shaft* predicate. On the other hand, an extensional description might be the set of the following facts:

$$shaft(s), \ equal(diameter(s), D), \ equal(length(s), L),$$

$$supported-by(s, b_1), \ supported-by(s, b_2), \ bearing(b_1), \ bearing(b_2).$$

In this example, an extensional description does not assume anything, e.g. *s* is just a name.

DM 17. *IDDL has two kinds of names: system names are internal and should be unique whereas user names are external and modifiable.*

In an extensional description we need to write numerous (and often very obvious) descriptions. However, modifying such a representation is just adding or deleting facts (thus incremental, cf. **DM** 3). On the other hand, an intensional description assumes a predefined scheme which might be difficult to change but shows high performance. For instance, it is easily predicted that the computation to determine two bearings supporting a shaft is reduced to an address calculation.

It is well-known that CAD applications request a flexible data description scheme which is easy to modify (Lorie 1982). From this point of view, the extensional description method is important in IIICAD because of the incremental nature of design processes. Independent (but still correlated) multiple views of design object demand independent small partitionings in the database. This is easily achieved by an extensional description method because we only have to pick up relevant facts. In an intensional description method this requires to create a totally new scheme.

As discussed in Section 3, there are variants which change during the design process. The extensional description can be used to describe these variants. To this end, we know that the logic programming paradigm (Kowalski 1978) is very useful to implement such description methods. On the other hand, there are invariants we use as building blocks for designing. They do not change their structural properties although values of their attributes might change. Invariants, therefore, can be represented in an intensional description method and their properties (or attributes) can be represented as covariants. Having intensional descriptions may also contribute to improving the performance.

DM 18. *IDDL should have both an extensional description method and an intensional one.*

DM 19. *Invariants in IDDL will be represented as objects, variants will be constructed on the predicate level, and covariants will be represented by functions.*

5. THEORY OF DESIGN OBJECTS

5.1. Theory of Machine Design

Design is regarded as a mapping from the function space onto the attribute space. This requires IDDL to have both attributive and functional representations. There are several issues in representing the attributive information (Tomiyama and Yoshikawa 1985; Tomiyama and ten Hagen 1987a). First, an attribute does not necessarily have a value. In the design process, it often happens that an attribute is only known to exist and its value is not yet decided. Hence IDDL should be based on three-valued logic including *unknown* (cf. **DM** 14). Second, attributive information refers to the structure of an entity. The structure of an entity might be characterized by existence of substructures and relationships among substructures. Information on the number of substructures is also needed.

DM 20. *IDDL should make a distinction between the facts that an entity has an attribute and that an attribute has a value.*

DM 21. *IDDL should be able to represent part-assembly relationships and relationships among parts in order to represent structures.*

DM 22. *IDDL should be able to deal with cardinalities (i.e., number of elements in a set).*

On the other hand, the representation of functions is a rather difficult issue. Unfortunately, it is not yet known in which language we can describe functions of e.g. machines. There is, however, a hope that functions can be represented in terms of physical phenomena that the machine exhibits (Tomiyama and Yoshikawa 1987). From this point of view, the representation of functions can be reduced to the representation of physical phenomena and qualitative reasoning (cf. Section 5.2).

At one time in the metamodel evolution model, the designer will focus at a particular part of the design process or the design object. When this focusing is taking place, the information about the rest of the design process or the design object should not be accessible and stay unaffected. (This is the principle of abstract data type languages.) In order to make focusing more effective, we must create a small world which represents it. For example, we must be able to control applicable predicates to particular classes of entities, although this inevitably asks for higher order predicate logic.

DM 23. *IDDL should have a focusing control mechanism which is able to create a small world where it is clearly defined what kind of information is accessible.*

We must also be able to see a particular part of the design object. When we are considering a particular object, we must be able to see its inner structure represented by variants. On the other hand, we may use that object as a building block when we are working on another object. This requests transition between different abstraction levels and the same information (e.g. an object at some level) must be seen differently (e.g. as a collection of predicates) (Fig. 18.3).

DM 24. *IDDL should be able to describe hierarchical enclosing controls.*

5.2. Naive Physics, Qualitative Reasoning, and Design

Naive physics observes that people are generally very good at functioning in the physical world and tries to develop a formal framework to serve as a basis to export this human capability to computers (Hayes 1985). As such, it constitutes a major part of what is known as *commonsense reasoning* in AI. Naive physics concepts are needed in design because in many cases design objects will have a physical existence and accordingly obey natural laws. If we want to create designs corresponding to physically realizable (read manufacturable) design objects then we will have to refer to naive physics primitives such as solids, space, motion, etc. Furthermore, if we want to reason about a design object in its destined environment (think of a pressure regulator to be installed in a nuclear reactor) we will need naive physics notions such as envisioning, simulation, diagnostics, etc.

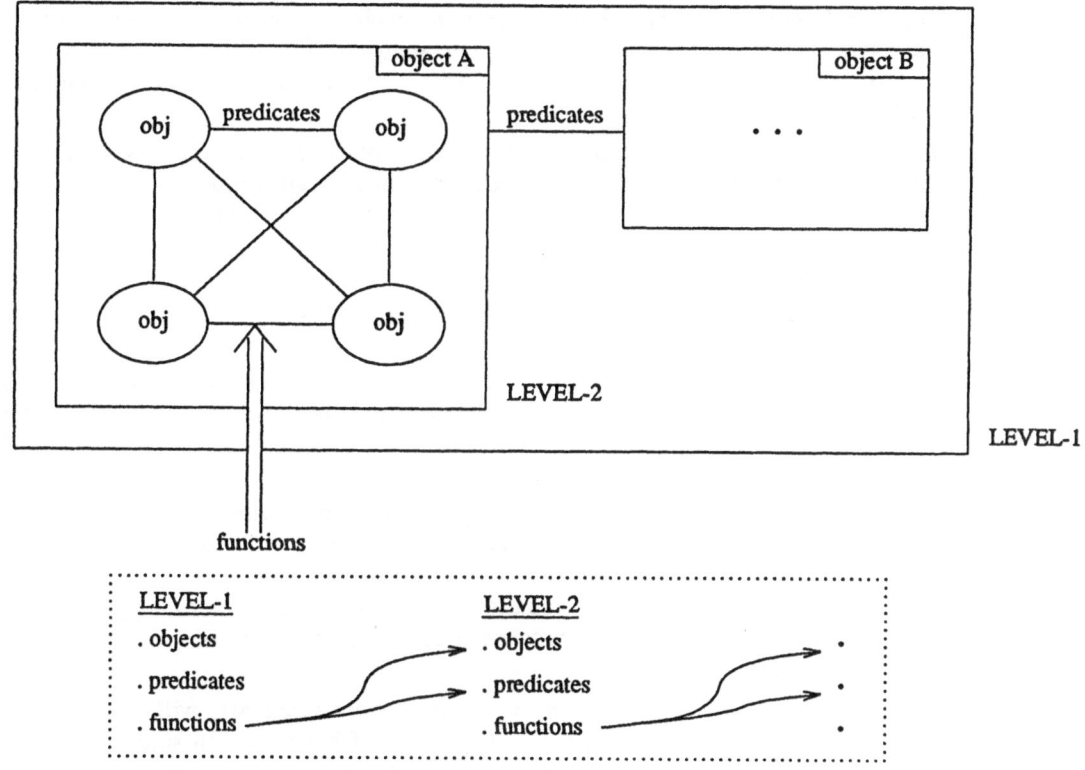

Fig. 18.3. Hierarchical enclosing control

A basic commonsense notion is *causation.* We seem to have no difficulty in grasping the relationship between two events (or states of affairs) such that the first brings about the second. Basic notions of modal (e.g. temporal) logic can be used to talk about causality. An intricacy is brought about by *teleological explanations* — i.e. certain phenomena seem to be best explained by intentions or purposes rather than by means of prior causes. This closely resembles to our way of looking at a design object from different viewpoints.

Classical physics seems not too useful in formalizing naive physics because it describes everything in exact terms. Starting with the introductory level textbooks, physics laws are based on the presupposition that the readers have a shared *prephysics* knowledge (de Kleer 1975). In fact, mathematical formalizations of physics, unless aided by verbal explanations, occasionally hide causality. What makes then a good formalization of naive physics? Both mathematical and computational criteria are important. Some guidelines may be given at this point:

- *Categorical angle:* Study objects through their universal properties which characterize them, rather than through their anatomical properties. See intriguing similarities and exploit abstractness to arrive at theories of utmost generality unattainable in other ways.

- *Mechanistic outlook and deductionism:* Regard physical situations as machines comprising individual components each of which contributing to the overall behavior of the machine. This means that the behavior of a physical structure will be completely accounted for by the way its physical constituents act. To a first order approximation, three kinds of constituents are enough (de Kleer and Brown 1984): materials (such as air, water), components (such as containers, wheels), and channels (such as electric cables, conduits).

- *No-function-in-structure:* This follows from the preceding guideline. Briefly, one has a catalog of components and these have associated laws which do not make assumptions about how they are employed in a certain context (de Kleer and Brown 1984).

- *Confluences:* These are better known as qualitative differential equations (Forbus 1984; Kuipers 1986). One first reduces continuous real-valued variables to discrete-valued variables taking only a small number of values, say $+$, $-$, and 0. This process maps differential equations to confluences. Normally, a single confluence will not be able to characterize the behavior of a component over its entire operation region.

- *Envisionment, simulation, and diagnosis:* In the former one starts with a structural description and determines all possible sequences of behavior. In simulation, one starts again with a structural description but this time he is given some initial conditions to determine a probable course of future behavior. In diagnosis, one starts with some specified behavior expected from a system and tries to see why the system is misbehaving.

- *Topological scene description:* This suggests that one may temporarily ignore the exact coordinates in some geometric situation and model it using topological notions like homotopies, isolation, etc.

- *Frame problem:* This is well-known. When some action takes place in a situational calculus-like representation, how does one tell what facts change and what facts stay unaffected? The answer is, one has to write explicit axioms that state what changes and what remains the same. One may avoid this problem at least partially by adopting *histories* (Hayes 1985) — descriptions extended through time but always spatially delimited (in contrast to situational calculus situations which are instants spatially unbounded). This reflects a choice to ban action at a distance (Forbus 1984).

- *Modal logic:* We find it handy to use logic with modes of truth, viz. modal logic with necessity and possibility. As explained earlier, here we have not only affirmations such as that proposition p is true, but also stronger ones such as that p is necessary, and weaker ones such as that p is possible. Modal logic is useful to code naive physics. Consider examples such as

$$below(obj,\ surface)\ \wedge not - glued(obj,\ surface)\ \supset\ \blacksquare\ fall(obj)$$

and

$$at - top(obj,\ inclined - surface)\ \wedge above(obj,\ inclined - surface)$$
$$\supset\ \square\ slide(obj)$$

where the latter possibility being dictated by our incomplete knowledge about e.g. friction.

According to these guidelines, **DMs** 9, 13, and 21 are relevant. In addition:

DM 25. *IDDL should be able to carry out simple algebraic manipulations — this is necessitated by e.g. qualitative reasoning with confluences.*

6. REQUIREMENTS TO IDDL AS A KERNEL LANGUAGE OF CAD

6.1. CAD Perspective

CAD is no longer considered as a tool for speeding up the creation of exact product definitions in the form of text and drawings. Because CAD needs a coordinated flow of information between system and user, the functional view of system design should consider the totality of design, not only those functions which will be carried out by a computer. It needs a language to represent the flow of design (cf. **DM** 1). The number of design rules which exhaust a realistic area within electrical, mechanical or civil engineering lies around tens of thousands. The amount of attributive data to be handled by the actual management of the design process could be around hundreds of megabytes, not mentioning the attributive data in the background databases of a design office.

DM 26. *IDDL should have a mechanism for structuring knowledge.*

DM 27. *As with design knowledge, design object representation calls for an encapsulation and structuring mechanism for it to be representable in reasonable form.*

Design produces intermediate results which are incomplete and even inconsistent during time spanning series of transactions. The design object's attributive representation must allow assumptions to be used for the evolution of the design object (cf. Section 3). Result of nonmonotonic reasoning must be checked for feasibility and consistency.

DM 28. *IDDL, using nonmonotonicity, should be able to retract assumed but later on unconsidered propositions.*

DM 29. *IDDL should be able to check consistency and completeness.*

From the viewpoint of interactive design, it is desirable for the human designer to be able to "mark" intermediate design stages, and later go back to them for examining or resuming from there.

DM 30. *The stages of design evolution must be representable on the level of IDDL.*

6.2. Software Perspective

From a software engineering point of view two types of requirements influence our language design. One is that the system to be built using IDDL will after all be a software system with high complexity. Therefore the language design must reflect considerations for managing complexity in software design (cf. Section 2). Especially, due to our inability of separating specification from experimentation, we would like to have an environment where the two can be done in parallel.

Excellent opportunities are found in object oriented style of programming (Stefik and Bobrow 1986). Object oriented programming delivers extensibility, flexible modifications of code, and reusability. Important issues in object oriented programming are data encapsulation and information hiding. Thus an object can be regarded as an independent program which knows everything about itself. Reuse of software is helped by the class inheritance mechanism. Other flexibilities of languages like Smalltalk-80 (Goldberg and Robson 1983) and Loops (Stefik, Bobrow, and Kahn 1986) are incremental compilation and dynamic binding. It is an additional benefit of object oriented programming systems that they offer rich system building tools and good user interfaces. Therefore object oriented programming is a good choice for creating IIICAD. On the other hand, logic programming is powerful for problem solving since it reflects the reasoning process most naturally and directly.

DM 31. *IDDL uses the logic programming paradigm to express the design process for manipulating design objects, whereas the object oriented programming paradigm is used to express design objects.*

DM 32. *In IDDL invariants, variants, and covariants will respectively be represented by objects, their internal and external relationships, and behavior of objects.*

The second aspect of language design is how representational tasks can be unified. A typical epistemological view (Brachman 1979) includes class − subclass, part − subpart, class − member-of-class, prototype-of-class − member-of-class, and functional abstraction.

DM 33. *IDDL should have mechanisms to represent inheritance.*

7. IDDL SPECIFICATIONS

7.1. From Design Policies to Specifications

In the preceding sections we have counted 33 design maxims. The derivation of IDDL specifications took place as follows. First we classified them into several categories so that we could see the relationships among them. We extracted a "keyword" from each of those categories, for example *encapsulation* was arrived at after considering **DMs** 10, 18, 19, 27, and 32; for *modalities* we considered **DMs** 13 and 28.

We then derived features considered to be essential for IDDL from those keywords. For instance, the feature *no automatic backtracking* was derived from the keyword *metaknowledge for control*. This is due to the expectation that if an automatic backtracking mechanism and metaknowledge for control are installed together then the control would become unmanageable.

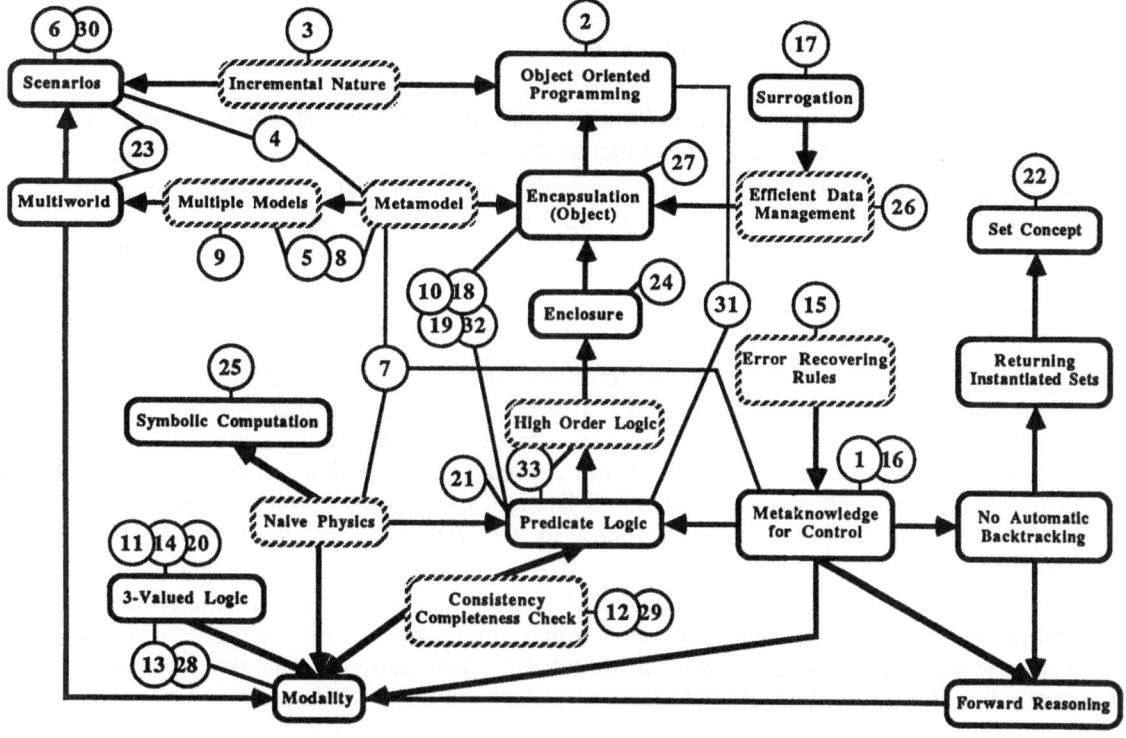

Fig. 18.4. Classification of IDDL design maxims

Figure 18.4 shows the relationships among design maxims, keywords, and features. In this figure, the small circles correspond to **DMs,** dotted boxes are keywords, and the solid boxes are features. Some nontrivial features are explained below.

- *Enclosure* is a mechanism to absorb the difference in abstraction levels (see Section 5.1).

- *Forward reasoning* is requested by the features *metaknowledge for control* and *no automatic backtracking.*

- *Metaknowledge for control* demands other features like predicate logic, no automatic backtracking, forward reasoning, and introduction of modality.

- *Modality* in the logic system is one of the main issues in IDDL, for the inference control will be dependent on it. We are thinking of incorporating necessity/possibility and known/unknown operators.

- *Multiworld* mechanism is used to describe metamodel evolution and evaluation of model. The scenario mechanism creates a completely isolated world which is independent but still capable of having relationships with other worlds.

- *Returning instantiated sets* to an inquiry is requested due to *no automatic backtracking* feature. Elements of IIICAD will return all possible instances to an inquiry, which does not require automatic backtracking for exhaustive search and gives possibilities for both depth- and breadth-first searches. This further begs for the introduction of *set concept*.

- *Scenarios* are the kernel mechanisms to realize multiworlds which are important to describe the stepwise nature of design processes.

- *Three valued logic* will play an important role in IDDL. It will however be introduced as operators rather than truth values to avoid unnecessary complexity in the inference algorithm.

7.2. Example from IDDL Prototype

Figure 18.5 shows a typical display from our prototype implementation of IDDL. Based on the discussions in Section 7.1 we have developed this version on our Smalltalk-80 system. In Fig. 18.5, design browsers are used to manipulate design information such as scenarios, constraints, predicates. In this version of IDDL, *constraints* correspond to the covariants of Section 3.1 and are meant to be the design specifications, while *predicates* correspond to the variants and are used for object description.

The two windows on the left of Fig. 18.5 are snapshots of the constraints for the bridge and its subparts. The other two overlapping windows on the right show the part-assembly relationships among substructures. The designer can explore various possibilities by communicating with the system through these windows.

8. CONCLUDING REMARKS

In this paper, we presented a unifying framework to describe design knowledge. Our starting point, theory of CAD, allowed us to formulate design maxims which were converted into IDDL specifications. Theory of CAD enabled us to understand, clarify, model, and formalize design processes and design knowledge in an intelligent CAD environment.

The development of IDDL was given priority to the development of other subsystems of IIICAD. The current goals of the project are as follows:

- To implement more powerful prototyping tools for IDDL development.
- To construct a more complete version of IDDL by using these tools.

Near future work includes:

- To develop subsystems of IIICAD such as SPV, API, and IUI.
- To incorporate existing CAD tools and "knowledge-based systems" in the IIICAD framework.
- To justify our methodology of developing IIICAD and eventually to prove the effectiveness of our theory of CAD by case studies.

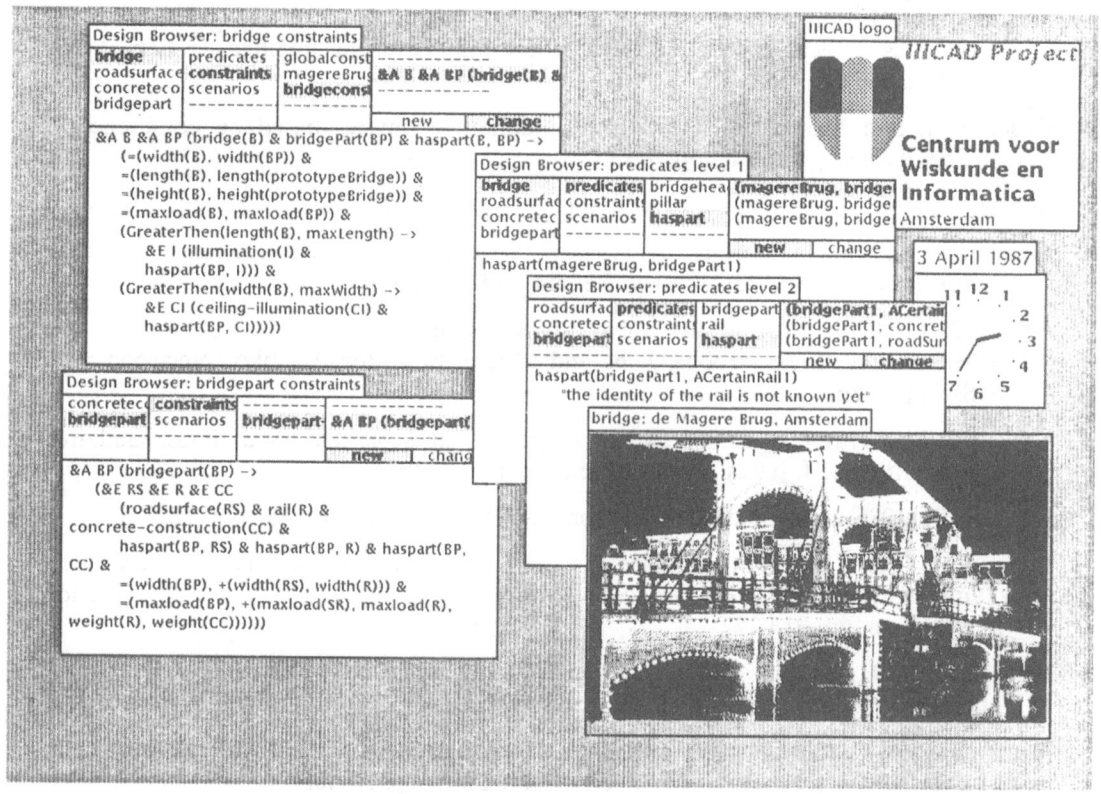

Fig. 18.5. Example bridge design with IDDL prototype

ACKNOWLEDGMENTS

We are grateful to M.M. Megens for her enthusiastic work in IDDL implementation and P.J.W. ten Hagen and Zs. Ruttkay for their critical remarks on the earlier version of this paper.

REFERENCES

Bobrow, D. G., Mittal, S., and Stefik, M. (September 1986): "Expert systems: Perils and promise," *Communications of the ACM*, **29**(9), pp. 880-894.

Brachman, R. J. (1979): "On the epistemological status of semantic networks," in *Associative Networks: Representation and Use of Knowledge by Computers*, Findler, N. V. (ed.), Academic Press, New York, pp. 3-50.

Brooks, F. (1975): *The Mythical Man-Month*, Addison-Wesley, Reading, MA, USA.

de Kleer, J. (December 1975): "Qualitative and quantitative knowledge in classical mechanics," Technical Report No. AI-TR-352, Artificial Intelligence Lab, MIT, Cambridge, MA, USA.

de Kleer, J. and Brown, J. S. (1984): "A qualitative physics based on confluences," *Artificial Intelligence*, **24** , pp. 7-83.

Forbus, K. (1984): "Qualitative process theory," *Artificial Intelligence*, **24** , pp. 85-168.

Goldberg, A. and Robson, D. (1983): *Smalltalk-80: The Language and its Implementation*, Addison-Wesley, Reading, MA, USA.

Hayes, P. J. (1985): "The second naive physics manifesto," in *Formal Theories of the Commonsense World*, Hobbs, J. R. and Moore, R. C. (eds.), Ablex, Norwood, NJ, USA, pp. 1-36.

Kowalski, R. (1978): "Logic for data description," in *Logic and Data Bases*, Gallaire, H. and Minker, J. (eds.), Plenum, London, pp. 77-103.

Kuipers, B. (1986): "Qualitative simulation," *Artificial Intelligence*, **29** , pp. 289-338.

Lorie, R. (1982): "Issues in database for design applications," in *File Structures and Data Bases for CAD*, Encarnacao, J. and Krause, F. L. (eds.), North-Holland, Amsterdam, pp. 213-222.

McDermott, D. (January 1982): "Nonmonotonic logic II: Nonmonotonic modal theories," *Journal of the ACM*, **29**(1), pp. 33-57.

Ramamoorthy, C., Shekhar, S., and Garg, V. (January 1987): "Software development support for AI programs," *IEEE Computer*, **20**(1), pp. 30-40.

Stefik, M. and Bobrow, D. G. (Winter 1986): "Object-oriented programming: Themes and variations," *AI Magazine*, **6**(4), pp. 40-62.

Stefik, M., Bobrow, D., and Kahn, K. (January 1986): "Integrating access oriented programming with a multiparadigm environment," *IEEE Software*, **3**(1), pp. 10-18.

Tomiyama, T. and Yoshikawa, H. (1985): "Requirements and principles for intelligent CAD systems," in *Knowledge Engineering in Computer-Aided Design, Proceedings of the IFIP Working Group 5.2 Working Conference 1984 (Budapest)*, Gero, J. S. (ed.), North-Holland, Amsterdam, pp. 1-23.

Tomiyama, T. and Yoshikawa, H. (1987): "Extended general design theory," in *Design Theory for CAD, Proceedings of the IFIP Working Group 5.2 Working Conference 1985 (Tokyo)*, Yoshikawa, H. and Warman, E. A. (eds.), North-Holland, Amsterdam, pp. 95-130.

Tomiyama, T. and ten Hagen, P. J. W. (April 1987): "The Concept of Intelligent Integrated Interactive CAD Systems," CWI Report No. CS-R8717, Centre for Mathematics and Computer Science, Amsterdam.

Tomiyama, T. and ten Hagen, P. J. W. (1987): "Organization of design knowledge in an intelligent CAD environment," in *Expert Systems in Computer-Aided Design, Proceedings of the IFIP W.G. 5.2 Working Conference 1987 (Sydney)*, Gero, J. S. (ed.), North-Holland, Amsterdam, pp. 119-147.

Tomiyama, T. and ten Hagen, P. J. W. (June 1987): "Representing Knowledge in Two Distinct Descriptions: Extensional vs. Intensional," CWI Report No. CS-R8728, Centre for Mathematics and Computer Science, Amsterdam.

Wegner, P. (July 1984): "Capital-intensive software technology," *IEEE Software*, **1**(3), pp. 7-45.

Yeomans, R., Choudry, A., and ten Hagen, P. J. W. (eds.) (1985): *Design Rules for a CIM System*, North-Holland, Amsterdam.

Yoshikawa, H. (1981): "General design theory and a CAD system," in *Man-Machine Communication in CAD/CAM, Proceedings of the IFIP Working Group 5.2 Working Conference 1980 (Tokyo)*, Sata, T. and Warman, E. A. (eds.), North-Holland, Amsterdam, pp. 35-58.

Report on Session 8

Chair: A. Bijl †
Cochair: P. Bernus ‡
Edited by: V. Akman ‡

The two papers in this session presented ambitious theories for CAD. Both presentations were nicely complementary.

Takala's talk was speculative, referring to a model of the physiology of the brain and using it as a basis for his design paradigm. He postulated the presence of a noise generator as contributing to creative ability. Intelligence was explained as the potential to resolve noisy guesses and to obtain verified truths. A strategy was proposed for linking an entropy generator to focusing and testing mechanisms. This may possibly lead to autonomous design generation in Takala's opinion.

Group Bart Veth's paper (presented by Tomiyama) dealt with a precise definition of a formalism for representing design as a transition from function space to attribute space. Functions refer to requirements and attributes refer to design objects that are described by their properties. The formalism was offered as a general environment in which design processes can be modelled. The formalism itself in not a model of any particular design process. Tomiyama's philosophy rests on a solid belief: Design knowledge and processes are formalisable, using standard and exotic logics innovatively, in a manner that can be implemented effectively. His presentation showed that a lot of fundamental work has been done towards realizing this goal.

The response from the audience was interesting in that participants appeared to be more comfortable with Takala's sketchy but evocative presentation, than with Tomiyama's more explicit (yet, unfortunately, somewhat abbreviated due to time limitations) presentation. Responses to the latter sought to evaluate General Design Theory at higher levels of application and answers (referring to textbooks) were received with frowns. The session chair admits that this perhaps is more of a comment on the audience's level and convictions than the presentation.

---

† EdCAAD, Department of Architecture, University of Edinburgh, 20 Chambers Street, Edinburgh EH1 1JZ, UK.
‡ Centre for Mathematics and Computer Science, Kruislaan 413, 1098 SJ Amsterdam, The Netherlands.

During the Q&A period, Letray asked Takala how he could explain the meaning of entropy generation. His answer was that this is sort of random (or pseudo-random) number generation. For example, fractals are generated using this technique.

Sunde posed the following question to Tomiyama: You said that design was a mapping from the function space to the attribute space. Where do the innovative and creative aspects of design appear in this mapping process? Tomiyama's reply was that in the beginning of a design one does not know the attributes, but only the function. At the end one knows all the attributes. In this sense, creativity enters the picture in between. Koegel wanted to know what other kinds of design processes could exist, for there may be ones that may not fit in the function-attribute mapping scheme. Tomiyama told Koegel that he is somewhat mistaken in assuming the latter. The kind of design depends on how you carry out the incremental steps. In the textbook case the steps are found in the book. If not, one, as a user, must think of what step to take next. This probably explains the diversity of design processes.

Hoeltzel asked if Tomiyama has different models for different design phases or he simply assumes some sort of common product model. Tomiyama's reply was that the metamodel will be described in one single knowledge base in his IIICAD system. On the other hand scenarios create worlds for given design situations and subscenarios can be used for having multiple worlds using the same metamodel. Since he uses an extensional description, it is possible to partition the metamodel even to represent inconsistency. Novak asked how one deals with actions that operate on entities. Tomiyama's argument was that actions are formulated as set operations on the extensional descriptions. Novak then said that there are valid and invalid operations, some you are allowed to carry out, and some you must not do because they are not meaningful. Tomiyama countered that caveat by stating that he intends to apply naive physics and qualitative reasoning for that purpose.

General Discussion 1

Report on General Discussion 1

Chair: T. Tomiyama †
Cochair: V. Akman †
Edited by: V. Akman †

General Discussion 1 was conducted to summarize the first day's presentations and to suggest a direction for evolving the workshop. First, the session chairs were asked to report about their sessions from the viewpoint of the following three aspects:

(A) What is *design*? (A manifesto was implicit in each presentation.)

(B) What is the role of (future) CAD systems?

(C) What is *intelligence* all about in CAD systems?

Their short summaries were formulated to highlight the distinct opinions among the papers. (Session chairs were ten Hagen, Takala, Mac an Airchinnigh, and Bernus.)

It turned out that these questions converged to a single issue. Under which philosophy is an intelligent CAD system constructed — an intelligent apprentice or a miracle autonomous equipment? In order to explain the various positions taken against this question, we should first describe the session in more detail. Schramel raised two rather all-encompassing questions:

(i) What is intelligence?

(ii) How intelligence can be used in AI or AI-based CAD?

(Cf. Milne's comment below for a reply to question (i).)

From a CAD point of view, as far as AI can be used one can take a utilitarian approach, Bernus said. Arbab pointed out that there could be two ways of building "intelligent CAD systems":

(1) the apprentice approach where the designer is supported by intelligent tools (as much as possible) yet the authority still resides with him, and

(2) the autonomous approach where the system fully automates the design process.

The latter obviously assumes that "Design is an activity that can be formalized." This seemed to be a central issue in the following discussion.

---

† Centre for Mathematics and Computer Science, Kruislaan 413, 1098 SJ Amsterdam, The Netherlands.

Kahn stated that design systems should not limit designers, thus supporting the apprentice approach. Bijl suggested that there is no substantial difference between positions (1) and (2), pointing out that an apprentice has to know too much about the designer and adding that "We do not simplify the problem by thinking of a machine as an assistant." Genin stated that CAD systems of future would aid the designers increasingly, thus leading to autonomous systems. He emphasized common sense reasoning as a desirable feature of intelligent CAD systems. On the other hand, we observed that the AI community usually declines to consider fundamental issues, such as the definition of intelligence, as leading to fruitful discussions. This was highlighted by Milne who said "If there exists a tool, we want to develop an AI-based product using it. We are interested in techniques to solve practical problems."

Following this statement, there was a discussion about the use of various reasoning paradigms such as deduction/induction, reasoning by analogy, etc. in CAD. This was raised by Schramel and Donker who especially mentioned the importance of induction.

Tomiyama summarized these discussions by reminding the audience about two opposing standpoints historically recognized in CAD:

(a) CAD as a tool to support designers.

(b) CAD as an automated package for a particular type of designing.

The above division has been long debated in the CAD community and it seems that these two "religions" surely will influence the development of intelligent CAD systems. Roughly speaking, (a) corresponds to the apprentice approach (1), and (b) corresponds to the autonomous approach (2) cited above.

General Discussion 2

Report on General Discussion 2

Chair: V. Akman †
Cochair: T. Tomiyama †
Edited by: V. Akman †

The Organizing Committee and the session chairs (Arbab, Kjellberg, Schramel, and Bijl) decided to ask the participants to state their views on subjects which emerged as the most arguable during the workshop. This aimed at clarifying the direction of the workshop and to stimulate technical debates towards further mutual understanding. Akman listed the items to be discussed:

(1) Inheritance vs. Delegation.

(2) Apprentice (Assistant) vs. Autonomous Design Systems.

(3) Multiple views of design objects.

(4) Data models for CAD.

Akman asked Arbab to give a short explanation of (1). According to Arbab, the inheritance mechanism of most object oriented programming languages has serious problems in the context of CAD. Among other things, the mechanism is very rigid and difficult to modify, making the inheritance hierarchy inflexible. On the other hand, delegation seems to be more appropriate for CAD applications since an object acts as a "representative" of a class. Novak suggested that this problem can be solved by employing dynamic class creation. Arbab then emphasized that modification of class hierarchy is not easy. Sunde supported the idea of having multiple structures of objects; i.e. an object is allowed to have a dedicated inner structure for a particular view. Bijl concluded this discussion by pointing out that all those features are not just inherent in choice (1) but refer to higher level considerations.

Akman reminded the audience that in General Discussion 1, the dichotomy in (2) became apparent and asked them to relate their positions. Milne drew attention to his belief that even when we need an autonomous system we should somehow start with an apprentice and build upon it. Boose countered the division given in (2) by stating that discussions like this do not really lead anywhere due to their circular nature. Duffy

---

† Centre for Mathematics and Computer Science, Kruislaan 413, 1098 SJ Amsterdam, The Netherlands.

joined him and added that intelligent CAD may profit from both approaches. To Bijl, the crucial issue seemed to be not this distinction per se but the question: What does any system have to know in order to serve the user? Bernus remarked that we should naturally let the system go and do the work as much as it can. However, ten Hagen cautioned that we should not forget that there is an interaction going on between the user and the system, and supported the apprentice approach. Mac an Airchinnigh dismissed the whole conversation and implied that a discussion on apprentice versus autonomous dichotomy is a waste of time. For Kahn, the autonomous approach is much less useful than the assistant approach, for there is no interaction. Kahn further added that even in automatic design the user should be given the opportunity to look inside the system (glass-box). Hoeltzel claimed that the system should simply know the design. Novak offered as a metaphor foreground/background job control in e.g. UNIX. He suggested: Let the system proceed until it gets stuck.

Akman then asked ten Hagen to define the issue in (3) which was raised by many speakers in terms of multiple views or representations of design objects. Kahn rephrased the problem: It is a matter of controlling attention during the design process; we do not necessarily need a master view but rather, independent dedicated models with commonality. She added that generic modeling is necessary. Bijl thought that what we have is multiple or partial views of single models rather than multiple representations. According to Bernus, the idea of a central model is an implementational issue different from multiple or metamodels (cf. Veth's position). Milne supported the idea of multiple views from the viewpoint of redundancy; the user simply prefers luxurious interfaces. Sunde and Novak both agreed with Bijl: Multiple views which even allow inconsistency are desirable. For Duffy, a core issue is that of how the user constructs the model through multiple views.

Finally, Akman moved to the last issue, (4), which was borne out of a question posed by Arbab during Koegel's presentation. Tomiyama, Kahn, and Arbab each gave a definition of data model for CAD applications. They all concurred in that a data model is an abstract formal description for organizing information. Mac an Airchinnigh underlined the mathematical character of formalization. For Kahn, the tool currently used for data modelling is VDM (Vienna Definition Method). Zimmer was very clear about the power of mathematics as the foundation of CAD yet he did not give too much thought to data modeling. Novak attempted to capture the experiences of Takala and Tomiyama about extensional and intensional distinctions. Takala borrows Tomiyama's concepts but uses them in a more philosophical way. Tomiyama is basically concerned with descriptions. Schramel questioned the necessity of data modeling in CAD. Bernus replied that data modeling as conceptual modeling clarifies conceptual dependencies: What kind of concepts can be embedded in CAD? Arbab supported Bernus by rephrasing the last question: What do you think are the important concepts and relationships that should be encapsulated in CAD? Mac an Airchinnigh repeated that formal mathematical descriptions are important, e.g. in current CAD systems there exist a lot of mathematics behind the system. He challenged the audience: Why don't you codify everything in a mathematical framework? Novak closed the discussion by saying that we still need to relate and associate concepts freely using open-ended data modeling.

Closing Discussion

Report on Closing Discussion

Chair: P.J.W. ten Hagen † and M. Mac an Airchinnigh ‡
Cochair: V. Akman †
Edited by: V. Akman †

The final discussion started with a quiz by Mac an Airchinnigh to convince the audience that we need a theory of CAD. (A bear walks 10 miles to the south, 10 miles to the east, and finally 10 miles to the north to arrive at his starting location. Question: What did he eat in the morning? It turns out that he was a polar bear and hence ate fish.) Novak pointed out that we need several theories and communication between them. (Bernus would later join him in the that respect.) Takala underlined a need for metaphorical representations. Akman pronounced the importance of common sense reasoning while Novak thought that analog mode of thinking is also a key factor. Novak further implied that you do not have to be rigorous to be theoretical.

The discussion then converged to the distinction and interrelationship between theory and model. Mac an Airchinnigh cited the well-known theories of geometry and their related models: i.e. Euclidean geometry and the plane, Riemannian geometry and the sphere, and elliptical geometry and the pseudo-sphere. Bernus raised a new concern by stating that theory may not be connected to real world. Takala commented on the use of theory. For him, intelligence is the ability to move from one theory to another. According to Mac an Airchinnigh, model has expressive forms such as pictorial, symbolic, and linguistic. Neelamkavil cautioned the audience about the assumptions behind theories: Our understanding of the world is constantly changing. Ten Hagen found the main use of theory in our ability to make quick decisions — one could achieve the same effect using only the model but that would be slow. Eshuis commented that the pictorial nature of expressive forms is crucial in design. Ten Hagen then asked: Is there a general theory of design? If yes, is it useful? (That is, isn't it too big and probably incomprehensible?) Schramel responded by citing an ability to extrapolate about the future as the basic strength of a theory. Tomiyama complained that one cannot simply say that design is equivalent to problem-solving. Kjellberg defined engineering as a combination of science and art, and insisted that we have to keep the designer in loop.

---

† Centre for Mathematics and Computer Science, Kruislaan 413, 1098 SJ Amsterdam, The Netherlands.
‡ Department of Computer Science, Trinity College Dublin, Dublin 2, Ireland.

Mac an Airchinnigh then asked the audience if they had a model of design in their head. Takala saw design as a pictorial and linguistic activity. For ten Hagen, design is a collection of things in search of "externalization." Bijl saw the design of a CAD machine as equivalent to the design of a "user." Mac an Airchinnigh then softened this serious discussion by making a funny remark: What do we care if it is Friday the 13th and you are the pilot of a 747 approaching to a flock of birds? The lesson of the remark is that we have no control over many things in real world. So why care about them anyway?

The discussion then turned to the definition of design. Koegel saw design as a process with stages whereas ten Hagen more liked a hierarchical decomposition view. Genin mentioned the existence of various possibilities at a given design stage and the need for going over and revising previously made decisions. Koegel required that models must state properties of design processes, e.g. are they deterministic, stochastic (in a controlled way, cf. Takala's paper), or what? Tomiyama summarized his position as declaring theory and model inseparable. Koegel then became specific and questioned the use of modal logic in IIICAD system. Eshuis answered by repeating that Tomiyama had no model and he could have chosen another formal tool. Akman said that IIICAD is an apprentice and regards design as an activity of people building their worlds.

At this point Mac an Airchinnigh returned to his 747 example again. He reminded: Theory may fail and hence, we have to validate our design. Mistry thought that a definition of the problem domain is required for that purpose. Takala said that IIICAD has no absolute specifications behind it, and only experience and testing will tell if it is a good CAD system. Tomiyama concurred on this view and further stated that "growing" specifications is not a bad thing. After all, this is why IIICAD is using rapid prototyping as a software development strategy.

As a good functional specification expected from a CAD system, Novak challenged the audience with this view: I want the system to respond to me as a group of designers would. Ten Hagen related his experience in programming language design and said that one ends at a focal point only after traveling through a maze long enough. But this is probably unavoidable.

Finally, Mac an Airchinnigh gave a short overview of the workshop. He said that we examined several views ranging from abstract (e.g. category theory) to concrete (e.g. expert systems), from theories (e.g. IIICAD) to phantasies (e.g. Takala's talk). He then posed the question: Where do we stand now? Wingard criticisized the workshop for being mostly on implementational issues while the emphasis was expected to be placed on intelligent CAD theory. Mac an Airchinnigh asked some members of the audience about their reactions. He had generally favorable responses and mostly positive views on the overall program. Mistry added that the lack of well-defined terms was confusing and asked for a glossary in the future workshops. Mac an Airchinnigh closed the Final Session by voicing his belief that we have done a good (although not perfect) job and reminded the audience that they should read the Proceedings. "Success depends on what you are going to do from now on," he concluded.

Bibliography

This bibliography was made by compiling all the bibliographical information, removing duplications. Each reference is cited by a set of authors' names and publication year. In the papers, if there are more than two articles by the identical authors in one year, they are distinguished by putting "a," "b," etc., at the end. For instance, (Tomiyama and ten Hagen 1987a) refers to the first occurrence of articles by Tomiyama and ten Hagen in 1987.

Adams, J. (1985): "Probabilistic reasoning and certainty factors," in *Rule-Based Expert Systems: The MYCIN Experiments of the Stanford Heuristic Programming Project*, Buchanan, B. and Shortliffe, E. (eds.), Addison-Wesley, Reading, MA, USA, pp. 263-271.

Adorni, G. and Massone, L. (1985): "Graphics and natural language: an integrated interface for man-machine communication," *IFAC Conference on Analysis, Design and Evaluation of Man-Machine Systems*, Varese, Italy.

Afsarmanesh, H., McLeod, D., Knapp, D., and Parker, A. (August 1985): "An extensible object-oriented approach to databases for VLSI/CAD," *11th International Conference on Very Large Data Bases*, Stockholm, pp. 13-24.

Allen, J. (1983): "Maintaining knowledge about temporal intervals," *Communication of the ACM*, **26**(11), pp. 832-843.

Allen, J. (1984): "Towards a general theory of action and time," *Artificial Intelligence*, **23**, pp. 123-154.

Arbab, F. and Wing, J. M. (1986): "Geometric reasoning: A new paradigm for processing geometric information," in *Design Theory for CAD, Proceedings of the IFIP W. G. 5.2 Working Conference 1985 (Tokyo)*, Yoshikawa, H. and Warman, E. A. (eds.), North-Holland, Amsterdam, pp. 107-121.

Aude, J. S. (1986): "Design Rule Representation within a Hardware Design System," Technical Report (Ph.D. thesis) No. UMCS-86-11-4, University of Manchester, Department of Computer Science, Manchester, UK.

Aude, J. S. and Kahn, H. J. (1986): "A design rule database system to support technology adaptable applications," *Proceedings of 23rd Design Automation Conference*, Las Vegas, USA, pp. 510-516.

Barr, A. and Feigenbaum, E. (eds.) (1981): *The Handbook of Artificial Intelligence, 3 Volumes*, William Kaufmann Press, Palo Alto, CA, USA.

Batory, D. S. and Buchmann, A. P. (August 1984): "Molecular objects, abstract data types, and data models: A framework," *Proceedings of the 10th International Conference on Very Large Data Bases*, Singapore, pp. 172-184.

Batory, D. S. and Kim, W. (September 1985): "Modeling concepts for VLSI CAD objects," *ACM Transactions on Database Systems*, **10**(3), pp. 322-346.

Behnke, H., Bachmann, F., Fladt, K., and Kunle, H. (1986): *Fundamentals of Mathematics Volume II: Geometry*, MIT Press, Cambridge, MA, USA. Translated by S.H. Gould.

Benneth, J. L. (1983): *Building Decision Support Systems*, Wiley, New York.

Berger, D., David, B. T., and Bernanose, C. (June 1986): "Le système MARS: Un système de modélisation architecturale à référence spatiale (The sytem MARS: A system for architectural modeling for spatial reference)," *Proceedings of International Joint Conference on CAD and Robotics in Architecture and Construction*, Marseille, France.

Bergmann, N. (1985): "Generalised CMOS — A technology independent CMOS IC design style," *Proceedings of 22nd Design Automation Conference*, Las Vegas, USA, pp. 273-278.

Bergstrom, M. (1986): "Mind-brain interaction: Consciousness as a neural macrostate," in *Finnish Artificial Intelligence Symposium (September, 1986), Vol. 1, AI and Philosophy*, Karjalainen et al. (ed.), Finnish Society of Information Processing Science.

Berkhout, E. E. (1971): "Optimizing in Spatial Design," Technical Report, OPM Group, Faculty of Building, Technical University of Delft, Delft, The Netherlands. In Dutch.

Berkhout, E. E., Dekker, A. P. M., and van Loon, P. P. (1979): "Methods, Techniques, Models," Lecture Notes, Faculty of Building, Technical University of Delft, Delft, The Netherlands. In Dutch.

Berkhout, E. E., Micheels, S., and van Loon, P. P. (1982): *Design and Planning Methodology*, Delft University Press, Delft, The Netherlands. In Dutch.

Berkhout, E. E. (1986): "Methodological Survey," Research Report, OPM Group, Faculty of Building, Technical University of Delft, Delft, The Netherlands. In Dutch.

Bernus, P. and Hatvany, J. (1979): "Computer aids to the design of integrated manufacturing systems," *Computers in Industry*, **1**(1), pp. 11-19.

Beyth-Marom, R. and Dekel, S. (1985): *An Elementary Approach to Thinking Under Uncertainty*, Lawrence Erlbaum Associates, London.

Bijl, A., Stone, D., and Rosenthal, D. S. H. (1979): "Integrated CAAD Systems," EdCAAD Report for the Department of the Environment , Edinburgh University, Edinburgh, UK.

Bijl, A. (1979): "Computer aided houses and site layout design," *Proceedings of PARC79 (Planning Architecture & Computer): International Conference on Application of Computers in Architecture, Building Design and Urban Planning*, Berlin, West Germany, pp. 283-292.

Bijl, A. (in preparation, 1988): *Architecture in Mind, Computer Discipline and Design Practice*, Wiley, London.

Birtwistle, G. M., Dahl, O. J., Myhrhaug, B., and Nygaard, K. (1973): *Simula Begin*, Van Nostrand Reinhold, New York.

Bobrow, D. G. and Stefik, M. J. (1983): "The Loops Manual," Technical Memo No. KBVLSI-81-13, Xerox Palo Alto Research Center, Palo Alto, CA, USA.

Bobrow, D. G. (November 1985): "If Prolog is the answer, what is the question? or What it takes to support AI programming paradigms," *IEEE Transaction on Software Engineering*, **SE-11**(11), pp. 1401-1408.

Bobrow, D. G. (November 1986): "CommonLoops: Merging lisp and object-oriented programming," *Special Issue of ACM SIGPLAN Notices Notices (Proceedings of Object-Oriented Programming Systems, Languages and Applications '86)*, **21**(11), pp. 17-29.

Bobrow, D. G., Mittal, S., and Stefik, M. (September 1986): "Expert systems: Perils and promise," *Communications of the ACM*, **29**(9), pp. 880-894.

Boden, M. (1981): *Artificial Intelligence and Natural Man*, Basic Books, New York.

Bolt, R. A. (1979): "Filing and retrieving in the future — Spatial data-management?," in *Man/Computer Communication, State of the Art Report*, Infortech, Maidenhead, UK, pp. 19-36.

Boose, J. H. (1985): "A knowledge acquisition program for expert systems based on personal construct psychology," *International Journal of Man-Machine Studies*, **23** , pp. 495-525.

Boose, J. H. (1986): *Expertise Transfer for Expert System Design*, Elsevier, New York.

Boose, J. H. (1987): "Rapid aquisition and combination of knowledge from multiple experts in the same domain," *Future Computing Systems*, **1**(2), pp. 191-216, Elsevier.

Boose, J. H. and Bradshaw, J. M. (in press, 1987): "Expertise transfer and complex problems: Using AQUINAS as a knowledge acquisition workbench for knowledge-based systems," *Special Issue on the AAAI Knowledge Acquisition for Knowledge-Based Systems Workshop, International Journal of Man-Machine Studies*, **25** .

Borning, A. (November 1986): "Classes versus prototypes in object oriented languages," in *Proceedings of Fall Joint Computer Conference*, ACM/IEEE, Dallas, Texas.

Boyse, J. W. and Gilchrist, J. E. (March 1982): "GMSolid: Interactive modeling for design and analysis of solids," *IEEE Computer Graphics and Applications*, **2**(2), pp. 27-40.

Brachman, R. J. (1979): "On the epistemological status of semantic networks," in *Associative Networks: Representation and Use of Knowledge by Computers*, Findler, N. V. (ed.), Academic Press, New York, pp. 3-50.

Brachman, R. J. and Levesque, H. J. (October 1983): "Krypton: A functional approach to knowledge representation," *IEEE Computer*, **16**(10), pp. 67-73.

Brooks, F. (1975): *The Mythical Man-Month*, Addison-Wesley, Reading, MA, USA.

Brown, C. M. (March 1982): "PADL-2: A technical summary," *IEEE Computer Graphics and Applications*, **2**(2), pp. 69-84.

Brown, D. C. and Chandrasekaran, B. (1985): "Expert systems for a class of mechanical design activity," in *Knowledge Engineering in Computer-Aided Design, Proceedings of the IFIP WG5.2 Working Conference 1984 (Budapest)*, Gero, J. S. (ed.), North-Holland, Amsterdam, pp. 259-290.

Brown, D. C. and Chandrasekaran, B. (July 1986): "Knowledge and control for a mechanical design expert system," *IEEE Computer (Special Issue on Expert Systems in Engineering)*, **19**(7), pp. 92-100.

Buchanan, B. G. and Shortliffe, E. H. (1984): *Rule-Based Expert Systems: The MYCIN Experiments of the Stanford Heuristic Programming Project*, Addison-Wesley, Reading, MA, USA.

Buchman, A. P. and Perez de Celis, C. (1985): "An architecture and data model for CAD databases," *Proceedings of the 11th International Conference on Very Large Data Bases*, Stockholm, pp. 105-114.

Cantone, R., Lander, W. B., Marrone, M. P., and Gaynor, W. (July 1985): "Automated knowledge acquisition in IN-ATE using component information and connectivity," *ACM SIGART Newsletter*, **93** , pp. 32-34.

Carpenter, A. F., Gosling, J. B., and Kahn, H. J. (1986): "An accurate hierarchical simulation engine," *Proceedings of 3rd Silicon Design Conference*, London, pp. 257-266.

Chang, E. (in press, 1987): "Participant systems," *Future Computing Systems*.

Charniak, E. and Wilks, Y. (1976): *Computational Semantics*, North-Holland, Amsterdam.

Charniak, E. and McDermott, D. (1985): *Introduction to Artificial Intelligence*, Addison-Wesley, Reading, MA, USA.

Chieng, W. H. and Hoeltzel, D. A. (in press, 1987): "An interactive hybrid (symbolic-numeric) system approach to near optimal design of mechanical components," *Engineering Optimization*, **11**(4), Gordon & Breach Science Pub.

Chikayama, T. (November 1984): "Unique features of ESP," *Proceedings of International Conference on Fifth Generation Computer Systems 1984*, Tokyo, pp. 292-298.

Chou, H. T. and Kim, W. (1986): "A unifying framework for version control in a CAD environment," *Proceedings of the 12th International Conference on Very Large Data Bases*, Kyoto, Japan, pp. 336-349.

Choy, J. K. and Agogino, A. M. (1986): "SYMON: Automated symbolic monotonicity analysis system for qualitative design optimization," *Proceedings of ASME International Computers in Engineering Conference*, Chicago, IL, USA, pp. 207-212.

Clancey, W. (1986): "Heuristic classification," in *Knowledge-Based Problem-Solving*, Kowalik, J. (ed.), Prentice-Hall, New York, pp. 1-67.

Clarke, E. M., Emerson, E. A., and Sistla, A. P. (1983): "Automatic Verification of Finite State Concurrent Systems Using Temporal Logic: A Practical Approach," Research Report No. CMU-CS-83-152, Carnegie Mellon University, Computer Science Department.

Clocksin, W. and Morgan, A. (July 1986): "Qualitative control," *Proceedings of ECAI-86*, Brighton, UK, pp. 350-356.

Copi, I. M. (1979): *Symbolic Logic*, Macmillan Publ. Co., New York.

Cugini, U., Devoti, C., and Galli, P. (1985): "System for parametric definition of engineering drawings," *Proceedings of MICAD'85*, pp. 50-64.

Dahl, O. J., Myhrhang, B., and Nygaard, K. (1970): "Simula 67 Common Base Language," Technical Report No. S-22, Norwegian Computing Center.

David, B. T. (1983): "Conceptual framework for CAD systems construction," in *CAD Systems Framework, Proceedings of the IFIP W.G 5.2 Working Conference 1982 (Roros, Norway)*, Bø, K. and Lillehagen, F. M. (eds.), North-Holland, Amsterdam, pp. 257-278.

Davis, R. and Shrobe, H. (July 1985): "The hardware troubleshooting group," *ACM SIGART Newsletter*, **93** , pp. 17-20.

de Kleer, J. (December 1975): "Qualitative and quantitative knowledge in classical mechanics," Technical Report No. AI-TR-352, Artificial Intelligence Lab, MIT, Cambridge, MA, USA.

de Kleer, J. (September 1979): "Causal and teleological reasoning in circuit recognition," Technical report No. TR-529, MIT AI Lab., Cambridge, MA, USA.

de Kleer, J. and Brown, J. S. (1984): "A qualitative physics based on confluences," *Artificial Intelligence*, **24** , pp. 7-83.

de Kleer, J. (August 1984): "Choices without backtracking," *Proceedings of the fourth National Conference on AI*, Austin, TX, USA, pp. 79-85.

de Kleer, J. and Williams, B. (August 1986): "Back to backtracking: Controlling the ATMS," *Proceedings of the Fifth National Conference on AI*, Philadelphia, PA, USA, pp. 910-917.

de Kleer, J. (1986): "Problem solving with the ATMS," *Artificial Intelligence*, **28**(2), pp. 197-224.

de Kleer, J. (1986): "Extending the ATMS," *Artificial Intelligence*, **28**(2), pp. 163-196.

de Kleer, J. (1986): "An assumption-based TMS," *Artificial Intelligence*, **28**(2), pp. 127-162.

de Man, H., Rabaey, J., Six, P., and Claesen, L. (December 1986): "Cathedral-II: A silicon compiler for digital signal processing," *IEEE Design and Test*, **3**(6), pp. 13-25.

de Man, H. (September 1986): "Evolution of CAD-tools towards third generation custom VLSI-design," *Digest European Conference on Solid-State Circuits, ESSCIRC*, Toulouse, France, pp. 256-256c.

Dejong, G. (1986): "An approach to learning from observation," in *Machine Learning: An Artificial Intelligence Approach* **II**, Morgan Kaufmann, Palo Alto, CA, USA, pp. 571-590.

Descartes, R. (1954): *Geometry*, Dover Publications, Inc., New York. Translated by D.E. Smith and M.L. Latham.

Descotte, Y. and Latombe, J. C. (1985): "Making compromises among antagonist constraints in a planner," *Artificial Intelligence*, **27** , pp. 183-217.

Dixon, J. R. and Simmons, M. K. (November 1983): "Computers that design: Expert systems for mechanical engineers," *Computers in Mechanical Engineering*, pp. 10-18.

Dixon, J. R., Howe, A., Cohen, P. R., and Simmons, M. K. (1986): "DOMINIC I: Progress towards independence in design by iterative redesign," *Proceedings of ASME International Computer in Engineering Conference*, Chicago, IL, USA, pp. 199-206.

Dodge, C. and Bahn, C. (June 1986): "Musical fractals," *BYTE*.

Doyle, J. (1979): "A truth maintenance system," *Artificial Intelligence*, **12** , pp. 231-272.

Doyle, J. (1983): "A Society of Mind," Technical Report No. CMU-CS-83-127, Carnegie-Mellon University, Department of Computer Science, Pittsburg, PA.

Dreyfus, H. L. (1979): *What Computers Can't Do — The Limits of Artificial Intelligence (revised edition)*, Harper Colophon Books, New York.

Duhovnik, J., Kimura, F., and Sata, T. (1983): "Contribution to methodic in CAD," in *Eurographics 83*, ten Hagen, P. J. W. (ed.), North-Holland, Amsterdam, pp. 113-132.

Duhovnik, J. (1984): *CAD, Report for Research Council Society of Slovenia*, Faculty for Mechanical Engineering (FS), University Edvarda Kardelja, Ljubljana, Yugoslavia.

Duhovnik, J. and Matičič, N. (1985): *Analitics Method in CAD, Lakos 11 Seminar*, Faculty for Mechanical Engineering (FS), University Edvarda Kardelja, Ljubljana, Yugoslavia.

Duhovnik, J., Ocepek, P., Matičič, N., and Prebil, I. (1986): "Expert system for mechanical elements," in *Knowledge Engineering and Computer Modelling in CAD*, Butterworth, London, pp. 158-169.

Eastman, C. M. and Preiss, K. (March 1984): "A review of solid shape modelling based on integrity verification," *Computer-Aided Design*, **16**(2), pp. 66-60.

Erman, L. D., Hayes-Roth, F., Lesser, V. R., and Reddy, D. R. (June 1980): "The HEARSAY-II speech understanding system: Integrating knowledge to resolve uncertainty," *Computing Surveys*, **12**(2), pp. 213-253.

Euclid (1933): *Elements*, J.M. Dent & Sons, London. I. Todhunter's edition of R. Simson's "Euclid".

Faludi, A. (1973): *Planning Theory*, Oxford University Press, Oxford, UK.

Faught, W. S. (July 1986): "Applications of AI in engineering," *IEEE Computer (Special Issue on Expert Systems in Engineering)*, **19**(7), pp. 17-27.

Feigenbaum, E. A. and Barr, A. (eds.) (1986): *The Handbook of Artificial Intelligence III*, Addison-Wesley, Reading, MA, USA.

Fitzgerald, W., Gracer, F., and Wolfe, R. (July 1981): "GRIN: Interactive graphics for modeling solids," *IBM Journal of Research and Development*, **25**(4), pp. 281-294.

Forbus, K. (1984): "Qualitative process theory," *Artificial Intelligence*, **24** , pp. 85-168.

Fourman, M. P. and Zimmer, R. M. (in preparation, 1987): An Algebraic Theory of Composition.

Fox, M. S. (1979): "On inheritance in knowledge representation," in *Proceedings of the Seventh International Joint Conference on Artificial Intelligence* **2**, Tokyo, pp. 282-284.

Fox, M. S. (August 1983): "Techniques for sensor-based diagnosis," *Proceedings of the Eighth International Joint Conference on Artificial Intelligence*, Karlsruhe, Germany, pp. 158-163.

Fox, M. S. and Baykan, C. A. (1985): "WRIGHT: An intelligent CAD system," *SIGART Newsletter*, **92** , pp. 61.

Fox, M. S. (to appear, 1987): "Industrial applications of AI," in *AI in Manufacturing*, Mowforth, P. (ed.), Springer-Verlag, Berlin, Heidelberg, New York, Tokyo.

Gaines, B. R. and Shaw, M. L. G. (1981): "New directions in the analysis and interactive elicitation of personal construct systems," in *Recent Advances in Personal Construct Technology*, Shaw, M. L. G. (ed.), Academic Press, New York, pp. 147-182.

Gaines, B. R. and Shaw, M. L. G. (August 1985): "Induction of inference rules for expert systems," *Fuzzy Sets and Systems*.

Gallaire, H. (July 1985): "Logic programming: Further developments," in *1985 Symposium on Logic Programming*, pp. 88-96.

Genesereth, M. (January 1984): "The use of design descriptions in automated diagnosis," Stanford Heuristic Programming Project Memo No. HPP-81-20, Stanford University, Stanford, CA, USA.

Gergely, T. and Szots, M. (1982): "About representation of semantics," in *Progress in Cybernetics and Systems Research, Proceedings of the Symposium of Austrian Society of Cybernetic Studies* **11**, Trappl, R., Findler, N. V., and Horn, W. (eds.), Hemisphere Publ. Co., Washington D.C., pp. 227-234.

Gero, J. S. and Coyne, R. (1984): "The place of expert systems in architecture," in *Proceedings of CAD84*, Wexler, J. (ed.), Butterworth, pp. 529-546.

Gero, J. S. (ed.) (1985): *Knowledge Engineering in Computer-Aided Design, Proceedings of the IFIP WG 5.2 Working Conference 1984 (Budapest)*, North-Holland, Amsterdam.

Gero, J. S. (1987): "Knowledge-based planning as a design paradigm," in *Design Theory for CAD, Proceedings of the IFIP W.G. 5.2 Working Conference 1985 (Tokyo)*, Yoshikawa, H. and Warman, E. A. (eds.), North-Holland, Amsterdam, pp. 339-379.

Giloi, W. K. (1986): Functionality and Architecture of Machine Vision Systems, Lecture Course (notes for the USLA computer vision course).

Goldberg, A. and Robson, D. (1983): *Smalltalk-80: The Language and its Implementation*, Addison-Wesley, Reading, MA, USA.

Gongaware, T., Hseih, L., McGuire, P., and Volgel, S. (1983): "Application of automated design process planning to electronics manufacturing," *CAM-I's 12th Annual Meeting & Technical Conference*, pp. 5.

Goossens, G., Rabaey, J., Vandewalle, J., and de Man, H. (November 1987): "An efficient microcode compiler for custom multiprocessor DSP-systems," *submitted to the IEEE ICCAD-87 (International Conference on Computer Aided Design)*, Santa Clara, CA, USA.

Gossard, D. C. and Lin, V. (1983): "Representation of part-families through variational geometry," in *Advances in CAD/CAM*, Ellis, T. M. R. and Semenkov, I. I. (eds.), North-Holland, Amsterdam, pp. 47-53.

Hankiss, E. (1985): *The Literary Work as a Complex Model*, Magveto, Budapest. (In Hungarian).

Hardwick, M. and Spooner, D. L. (March 1987): "Comparisons of some datamodels for engineering objects," *IEEE Computer Graphics and Applications*, 7(3), pp. 56-66.

Harmon, P. (August 1986): *Expert Systems Strategies*, 2(8), Cutter Information Corp.

Hatvany, J. (1984): "CAD — State of the art and a tentative forecast," *Robotics and Computer-Integrated Manufacturing*, **1**(1), pp. 61-64.

Hayes, P. J. (1985): "The second naive physics manifesto," in *Formal Theories of the Commonsense World*, Hobbs, J. R. and Moore, R. C. (eds.), Ablex, Norwood, NJ, USA, pp. 1-36.

Hayes-Roth, F. and Waterman, D. A. (1986): *Building Expert Systems*, Addison-Wesley, Reading, MA, USA.

Herot, C. F. and et. al (1980): "A Prototype Spatial Data Base Management System," *Computer Graphics*, **14**(3), pp. 63-70.

Hilbert, D. (1965): *The Foundations of Geometry*, The Open Court Publishing Company, La Salle, IL, USA. Authorized English translation by E.J. Townsend.

Hilfinger, P. (May 1985): "A high-level language and silicon compiler for digital signal processing," *IEEE CICC Conference*, Portland, OR, USA, pp. 213-216.

Hillyard, R. C. and Braid, I. C. (May 1978): "Analysis of dimensions and tolerances in computer-aided mechanical design," *Computer-Aided Design*, **10**(3), pp. 161-166.

Hinkle, D. N. (1965): "The Change of Personal Constructs from the Viewpoint of a Theory of Implications," Ph.D. dissertation, Psychology Department, Ohio State University, OH, USA.

ISO (May 1983): "Open Systems Interconnection, Basic Reference Model," ISO International Standard 7498.

ISO (August 1985): "Information Processing Systems — Computer Graphics — Graphical Kernel System (GKS) Functional Description," ISO International Standard 7942.

Israel, D. J. (October 1983): "The role of logic in knowledge representation," *IEEE Computer*, **16**(10), pp. 37-41.

Israel, D. J. and Brachman, R. J. (1984): "Some remarks on the semantics of representation languages," in *On Conceptual Modeling: Perspectives from Artificial Intelligence, Databases and Programming Languages*, Brodie, M. L., Mylopoulos, J., and Schmidt, J. W. (eds.), Springer-Verlag, Berlin, Heidelberg, New York, Tokyo, pp. 119-146.

Kahn, H. J. and Filer, N. P. (1985): "An application of knowledge-based techniques to VLSI design," *Proceedings of Expert Systems 85*, Warwick, UK, pp. 307-321.

Kahn, H. J. (1986): "A multi-view data model for VLSI," *IEE Colloquium on Design Databases, Digest No. 1986/29*, London, pp. 8/1-8/4.

Katz, R. H., Anwarrudin, M., and Chang, E. (1986): "Organizing a design database across time," in *On Knowledge Base Management Systems*, Brodie, M. L. and Mylopoulos, J. (eds.), Springer-Verlag, Berlin, Heidelberg, New York, Tokyo, pp. 287-295.

Kawagoe, K. and Managaki, M. (January 1984): "Parametric Object Model: A Geometric Data Model for Computer-Aided Engineering," NEC Research & Development No. 72, Nippon Electric Co., Tokyo.

Kelly, G. A. (1955): *The Psychology of Personal Constructs*, Norton, New York.

Ketabchi, M. A., Berzins, V., and March, S. (1986): "ODM: An object-oriented data model for design databases," *1986 ACM Computer Science Conference*, Cincinnati, OH, USA, pp. 261-269.

Kimura, F., Suzuki, H., and Wingard, L. (1986): "A uniform approach to dimensioning and tolerancing in product modelling," in *Proceedings of CAPE '86, Second International Conference on Computer Applications in Production and Engineering*, Bo, K., Estensen, L., and Warman, E. A. (eds.), Copenhagen, pp. 165-171.

Kitto, C. and Boose, J. H. (in press, 1987): "Heuristics for expertise transfer: The automatic management of complex knowledge acquisition dialogs," *Special Issue on the AAAI Knowledge Acquisition for Knowledge-Based Systems Workshop, International Journal of Man-Machine Studies*, **25** .

Koegel, J. (January 1987): "POOL: Parallel object-oriented logic," Technical Report No. MS-R-8701, University of Denver, Department of Computer Science, Denver, CO, USA.

Koegel, J. (March 1987): "Molecular Objects in POOL," Technical Report No. MS-R-8702, University of Denver, Department of Computer Science.

Koegel, J. (June 1987): "POOL: Parallel object-oriented logic," *Proceedings of the 1987 Rocky Mountain AI Conference*, Boulder, CO, USA.

Kohonen, T. (1984): *Self-Organization and Associative Memory*, Springer-Verlag, Berlin, Heidelberg, New York, Tokyo.

Koller, R. (1985): *Konstruktion Lehre für den Machinenbau (Zweiter Auflage)*, Springer-Verlag, Berlin, Heidelberg, New York, Tokyo.

Kowalski, R. (1978): "Logic for data description," in *Logic and Data Bases*, Gallaire, H. and Minker, J. (eds.), Plenum, London, pp. 77-103.

Kowalski, R. (1979): *Logic for Problem Solving (Artificial Intelligence Series 7)*, North-Holland, Amsterdam.

Krekelberg, D. G., Shragowitz, E., Sobelman, G. E., and Lin, L. S. (1986): "Automated layout synthesis in the YASC silicon compiler," *Proceedings of 23rd Design Automation Conference*, Las Vegas, USA, pp. 447-453.

Krishnamurti, R. (1987): "Representing Design Knowledge," *(submitted to) Environment and Planning B*, Planning & Design, UK.

Kuhn, T. (1962): *The Structure of Scientific Revolutions*, Chicago University Press, Chicago, IL, USA.

Kuipers, B. (1986): "Qualitative simulation," *Artificial Intelligence*, **29** , pp. 289-338.

Laithwaite, R. N. W. (1986): "An expert system to aid placement on gate arrays," *Proceedings of 3rd Silicon Design Conference*, London, pp. 305-313.

Lakoff, G. (1987): *Women, Fire, and Dangerous Things: What Categories Reveal about the Mind*, Chicago University Press, Chicago, IL, USA.

Lansdown, J. (1986): "Do CAD systems fulfill designers' needs?," *Computer Aided Design*, 18(6), pp. 299-300.

Lebowitz, M. (1986): "Concept learning in a rich input domain: Generalization-based memory," in *Machine Learning: An Artificial Intelligence Approach* II, Morgan Kaufmann, Palo Alto, CA, USA, pp. 193-214.

Lee Jr., D. B. (May 1973): "Requiem for large scale models," *Journal of the American Institute of Planners*.

Lenat, D. (1983): "The nature of heuristics (Part 1)," *Artificial Intelligence*, 19 .

Lenat, D. (1983): "The nature of heuristics (Part 2)," *Artificial Intelligence*, 21 .

Lieberman, H. (1986): "Delegation and inheritance: Two mechanisms for sharing knowledge in object oriented systems," in *3eme Journees d'Etudes Langages Orientes Objets*, Bezivin, J. and Cointe, P. (eds.), AFCET, Paris, France.

Lieberman, H. (November 1986): "Using prototypical objects to implement shared behavior in object-oriented systems," *Special Issue of ACM SIGPLAN Notices Notices (Proceedings of Object-Oriented Programming Systems, Languages and Applications '86)*, 21(11), pp. 214-223.

Light, R. A. and Gossard, D. C. (1982): "Modification of geometric models through variational geometry," *Computer Aided Design*, 14(4), pp. 209-214.

Lin, V. C., Gossard, D. C., and Light, R. A. (1981): "Variational geometry in computer-aided design," *Computer Graphics*, 15(3), pp. 171-178.

Liskov, B., Snyder, A., Atkinson, R., and Schaffert, C. (August 1977): "Abstraction mechanisms in CLU," *Communications of the ACM*, 20(8), pp. 564-576.

Liskov, B., Atkinson, R., Bloom, T., Moss, E., Schaffert, C., Scheifler, R., and Snyder, A. (1981): *CLU Reference Manual*, Springer-Verlag, Berlin, Heidelberg, New York, Tokyo.

Lorie, R. (1982): "Issues in database for design applications," in *File Structures and Data Bases for CAD*, Encarnacao, J. and Krause, F. L. (eds.), North-Holland, Amsterdam, pp. 213-222.

Love, S. F. (1980): *Planning and Creating Successful Engineering Designs*, Van Nostrand Reinhold, New York.

Machlup, F. and Mansfield, U. (eds.) (1983): *The Study of Information*, Wiley, New York.

Marcus, S., McDermott, J., and Wang, T. (August 1985): "Knowledge acquisition for constructive systems," *Proceedings of the Ninth Joint Conference on Artificial Intelligence*, Los Angeles, CA, USA, pp. 637-639.

Marcus, S. and McDermott, J. (1986): "SALT: A Knowledge Acquisition Tool for Propose-and-Revise Systems," Technical Report, Pittsburgh, PA, USA.

Marcus, S. (in press, 1987): "Taking backtracking with a grain of SALT," *Special Issue on the AAAI Knowledge Acquisition for Knowledge-Based Systems Workshop, International Journal of Man-Machine Studies,* **25** .

McAllester, D. and Zabih, R. (November 1986): "Boolean classes," *Special Issue of ACM SIGPLAN Notices Notices (Proceedings of Object-Oriented Programming Systems, Languages and Applications '86),* **21**(11), pp. 417-423.

McDermott, D. (January 1982): "Nonmonotonic logic II: Nonmonotonic modal theories," *Journal of the ACM,* **29**(1), pp. 33-57.

Melkanoff, M. A. and Kamvar, E. (1985): "A manufacturing-oriented intelligent CAD-system," *Annals of the CIRP,* **34**(1), pp. 159-162.

Middendorf, W. H. (1986): *Design of Devices and Systems,* Marcel Dekker Pub.

Milne, R. and Chandrasekaran, B. (eds.) (July 1985): "Reasoning about structure, behavior and function," *ACM SIGART Newsletter,* **93** , pp. 4-59.

Milne, R. (July 1985): "The theory of responsibilities," *ACM SIGART Newsletter,* **93** , pp. 25-30.

Milne, R. (October 1985): "Diagnosing faults through responsibility," *1985 ACM Annual Conference,* Denver, CO, USA, pp. 88-91.

Milne, R. (August 1985): "Fault diagnosis through responsibility," *Proceedings of the Ninth International Joint Conference on Artificial Intelligence,* Los Angeles, CA, USA, pp. 424-427.

Milne, R. (May 1986): "Artificial intelligence applied to condition health monitoring," *Chartered Mechanical Engineer,* **33**(5), pp. 45-46, Mechanical Engineering Publications Ltd.

Milne, R. (April 1986): *Applications of Artificial Intelligence in Engineering Problems, Proceedings of 1st International Conference (1986),* Southampton University, UK, Springer-Verlag, Berlin, Heidelberg, New York, Tokyo.

Milne, R. (April 1986): "Fault diagnosis & expert systems," *The 6th International Workshop on Expert Systems & Their Applications,* Avignon, France, pp. 603-612.

Milne, R. (to appear, May 1987): "Strategies for Diagnosis," *IEEE Systems, Man and Cybernetics,* **17**(3).

Minsky, M. (1975): *The Psychology of Computer Vision,* McGraw-Hill, New York.

Minsky, M. (1980): "K-lines: A theory of memory," *Cognitive Science,* **4** , pp. 117-133.

Mitchell, W. (1977): *Computer Aided Architectural Design,* Petrocelli Charter.

Mittal, S., Bobrow, D. G., and Kahn, K. (November 1986): "Virtual copies: At the boundary between classes and instances," *SIGGPLAN Notices, Proceedings of the Object-Oriented Programming Systems, Languages, and Applications Workshop, Portland, Oregon,* **21**(11).

Moise, E. E. (1974): *Elementary Geometry from an Advanced Standpoint*, Addison-Wesley, Reading, MA, USA.

Moore, R. (April 1986): "Expert systems in process control: Applications experience," in *Applications of Artificial Intelligence in Engineering Problems, Proceedings of 1st International Conference (1986), Southampton University, UK*, Sriram, D. and Adey, R. (eds.), Springer-Verlag, Berlin, Heidelberg, New York, Tokyo, pp. 21-30.

Morris, P. and Nado, R. (August 1986): "Representing actions with an assumption-based truth maintenance system," *Proceedings of the AAAI-86*, Philadelphia, PA, USA, pp. 13-17.

Mortenson, M. E. (1985): *Geometric Modeling*, John Wiley & Sons, New York.

Mostow, J. (Spring 1985): "Towards Better Models of the Design Process," *AI Magazine*, **6**(1), pp. 44-56.

NBS (1983): "Initial Graphics Exchange Specifications (IGES), Version 2.0.," NBSIR-82-2631, National Bureau of Standard.

Negroponte, N. (1970): *The Architecture Machine*, MIT Press, Cambridge, MA, USA.

Nilsson, N. (1971): *Problem Solving Methods in Artificial Intelligence*, McGraw-Hill, New York.

Ogawa, Y., Shima, K., Sugawara, T., and Takagi, S. (November 1984): "Knowledge Representation and INference Environment: KRINE, — An Approach to the Integration of Frames, Prolog, and Graphics," *Proceedings of International Conference on Fifth Generation Computer Systems 1984*, Tokyo, pp. 643-651.

Pahl, G. and Beitz, W. (1984): *Engineering Design*, Springer-Verlag, Berlin, Heidelberg, New York, Tokyo.

Parden, G. and Newell, R. G. (April 1984): "A dimension-based parametric design system," *Proceedings of CAD84*, Brighton, UK, pp. 252-259.

Pask, G. (1975): *Conversation, Cognition and Learning*, Elsevier, Amsterdam.

Pask, G. (November 1980): "Developments in conversation theory," *International Journal of Man Machine Studies*, **13**(4), pp. 357-411.

Pask, G. (1982): "Concepts, coherence and language," in *Progress in Cybernetics and Systems Research, Proceedings of the Symposium of Austrian Society of Cybernetic Studies* **11**, Trappl, R., Findler, N. V., and Horn, W. (eds.), Hemisphere, Washington D.C., pp. 421-427.

Pereira, F. C. (June 1983): "Can Drawings Be Liberated from the von Neumann Style?," Technical Note No. 282, SRI International, Palo Alto, CA, USA.

Popplestone, R., Smithers, T., Corney, J., Koutsou, A., Millington, K., and Sahar, G. (1986): "Engineering Design Support Systems," Report No. DTOP/EXT/EDAI/09/1, Edinburgh University, Department of Artificial Intelligence, Edinburgh, UK.

Preiss, K. and Kaplansky, E. (September 1983): "Solving CAD/CAM problems by heuristic programming," *Computers in Mechanical Engineering*, pp. 56-60.

Quinlan, J. R. (1982): "Semi-autonomous acquisition of pattern-based knowledge," in *Introductory Reading in Expert Systems*, Michie, D. (ed.), pp. 192-207.

Quinlan, J. R. (1983): "Learning efficient classification procedures and their application to chess endgames," in *Machine Learning — An Artificial Intelligence Approach* **1**, Michalski, R. S., Carbonell, J. G., and Mitchell, T. M. (eds.), Tioga, Palo Alto, CA, USA, pp. 463-482.

Rabaey, J. and de Man, H. (May 1987): "Cathedral II: Computer aided synthesis of digital signal processing systems," *IEEE CICC-87 (Custom Integrated Circuits Conference)*, Portland, OR, USA.

Ramamoorthy, C., Shekhar, S., and Garg, V. (January 1987): "Software development support for AI programs," *IEEE Computer*, **20**(1), pp. 30-40.

Requicha, A. A. G. (May 1977): "Part and assembly description languages: dimensioning and tolerancing," Technical Memo No. 19, Production Automation Project, University of Rochester, Rochester, NY, USA.

Requicha, A. A. G. (December 1980): "Representations for rigid solids: Theory, methods, and systems," *ACM Computing Surveys*, **12**(4), pp. 437-464.

Requicha, A. A. G. (1980): "Representation of Rigid Solid Objects," in *Computer Aided Design — Modelling, Systems Engineering, CAD-Systems (Lecture Notes in Computer Science)* **89**, Encarnacao, J. (ed.), Springer-Verlag, Berlin, Heidelberg, New York, Tokyo, pp. 2-78.

Rivero, V. (June 1977): "Une Contribution à la Conception Architecturale Assistée par Ordinateur (A Contribution to Computer Aided Architecture Design)," Thèse de Docteur-Ingénieur (Doctorial thesis), Université Scientifique et Médicale de Grenoble, Institut National Polytechnique de Grenoble, Grenoble, France.

Robin, H. (1978): "Dimensions and tolerances in shape design," Technical Report No. 8, University of Cambridge Computer Laboratory, Cambridge, UK.

Rowe, P. (1987): *Design Thinking*, MIT Press, Cambridge, MA, USA.

Royal, N., Hunter, J., and Buchanan, I. (1985): "A case study in process independence," *Proceedings of 22nd Design Automation Conference*, Las Vegas, USA, pp. 591-596.

SL-2000 (1985): *SL-2000*, Silvar Lisco Corporation, Menlo Park, CA, USA.

Saaty, T. L. (1980): *The Analytic Hierarchy Process*, McGraw-Hill, New York.

Sacerdoti, E. D. (1974): "Planning in a hierarchy of abstraction spaces," *Artificial Intelligence*, **5** , pp. 115-135.

Sacerdoti, E. D. (1975): "A Structure for Plans and Behavior," Technical Note No. 109, SRI International.

Sacerdoti, E. (1977): *A Structure for Plans and Behavior*, North-Holland, Amsterdam.

Sata, T., Kimura, F., Hiraoka, H., Suzuki, H., and Fujita, T. (1986): "Comprehensive modelling of a machine assembly for off-line programming of industrial robots," *IFIP Working Conference on Off-line-programming of Industrial Robots*, Stuttgart, West Germany.

Sathi, A., Fox, M. S., and Greenberg, M. (September 1985): "Representation of activity knowledge for project management," *IEEE Transaction on Pattern Analysis and Machine Intelligence*, **PAMI-7**(5), pp. 531-552.

Scarl, E. A., Jamieson, J. R., and Delaune, C. I. (July 1985): "Process monitoring and fault location at the Kennedy Space Center," *ACM SIGART Newsletter*, **93** , pp. 38-44.

Searle, J. (1984): *Minds, Brains and Science (1984 Reith Lectures)*, BBC Publication, London.

Shafer, G. and Tversky, A. (1985): "Languages and designs for probability judgment," *Cognitive Science*, **9** , pp. 309-339.

Shannon, C. E. and Weaver, W. (1949): *The Mathematical Theory of Communication*, Urbana, IL, USA.

Shaw, M. L. G. and Gaines, B. R. (in press, 1987): "PLANET: A computer-based system for personal learning, analysis, negotiation and elicitation techniques," in *Cognition and Personal Structure: Computer Access and Analysis,*, Mancuso, J. C. and Shaw, M. L. G. (eds.), Praeger Press.

Shaw, M. L. G. and Gaines, B. R. (in press, 1987): "Techniques for knowledge acquisition and transfer," *Special Issue on the AAAI Knowledge Acquisition for Knowledge-Based Systems Workshop, International Journal of Man-Machine Studies*, **25** .

Shinohara, K., Hanakawa, R., Kawagoe, K., and Managaki, M. (1985): "Parametric object modeler: A conceptual object modeler for computer aided engineering," in *Design and Synthesis, Proceedings of the International Symposium on Design and Synthesis, Tokyo, Japan, July 11-13, 1984*, Yoshikawa, H. (ed.), North-Holland, Amsterdam, pp. 477-482.

Sibert, J., Hurley, W., and Blesser, T. (1986): "An object-oriented user interface management system," *ACM Computer Graphics (SIGGRAPH '86)*, **20**(4), pp. 259-269.

Simmons, R. F. (1984): *Computations from the English*, Prentice-Hall.

Smith (1984): "Plants, fractals and formal languages," *ACM Computer Graphics (SIGGRAPH'84 Proceedings)*, **18**(3).

Sowa, J. F. (1983): *Conceptual Structures*, Wiley, New York.

Spink, M. A. and Kahn, H. J. (1986): "Hierarchical design manipulation within a general silicon compiler," *Proceedings of 3rd Silicon Design Conference*, London, pp. 97-104.

Sriram, D. and Rychener, M. D. (1986): "Expert systems for engineering applications," *IEEE Software*, **3**(2), pp. 3-5.

Sriram, D. S. and Maher, M. L. (April 1986): "The representation and use of constraints in structural design," in *Applications of Artificial Intelligence in Engineering Problems, Proceedings of 1st International Conference (1986), Southampton University, UK*, Sriram, D. and Adey, R. (eds.), Springer-Verlag, Berlin, Heidelberg, New York, Tokyo, pp. 355-368.

Steele, B. D. (1981): "EXPERT — The Implementation of Data-Independent Expert Systems with Quasi Natural Language Information Input," M.Sc. thesis, Imperial College, Department of Computing Science, London.

Stefik, M. (1981): "Planning and meta-planning (MOLGEN: Part 1 and 2)," *Artificial Intelligence*, **19** , pp. 111-139 and 141-169.

Stefik, M. and Bobrow, D. G. (Winter 1986): "Object-oriented programming: Themes and variations," *AI Magazine*, **6**(4), pp. 40-62.

Stefik, M., Bobrow, D., and Kahn, K. (January 1986): "Integrating access oriented programming with a multiparadigm environment," *IEEE Software*, **3**(1), pp. 10-18.

Stiny, G., "Introduction to shape and shape grammars," *Environment and Planning*, **7** , pp. 343-351.

Sunde, G. (May 1986): "Specification of shape by dimensions and other geometric constraints," *Proceedings of the IFIP WG 5.2 Workshop on Geometric Modeling*, Rensselaerville, NY.

Sussman, G. J. (1975): *A Computer Model of Skill Acquisition (Artificial Intelligence Series 1)*, North-Holland, Amsterdam.

Sussman, G. J. and Steele, Jr., G. L. (1980): "Constraints — A language for expressing almost-hierarchical descriptions," *Artificial Intelligence*, **14** , pp. 1-39.

Sutherland, I. E. (May 1965): "Sketchpad: A man-machine graphical communication system," MIT Lincoln Laboratory Report No. 296, Cambridge, MA, USA.

Suzuki, H., Kimura, F., and Sata, T. (1985): "Treatment of dimensions on product modeling concept," in *Design and Synthesis, Proceedings of the International Symposium on Design and Synthesis, Tokyo, Japan, July 11-13, 1984*, Yoshikawa, H. (ed.), North-Holland, Amsterdam, pp. 491-496.

Szalapaj, P. J. and Bijl, A. (1984): "Knowing where to draw the line," in *Knowledge Engineering in Computer-Aided Design, Proceedings of the IFIP WG 5.2 Working Conference 1984 (Budapest)*, Gero, J. S. and Gero, J. S. (eds.), North-Holland, Amsterdam, pp. 149-169.

Takala, T. (1987): "Theoretical framework for computer aided innovative design," in *Design Theory for CAD, Proceedings of the IFIP W.G. 5.2 Working Conference 1985 (Tokyo)*, Yoshikawa, H. and Warman, E. A. (eds.), North-Holland, Amsterdam, pp. 323-338.

Takala, T. (in preparation, 1987): "Essays on Design Theory," CWI report, Centre for Mathematics and Computer Science, Amsterdam.

Tomiyama, T. and Yoshikawa, H. (1985): "Requirements and principles for intelligent CAD systems," in *Knowledge Engineering in Computer-Aided Design, Proceedings of the IFIP Working Group 5.2 Working Conference 1984 (Budapest)*, Gero, J. S. (ed.), North-Holland, Amsterdam, pp. 1-23.

Tomiyama, T. and Yoshikawa, H. (June 1985): "Knowledge engineering and CAD," *Future Generations Computer Systems*, 1(4), pp. 237-243, North-Holland.

Tomiyama, T. and ten Hagen, P. J. W. (June 1987): "Representing Knowledge in Two Distinct Descriptions: Extensional vs. Intensional," CWI Report No. CS-R8728, Centre for Mathematics and Computer Science, Amsterdam.

Tomiyama, T. and ten Hagen, P. J. W. (1987): "Organization of design knowledge in an intelligent CAD environment," in *Expert Systems in Computer-Aided Design, Proceedings of the IFIP W.G. 5.2 Working Conference 1987 (Sydney)*, Gero, J. S. (ed.), North-Holland, Amsterdam, pp. 119-147.

Tomiyama, T. and Yoshikawa, H. (1987): "Extended general design theory," in *Design Theory for CAD, Proceedings of the IFIP Working Group 5.2 Working Conference 1985 (Tokyo)*, Yoshikawa, H. and Warman, E. A. (eds.), North-Holland, Amsterdam, pp. 95-130.

Tomiyama, T. and ten Hagen, P. J. W. (April 1987): "The Concept of Intelligent Integrated Interactive CAD Systems," CWI Report No. CS-R8717, Centre for Mathematics and Computer Science, Amsterdam.

ULACAD (1986): *ULACAD Design System*, Ferranti Electronics Ltd., Manchester, UK.

Ullman, D. G. and Dietterich, T. A. (1986): "Mechanical design methodology: Implementation on future developments of computer-aided design and knowledge-based systems," *Proceedings of ASME International Computers in Engineering Conference*, 1 , pp. 173-180.

VDI (1979): *VDI Richtlinie 2222: Konstruktionsmethodik, Koncepieren Technischer Produkte*, VDI-Verlag, Düsseldorf, West Germany.

van Brunt, N. (1983): "The ZYCAD logic evaluator and its application to modern system design," *Proceedings of IEEE ICCD 83*, Port Chester, NY, USA, pp. 232-233.

Vanderplaats, G. N. (June 1985): *COPES/ADS A Fortran Control Program For Engineering Synthesis Using The ADS Optimization Program*, Engineering Design Optimization (EDO) Inc.

Vanhoof, J., Rabaey, J., and de Man, H. (August 1987): "A knowledge-based CAD system for synthesis of multi-processor digital signal processing chips," *VLSI-87*, Vancouver, Canada.

Veth, B. (1987): "An integrated data description language for coding design knowledge," in *this volume*, Springer-Verlag.

Wegner, P. (July 1984): "Capital-intensive software technology," *IEEE Software*, **1**(3), pp. 7-45.

Weizenbaum, J. (1976): *Computer Power and Human Reason*, Freeman, San Francisco, CA, USA.

Wesley, M. A., Lozano-Perez, T., L.Lieberman, Lavin, M. A., and Grossman, D. D. (January 1980): "A geometric modeling system for automated mechanical assembly," *IBM Journal of Research and Development*, **24**(1), pp. 64-74.

White, B. Y. and Fredericksen, J. R. (July 1985): "QUEST: Qualitative understanding of electrical system troubleshooting," *ACM SIGART Newsletter*, **93** , pp. 34-37.

Wiener, N. (1948): *Cybernetics*, Wiley, New York.

Wilkins, D. E. (1984): "Domain-independent planning: Representation and plan generation," *Artificial Intelligence*, **22** , pp. 269-301.

Winskel, G. (1984): "A New Definition of Morphism on Petri Nets," *Lecture Notes on Computer Science*, **166** , Springer-Verlag.

Winston, P. H. (1980): "Learning and reasoning by analogy," *Communication of ACM*, **23** , pp. 689-702.

Winston, P. H. (1982): "Learning new principles from precedents and exercises," *Artificial Intelligence*, **19** , pp. 321-350.

Winston, P. H. (1984): *Artificial Intelligence*, Addison-Wesley, Reading, MA, USA. 2nd ed.

Woods, A. W. (1975): "What's in a link: Foundations for semantic networks," in *Representations and Understanding*, Bobrow, D. G. and Collins, A. M. (eds.), Academic Press, New York, pp. 35-82.

Yeomans, R., Choudry, A., and ten Hagen, P. J. W. (eds.) (1985): *Design Rules for a CIM System*, North-Holland, Amsterdam.

Yoshikawa, H. (1981): "General design theory and a CAD system," in *Man-Machine Communication in CAD/CAM, Proceedings of the IFIP Working Group 5.2 Working Conference 1980 (Tokyo)*, Sata, T. and Warman, E. A. (eds.), North-Holland, Amsterdam, pp. 35-58.

Yoshikawa, H. (1985): *General Design Theory: Theory and Application*, Department of Precision Machinery Engineering, The University of Tokyo, Tokyo, Japan.

Yoshikawa, H. and Warman, E. A. (eds.) (1987): *Design Theory for CAD, Proceedings of the IFIP WG 5.2 Working Conference 1985 (Tokyo)*, North-Holland, Amsterdam.

Yovits, M. C. and Cameron, S. (eds.) (1960): *Self Organizing Systems*, Pergamon Press, London.

Zadeh, L. A. (October 1983): "Commonsense knowledge representation based on fuzzy logic," *IEEE Computer*, **16**(10), pp. 61-65.

Zaniolo, C. (February 1984): "Object-oriented programming in prolog," *1984 International Symposium on Logic Programming*, Atlantic City, NJ, USA, pp. 265-271.

Zimmermann, H. (April 1980): "OSI reference model — The ISO model of architecture for open systems interconnection," *IEEE Transaction on Communication*, **COM-28**(4), pp. 425-432.

Biographies of Participants

Varol Akman is a researcher in the Interactive Systems Department of the Centre for Mathematics and Computer Science (CWI), Amsterdam. For the 1985-86 academic year he was a guest docent with the Department of Computer Science at the University of Utrecht. His current research is concentrated on the theory and implementation of an intelligent computer-aided design system based on artificial intelligence techniques and advanced graphics. His other interests include geometry, theoretical computer science, and robotics.

Akman received his BSc (1979, honors) in electrical engineering and the MSc (1980, honors) in computer engineering from the Middle East Technical University, Ankara. From 1980 to 1985 he was a Fulbright scholar and a research assistant at Rensselaer Polytechnic Institute, New York, studying computer graphics at the Center for Interactive Computer Graphics, and later at the Image Processing Laboratory, and received his PhD in computer and systems engineering. He is a member of ACM, IEEE, and SIAM.

(Dr. V. Akman, Centre for Mathematics and Computer Science, Kruislaan 413, 1098 SJ Amsterdam, The Netherlands)

Farhad Arbab joined the Computer Science faculty at the University of Southern California in January 1984. He received his PhD in computer science from the University of California, Los Angeles in 1982.

Dr. Arbab has worked for IBM World Trade Corporation, Control Data Corporation, IBM Scientific Center, and has been a consultant in computer graphics and CAD/CAM. In 1983 he was a visiting assistant professor with the Manufacturing Engineering Program at UCLA. Dr. Arbab's fields of interest include geometric reasoning, intelligent CAD systems, geometric Models of Solid Objects, and CAD/CAM databases. His current research activity in geometric reasoning is funded by the National Science Foundation.

(Prof. F. Arbab, Computer Science Department, University of Southern California, SAL 200, MC 0782, Los Angeles, CA 90089, USA)

E.E. Berkhout Educational background: TU-Delft, methodology of building- and environmental design. Research background: applied decision theory and OR-methods in building- and environmental design; development of a formal design method ("Goal Processing" (GP)) based on OR-methods and decision theory; development of a preliminary — experts — interface to GP, mainly intended for interest groups, involved in intelligent bargaining. Evaluated in 20-30 real life design and development projects; developing an intelligent interface to GP, mainly intended for environmental design; developing a formal design method for architecture and building.

(Ir. E.E. Berkhout, Technical University Delft, Faculty of Building and Architecture, Berlageweg 1, 2628 CR Delft, The Netherlands)

Peter Bernus graduated from Budapest Technical University as an engineer in electronic technology. Worked in several fields of CAD, including graphics software and circuit analysis. Affiliated with the Computer and Automation Institute of the Hungarian Academy of Sciences. He has worked ten years in the research of requirements specification methodologies for CAD/CAM, and led the research group developing the SATT methodology and its software support, using computer graphics. He holds the Candidate of Sciences degree from the Hungarian Academy of Sciences, based on his thesis written on the functional analysis and synthesis of CAD/CAM systems. He is currently working in the IIICAD Group of the Interactive Systems Department at the Centre for Mathematics and Computer Science. AI background: visiting scholar at MIT AI Lab and Computer Science Department of Brown University, work on the semantic foundations of SADT (Structured Analysis and Design Technique) — leading to the development of SATT (Structured Analysis Techniques and Technology).

(Dr. P. Bernus, Centre for Mathematics and Computer Science, Kruislaan 413, 1098 SJ Amsterdam, The Netherlands)

John H. Boose (biography not available)

(Dr. J.H. Boose, Knowledge System Laboratory, Boeing Advanced Technology Center, Boeing Computer Services, P.O. Box 24346, Seattle, WA 98124, USA)

Frank W.G.M. de Bruyn started studying at the Eindhoven University of Technology in 1979. Although not majoring in computer science he took active interest in operations research, CAD, and computers in general. He joined Prof. Gero's Computer Applications Research Unit at the Sydney University. Most of that period in 1984 was devoted to studying and researching the potential of artificial intelligence, knowledge based systems, and logic programming for design environments.

Having come back to Eindhoven in 1985, he joined a CAD/CAM consultancy and software developer to accommodate industrial applications of knowledge based systems. Currently he holds a position in the Applied Artificial Intelligence Unit at Philips Eindhoven. His function at Philips involves him in AI consultancy and knowledge based systems design/development.

He has written and published several articles and reports on knowledge based systems, and authored a book on the Prolog language. In 1985 and 1986 he taught a Prolog course at Eindhoven Tech's Computing Centre and still functions as their "consultant" for logic programming. By mid 1987 he expects to complete his studies at Eindhoven Tech with a MS thesis on knowledge based resource scheduling for project management.

(Mr. F.W.G.M. de Bruyn, Philips Corp. ISA/CAD Centre, Advanced Developments/Applied AI, P.O. Box 218, 5600 MD Eindhoven, The Netherlands)

Aart Bijl, reader in charge of EdCAAD (Edinburgh Computer Aided Design), Department of Architecture, University of Edinburgh. He started his career as a practising designer, followed by 20 years of research into computer aided design progressing from ambitious and successful computer applications, towards deeper issues that are presented in the fields of artificial intelligence and cognitive science. His principal interest is to explore relationships between formal knowledge that supports technological advances, and the world of other non-specialist people, the world in which technology is employed.

Mr. Bijl was born in the Netherlands, grew up in South Africa and the UK, and graduated from the University of Cape Town as an architect. After practising in South Africa, the Netherlands and the UK, he took an appointment at Edinburgh in 1965, where he has continued his research to the present time. His work has included major funded projects on applications and speculative research, with a consistent focus on the link between graphics and design knowledge.

(Mr. A. Bijl, EdCAAD, Department of Architecture, University of Edinburgh, 20 Chambers Street, Edinburgh EH1 1JZ, United Kingdom)

Bertrand T. David graduated in Computer Science from the University of Grenoble in 1971. From 1974 to 1983 he worked at CNRS (French National Centre for Scientific Research), IMAG lab. in Grenoble. His main activity was on integrated CAD systems construction. He joined Ecole Centrale de Lyon (Lyon Graduate Scientific School) in 1983 as Professor of Computer Science and head of Mathematics — Computer Science — Automation Department. B.T. David is primary interested in Software Engineering, User Interface Management Systems, CAD systems construction methodology and knowledge-based approaches to these topics. He is a member of AFCET, IFIP WG 5.2 and Eurographics.

(Prof. B.T. David, Ecole Central de Lyon, Dept. Mathematiques-Informatique Systemes, B.P. 163, 69131 Ecylly Cedex, France)

Anneke Donker Present position: Research engineer, Informatics Division of the National Aerospace Laboratory (NLR), Department of Numerical Mathematics and Applications Programming.

Experience: 1983, at Technical University Delft, thesis treating the theoretical and numerical modelling of direct and inverse seismic problems with ray tracing techniques. From 1984 onwards, at NLR: contribution to the design and implementation of a graphics workstation as part of a nationwide military command control information system; responsible for building expert system development tools at NLR, including NEXT (the NLR Engineering X-pert system Toolkit).

Present activities: AI group leader within NLR Informatics Division; member of the board of the CIAD section "Knowledge Based Systems"; development of a theoretical model for reasoning with incomplete or uncertain data and knowledge; research on aircraft design theory and practice; study on application of AI-techniques to the software development life cycle; project on application of AI techniques to robot operations in space.

(Ir. J.C. Donker, National Aerospace Laboratory (NLR), Anthony Fokkerweg 2, 1059 CM Amsterdam, The Netherlands)

Alex Duffy spent over six years in industry before he became a student in naval architecture at the University of Strathclyde. After graduating in 1982 he began research work on a knowledge based system for numerical design. Based on this work I completed my PhD in 1986 and have for the past year been a lecturer in the CAD Centre at Strathclyde. My current research intertest is centered around the development of an intelligent design environment using a combination of knowledge based and conventional tools.

(Dr. A. Duffy, CAD Centre, University of Strathclyde, 131 Rottenrow, Glasgow G4 0NG, United Kingdom)

Jozé Duhovnik graduated in 1971; master degree 1974; 1978-1979 he was leader of a project on the uses of mechanical engineering in industry; doctorate 1980; post-doctorate study at the University of Tokyo (1982/83); 1981 was an Assistant Professor, and has taught postgraduate seminars about geometrical modelling and theory of CAD; articles and publications — CAD and driving techniques; realized projects and research results applied in industry at the University Edvarda Kardelja, Ljubljana (Department of Science for Mechanical Design). His research work is on expert systems and the theory of systematic design in the mechanical engineering field, and he is a leader of the project "Systematic Design" which is funded from the RSS (Research Society of Slovenia). He is a member of VDI (Society of Engineers in West Germany), JUDEKO (Yugoslav Society for Mechanical Elements), and SIT (Slovenian Society for Mechanical Engineering).

(Dr. J. Duhovnik, University Edvarda Kardelja Department of Science for Mechanical Design, Murnikova 6, 61000 Ljubljana, Yugoslavia)

Wim Eshuis 1972: Education in the foundation of mathematics, logic and general linguistics; 1982: Laboratory of Atomic and Molecular Physics: pattern recognition on pyrolysis mass spectra of mixtures of bio-materials; 1985: Technical University Delft: pattern recognition ; hence: real-time experts; presently at the Centre for Mathematics and Computer Science in the IIICAD Group. He is involved with several projects: intelligent CAD systems; participating in several projects, as well as a project for production planning systems; method bases, user interfaces and intelligent workstations.

(Drs. W. Eshuis, Centre for Mathematics and Computer Science, Kruislaan 413, 1098 SJ Amsterdam, The Netherlands)

Dominique Genin received his master in software engineering from the University in Liege in 1980. He was awarded the "Montefiore Prize" for his master thesis. In 1980 he joined the research center of Bell Telephone in Antwerp. In 1983 he was responsible for the expert system group. Prototypes of expert systems for the configuration of telephone switching systems were built, while the main interest of the group was to integrate different CAD tools in a knowledge based system for VLSI design and verification.

In 1986 he joined Tektronix. After a few months spent in the Tektronix Research Center in Beaverton, Oregon, he came back to Belgium to become the head of the Tektronix AI Research Center in Leuven. His main interests are in knowledge engineering, cognitive science, and VLSI design.

(Dr. D.R. Genin, AI Research Center, Tektronix, Kapeldreef 75, 3030 Heverlee-Leuven, Belgium)

Paul J.W. ten Hagen graduated in computer science from the Free University, Amsterdam in 1972. Since 1965 he has been working at the Centre for Mathematics and Computer Science, Amsterdam, where he now acts as head of the Department of Interactive Systems. From 1979 - 1983 he was convener of ISO/TC97/SC5/WG2 on Graphics Standardisation. This working group produced the international standard for graphics, GKS. He has been co-founder of the European Association for Computer Graphics and chairman for the years 1985 and 1986. As a member of the executive board he started the journal Computer Graphics Forum. He is conference chairman of the Eurographics '87 conference, to be held in Amsterdam.

As a member of IFIP WG 5.2 since 1977, he has participated and contributed to several conferences on computer aided design and workshops. He acted as the organiser of SEILLAC II Workshop on Methodology of Interaction (1978) and also for the Workshop on User Interface Management (1982). IFIP WG 5.2 is currently involved in knowledge engineering for CAD and aspects of pictorial databases. Both topics are within the scope of his research plans for the future.

(Drs. P.J.W. ten Hagen, Centre for Mathematics and Computer Science, Kruislaan 413, 1098 SJ Amsterdam, The Netherlands)

Geir Hasle received his M.Sc. in Computer Science at the University of Oslo in 1978. Since 1979 he has been involved in various projects of the Center for Industrial Research (SI): Internal project leader SPICS (Standard Program Libraries in Interactive CAD Systems); project leader IDDS (Interactive Design and Drafting System); project leader for development of problem oriented CAD system for design of wooden houses (DAKHUS); project leader "distributed engineering"; specification of system development tools for CAD systems; and leader of the "DABKON" project where two prototype KBS systems in the area of bearing type selection were developed.

(Dr. G. Hasle, Center for Industrial Research, Forskningsveien 1, 0371 Oslo 3, Norway)

David A. Hoeltzel: Assistant Professor of Mechanical Enginering Columbia University and Director of the Laboratory for Intelligent Mechanical Design. Research areas include the development of intelligent strategies from mechanism design, statistical machine learning from the cognitive selection of nonlinear programming methods in design optimization, symbolic methodologies for hidden element removal in computer graphics and knowledge-based user interfaces for 3-D object digitization.

(Prof. D.A. Hoeltzel, Department of Mechanical Engineering, Columbia University, New York, NY 10027, USA)

Josef Hofer-Alfeis Diplom-ingenieur in electrical engineering 1974, Technical University of Munich; 1974-1984 research in image processing and pattern recognition and assistant lecturer at the Institute of Communications Sciences at the Technical University of München. 1982 Dr.-Ing., thesis on image reconstruction from projections. Since 1984 with Siemens AG, Corporate Research and Technology, Corporate Production Engineering, Computer Aided Design and Manufacturing.

Main fields of research and development: intelligent input of existing engineering drawings to CAD systems (drawing conversion); methods and systems for the first design stages (rough design hand sketch input) and for variational design and optimization; user interface and graphics for model representation, especially 3D models for CAD systems.

(Dr. J. Hofer-Alfeis, Siemens AG, ZFA AUT 42, Otto-Hahnring 6, 8000 München, West Germany)

Lex James is a system designer of process planning systems at the University of Twente.

(Ir. L. James, University of Twente, Postbus 217, 7500 AE Enschede, The Netherlands)

L.S. de Jong studied applied mathematics at the University of Groningen and received his degree in 1975 at the Technical University, Eindhoven. Thesis: Numerical Aspects of Realisation Algorithms. For the next 10 years he worked at the Computing Center of the Technical University Eindhoven and was engaged in simulation of and numerical methods for evolutionary equations.

Since 1985 he has been with the Advanced Development division of ISA/CAD Centre at Philips Eindhoven and is engaged in research on software engineering and artificial intelligence.

(Dr. L.S. de Jong, Philips Corp. ISA/CAD Centre, Advanced Developments/Applied AI, P.O. Box 218, 5600 MD Eindhoven, The Netherlands)

Hilary Kahn: Graduate in Latin and Greek from University of London. Postgraduate work in Computer Science at University of Newcastle upon Tyne. Has been a lecturer in Computer Science at the University of Manchester since 1968. Involved in design and implementation of many practical CAD systems for electronic system design. Research interests include data modelling for electronic CAD, simulation, layout, design interchange standards and the applications of knowledge based techniques to CAD. Currently heads a research group of about 16 people working on a number of industrially/government funded CAD research projects.

(Ms. H.J. Kahn, Department of Computer Science, University of Manchester, Oxford Road, Manchester M13 9PL, United Kingdom)

Torsten Kjellberg is the head of a research group of 15 people working to large extent on a national program on CAD/CAM based on product modelling and artificial intelligence. Has been the Swedish project leader of the Internordic project "Geometric Product Models," GPM. Associate professor in computer systems for design and manufacturing at the Department of Manufacturing Systems of the Royal Institute of Technology. Doing research in product modelling-knowledge bases (PM-KB), and in engineering applications based on PM-KB techniques and AI in general. Has written a number of papers.

(Prof. T. Kjellberg, Department of Manufacturing Systems, Royal Institute of Technology, Brinellvägen 81, 100 44 Stockholm, Sweden)

John F. Koegel is a full-time lecturer in computer science at the University of Denver where he has taught since 1983. He holds the MSEE and BSEE from MIT and is a candidate for the PhD from Technical University Graz (Austria) which he expects to receive in the Fall of 1987. From 1980-1982 he was employed by IBM Corp. He has developed a number of graphics applications including an online weather information system for private pilots, an editor for petroleum production decline curves, and a graphics package for an ultrasonic imaging system. His research in expert systems and expert system tools led to his current work in intelligent CAD systems. This work includes: a frame-based production system (OPS86), a parallel object-oriented logic programming language (POOL), an animation system for Prolog (Anilog), a data model for CAD, and a programmable graphics editor.

(Prof. J.F. Koegel, Department of Mathematics and Computer Science, University of Denver, 2360 S. Gaylord Denver, CO 80208-0189, USA)

Frank v.d. Kruk is studying architecture at the Technical University of Eindhoven, and is currently doing practical research at I3P where he is working on a graduation project on techniques and aids from the knowledge-based systems (as part of AI). These aids and techniques will be used for modelling the design process of a specific type of shop (configuration or redesign). The objective is the design of an expertise model for redesigning a specific type of shops. This model is the base of a shop configuration (designing) expert system.

(Mr. F. van der Kruk, I3P Raadgevend Ingenieursbureau B.V., P.O. Box 2623, 60266 ZG Maarheeze, The Netherlands)

Zoltan Létray graduated in 1974 from the Technical University of Budapest as an electrical engineer. He also studied instructional technology and formal logic on which had been lecturing at the Györ Polytechnic for Transport and Telecommunication. He was also involved in designing and developing psychological diagnosis instruments. He wrote his doctoral thesis about the teaching process supported by personal computers. In 1984, he joined the Mechanical Engineering Automation Division of the Computer and Automation Institute at the Hungarian Academy of Sciences. He has worked in the field of large scale systems design methodologies and is also interested in knowledge representation techniques for intelligent CAD/CAM systems.

(Dr. Z. Letray, Computer and Automation Institute, Hungarian Academy of Sciences, Kende ut. 13-17, 1111 Budapest, Hungary)

Michael Lutz studied mathematics, computer science, and electrical engineering at the TH Darmstadt, West Germany, and received his degree as "Diplom Mathematiker" in May 1986. Since then he is an assistant of Professor Encarnacao at the TH Darmstadt, Interactive Graphics Systems. At the moment he is working with Prolog and GDDM, which is a graphic system to investigate knowledge representation and to support graphical explanations of knowledge.

(Mr. M. Lutz, Technical University Darmstadt, Fachbereich 20, GRIS, Alexanderstrasse 24, 6100 Darmstadt, West Germany)

Michael Mac an Airchinnigh Degrees: B.Sc. (pure mathematics), University of London, 1978; M.Sc. (computer science), University of Dublin, 1981; M.A. (jure officii), University of Dublin, 1985. Research: Ph.D. Dissertation on Conceptual Models and Computing: Theory and Practice (Summer 1987), University of Dublin; formal methods (VDM, Z, CSP); executable specifications (R-Meta-IV in Prolog); intelligent CAD systems (conceptual modelling and UIMS); software engineering methodology (Ada Software Methodology). Position: Lecturer in computer science at the University of Dublin, Trinity College Dublin, 1980 — (Problem Solving and Computing, Computer Graphics, CAD, UIMS); Technical Director of Generics (Software) Ltd., 1984 — (computer integrated manufacturing, Ada methodology, computer graphics, VDM, and Prolog). Member: ACM (SIGADA, SIGCHI, SIGGRAPH, SIGPLAN); IEEE (Computer Society); Eurographics; Ada-Europe (Chairman Education WG); VDM-Europe (Treasurer); IFIP WG2.7 (Operating System Interfaces); AAAI. Programming languages: Ada, Simula-67, Pascal, Lisp, Prolog; APL.

M. Mac an Airchinnigh is primarily interested in two interrelated areas: User's conceptual model (UCM) and user interface management systems (UIMS). Early work in UCM led to the M-R-C notation. Sowa's Conceptual Graph Theory was employed in 1984/85 for knowledge representation in CAD Systems and a prototype was developed in Ada. Current work involves unification of the conceptual foundations of algebraic structures and their morphisms with the theory of algorithms and formal specifications, showing that certain executable specifications are equivalent to homorphisms of structures. To support this work Relational Meta-IV (R-Meta-IV) was developed. M. Mac an Airchinnigh led the Irish team (December 1984-May 1986) in the Esprit project 496: Papillon — A Configurable Graphics Subsystem for CIM. Formal specifications in VDM Meta-IV and executable specifications in R-Meta-IV preceded the eventual encoding in Ada.

(Mr. Michael Mac an Airchinnigh, Department of Computer Science, Trinity College Dublin, Dublin 2, Ireland)

Zdravko I. Markov in 1981, received M.Sc. degree in the Faculty of Mathematics of the Sofia University and joined the Institute of Engineering Cybernetics and Robotics of the Bulgarian Academy of Sciences. His initial activities were in microcomputer programming. In the last four years he has been working on Prolog and has implemented some new versions of Prolog for microcomputers and used them for experiments in computer graphics programming. He had some results concerning the integration of Prolog with some graphics packages, especially the integration of Prolog and GKS. In 1985 he was a visiting researcher in GRIS (TU Darmstadt, West Germany), where he worked in the group of Professor Encarnacao for a few months. During his stay in GRIS he worked on GKS based programming in Prolog and on the design of an expert system implemented in Prolog. His latest interests are in the use of expert systems in CAD, especially using Prolog for knowledge representation in CAD and for geometric modelling.

(Mr. Z.I. Markov, Institute of Engineering Cybernetics and Robotics, Bulgarian Academy of Sciences, Acad. G. Bonchev str. Block 2, 1113 Sofia, Bulgaria)

Monique Megens is studying computer science at the University of Amsterdam. At the moment she is finishing her study working on IIICAD systems at the Centre for Mathematics and Computer Science in Amsterdam.

(Ms M.M. Megens, Centre for Mathematics and Computer Science, Kruislaan 413, 1098 SJ Amsterdam, The Netherlands)

Robert W. Milne has a BSc from MIT in electrical engineering and computer science and a PhD from the University of Edinburgh on artificial intelligence. He is the founder and chief executive of Intelligent Applications Limited. Before the creation of Intelligent Applications Ltd., Dr. Milne was the chief scientist at the United States Army Artificial Intelligence Centre in The Pentagon. He also taught as a professor in the United States Air Force Institute of Technology. He has over five years experience in the applications of artificial intelligence, primarily in the domain of electronic testing and troubleshooting, and coupling expert systems with automatic test equipment. He has published over 30 research papers. Intelligent Applications Ltd. specializes in the applications of artificial intelligence to electronic fault diagnosis and mechanical health monitoring.

(Dr. R. Milne, Intelligent Applications Ltd, Kirkton Business Centre, Kirk Lane, Livingston Village, West Lothian EH54 7AY, United Kingdom)

Rashmi Mistry graduated from Brunel University in 1981 with an honours degree in mathematics and computer science. He then went on to do an M.Sc in numerical analysis at Brunel which he completed in 1982. He joined Quest C.A.E. and worked at Quest on a 2D mechanical drafting system called Quadrant and later undertook a project for integration of PADL., the 3D modelling system, with Quadrant. He moved from Quest to join GMWComputers in 1983. At GMWC he worked on Rucaps, the building modelling system. The system has been developed for various users of CAD, such as architects, building services engineers, structural engineers, interior designers, retailers, and property managers. A fundamental part of the Rucaps project, is the Rucaps intelligent component system. As a senior software engineer he undertook the project of design and implementation of Rucaps intelligent component system. The Rucap intelligent component system allows the user to design his components with user defined prompts and defaults. The user is also able to work with complex algebraic expressions which will be executed at the building assembly stage as known in Rucaps.

(Mr. R. Mistry, GMWComputers Ltd, Castle Mill, Lower Kings Road, Berkhamsted, Herts. HP4 2AD, United Kingdom)

Mihai Nadin applies original theoretic knowledge to various areas, such as art, media, communication, computer technology, and artificial intelligence. He was one of the first to work in computer graphics in Europe. Since living in the West, he has acted as consultant to European and American firms dealing with electronics and computers and has advised several institutes of higher learning in applying computers to art and design and other academic areas. His professional activity in the United States started at the Rhode Island School of Design, where he introduced courses in computer graphics. His association with the Computer Science Department at Brown University involved him in problems concerning the electronic book. Due to his background in technology, semiotics, and aesthetics, he was able to suggest ways of formalizing design knowledge, an important preliminary for intelligent CAD. Acting on his proposal, Ohio State University has established the Advanced Computing Center for Arts and Design. Among his publications (covering mathematics, logic, semiotics, the arts, computers) the most recent are: Interface Design: A Semiotic Paradigm; Twelve Theses Regarding the Semiotics of the Visual; Computers in Design Education: a Case Study. As a lecturer, he has addressed audiences regarding the aesthetics of computer graphics; Can the analytical engine run the art machine?; Design as interface; The semiotic model of artificial intelligence — a shift in paradigm. His involvement in intelligent CAD led to a conceptual model of design and to the concept of the Design Machine, in which parallel computing and AI models are brought together.

(Prof.Dr. M. Nadin, Department of Art and Design Technology, Ohio State University, 1501 Neil Avenue, Columbus, OH 43201, USA)

Francis Neelamkavil is senior lecturer in computer science at the Department of Computer Science, Trinity College, Dublin. His research interests include intelligent CAD systems, interactive graphics, user interface management systems (UIMS), and expert simulation systems. He has published several papers on computer graphics, CAD optimization, modelling and simulation and authored a book entitled "Computer Simulation and Modelling," published by John Wiley and Sons in 1987. Dr. Neelamkavil took his Ph.D. degree from Trinity College Dublin and currently he is in charge of CAD and Simulation courses there. He is the director of ESPRIT-496 (configurable graphics subsystem for CIM) project in Trinity and our research is oriented towards the development of UIMS; we use VDM for specification, Prolog for prototyping and Ada for implementation. Dr. Neelamkavil is a member of ACM, IEEE, IEI and Eurographics and he was a visiting professor of computer science at the University of Kansas during 1984-85.

(Dr. F. Neelamkavil, Department of Computer Science, Trinity College Dublin, Dublin 2, Ireland)

Marcos Novak (biography not available)

(Prof. M. Novak, Department of Art and Design Technology, Ohio State University, 1501 Neil Avenue, Columbus, OH 43201, USA)

Bernd Pätzold is a scientific researcher at the Department of Computer Aided Design and Manufacturing at the University of Karlsruhe. He worked in the Product Modelling Group of the Department with special interests in designing by geometrical constraints and parametric parts.

(Mr. B. Pätzold, Department of Computer Aided Design and Manufacturing (CAD/CAM), University of Karlsruhe, D-7500 Karlsruhe, West Germany)

K.A. Popal is a student in computer science at the University of Dortmund and is engaged in intelligent CAD systems.

(Mr. K.A. Popal, (University of Dortmund) Gärtnerstrasse 35, D-4300 Essen 1, West Germany)

Jan Rogier Scientific engineer University of Technology Delft, Depart. Building 1983-1986. Scientific engineer TNO-ITT, 1986-present. Member of the IIICAD project group CWI, 1986-present.

(Ir. J. Rogier, TNO-ITT, Schoemakerstraat 97, 2628 VK Delft, The Netherlands)

Zsófia Ruttkay is a scientific researcher in the Computer and Automation Institute of the Hungarian Academy of Sciences. She attended Eötvös Lóránd University at Budapest, where she received her M.S. degree in applied mathematics in 1979. After having worked as a software engineer on linear programming and scheduling applications, she joined the Institute in 1985. First she worked on problems of machine translation and natural language database questioning, and in 1986 she joined the newly formed Intelligent Systems Department. She has been working in the project on practical applications of AI tools and techniques in manufacturing. Her recent research interests are intelligent interfaces and adequate knowledge representation for CAD systems. She is also involved in an intelligent fixture design system developed for a bus factory. She has been coauthor of four articles about issues and experiences in applying AI in computer integrated manufacturing.

(Ms Zs. Ruttkay, Computer and Automation Institute, Hungarian Academy of Sciences, Kende ut. 13-17, 1111 Budapest, Hungary)

F.J. Schramel is senior director of Philips International B.V., Eindhoven.

(Prof.Dr. F.J. Schramel, Philips International B.V., Dept. Corp. ISA, Building VN 305, P.O. Box 218, 5600 MD Eindhoven, The Netherlands)

Geir Sunde M.Sc., The Norwegian Institute of Technology (NTH), Trondheim, 1983, Mechanical Engineering (including CAD/CAM and Computer Science). Mainly been working with different aspects of geometric modelling, particularly with a new form for specification of the geometric shapes, specifications with dimensions and other geometric constraints.

(Mr. G. Sunde, Center for Industrial Research, Forskningsveien 1, 0371 Oslo 3, Norway)

Tapio Takala received the M.Sc.tech. in computer science and electronics in 1977, and the Lic.Tech. in computer science in 1983, both from the Helsinki University of Technology, Finland. Since 1980 he has been working as a researcher in CAD projects and as a senior instructor with the Laboratory of Information Processing Science at Helsinki University of Technology. In 1986 he spent three months as visiting researcher with the IIICAD project at CWI, Amsterdam. Currently he is leading a project for developing an intelligent CAD system for industrial designers. His special interests include CAD, computer graphics, modelling, user interface management, and design theory.

(Dr. T. Takala, Laboratory of Information Processing Science, Helsinki University of Technology, Otakaari 1, 02150 Espoo, Finland)

Frits Tolman is head of the Software Engineering Department of TNO-IBBC and head of the CAD/CAM Department of TNO-ITI. Working in "computers in industry" since 1968. Current research on "integration concepts" in CIM, Product Modelling, Finite Element Method software, Information Engineering in industry, User Interface

Management systems, AI-applications in industry and special purpose microprocessor systems and robotics.

(Ir. F. Tolman, Head, Software Engineering Department, TNO-IBBC, Lange Kleiweg 31, 2611 NX Delft, The Netherlands)

Tetsuo Tomiyama graduated from the Department of Precision Machinery Engineering, Faculty of Engineering, the University of Tokyo, in 1980. In 1985, he received his doctor's degree in precision machinery engineering from the graduate school of the University of Tokyo. The title of his thesis was "Theory of CAD." From 1979 to 1985, he worked with Professor Hiroyuki Yoshikawa at the University of Tokyo where he was concerned with design theory, computer aided design, and application of knowledge engineering to machine design. Since July 1985, he has been working at Centre for Mathematics and Computer Science), Amsterdam. He is currently the project leader for the "Intelligent Integrated Interactive CAD Systems" (IIICAD) project at the Department of Interactive Systems headed by Paul J.W. ten Hagen. He is a member of ACM (Association for Computing Machinery), AAAI (American Association for Artificial Intelligence), Eurographics, JSPE (Japan Society of Precision Engineers), JSME (Japan Society for Mechanical Engineering, and IPSJ (Information Processing Society of Japan).

(Dr. T. Tomiyama, Centre for Mathematics and Computer Science, Kruislaan 413, 1098 SJ Amsterdam, The Netherlands)

Paul Veerkamp studied mathematics at the University of Amsterdam (1978-1981) and then computer science at the University of Amsterdam (1981-1986). He graduated from the computer science department of the University of Amsterdam in 1986. From 1985-1986 he was a scientific assistant at CWI and worked on 3D-GKS. Since 1986 he has been working on the IIICAD project at CWI.

(Drs. P. Veerkamp, Centre for Mathematics and Computer Science, Kruislaan 413, 1098 SJ Amsterdam, The Netherlands)

Eric Weijers is studying computer science at the University of Amsterdam. At the moment she is finishing her study working on IIICAD systems at the Centre for Mathematics and Computer Science in Amsterdam.

(Mr. E.J. Weijers Centre for Mathematics and Computer Science, Kruislaan 413, 1098 SJ Amsterdam, The Netherlands)

Lars Wingard MSc in engineering physics, majoring in applied mathematics and computer science, in 1981. Working at the Royal Institute of Technology since 1981, with geometric and products modelling. From October 1986 to March 1986 research scholar at the University of Tokyo, working with dimensions and tolerances in CAD. Currently working with Development of Product Modelling using Lisp and artificial intelligence techniques in an environment with Symbolics Lisp-machines and Sun workstations connected in a network to a VAX11-780 computer.

(Mr. L. Wingard, Department of Manufacturing Systems, Royal Institute of Technology, Brinellvägen 81, 100 44 Stockholm, Sweden)

Shin-Ting Wu has received her degree of electrical engineering from University of Minas Gerais (UFMG) Brazil in 1982, and her MS degree in engineering (specialisation: automation) from UNICAMP, Brazil in 1984. During this time she attended the graduate school Philosophy Department in UNICAMP and participated in the implementation of GKS-2D and CORAS-Unicamp. In 1985 she was employed by the CTI (Technological Centre for Informatic), Brazil, where she has participated in the implementation of CAD for Control System (CADCLA). Now she is a doctoral student at the University of Darmstadt (THD-GRIS), West Germany, and is working on the automation of the generation of graphical scenes from high-level object descriptions (conceptual model). She has already implemented an interface between GKS-3/Prolog and automatic generation of tree structure. Her present interests include automatic synthesis of graphical scene descriptions, concept and graphical object modelling, classification of different levels of knowledge and their relationship, logical deduction and acquisition of knowledge.

(Ms Shin-Ting Wu, Fachbereich 20, GRIS, Technical University Darmstadt, D-6100 Darmstadt, West Germany)

Bart Veth it has been rumoured that this name has something to do with a group of researchers at the Centre for Mathematics and Computer Science, Amsterdam. These allegations have never been proved.

(Bart Veth, Centre for Mathematics and Computer Science, Kruislaan 413, 1098 SJ Amsterdam, The Netherlands)

Robert Zimmer received his PhD in automation theory, supervised by Samuel Eilenberg, from the Mathematics Department of Columbia University in 1985. Since then he has been with the Abstract Hardware Workshop at Brunel University. The Abstract Hardware Workshop is studying mathematical models of hardware and ways of incorporating these in a CAD system. The Workshop is also interested in hardware verification.

(Dr. R. Zimmer, Abstract Hardware Workshop, Electrical Engineering Department, Brunel University, Howell Building, Cleveland Road, Uxbridge, Middx. UB8 3PH, United Kingdom)